Springer Texts in Social Sciences

This textbook series delivers high-quality instructional content for graduates and advanced graduates in the social sciences. It comprises self-contained edited or authored books with comprehensive international coverage that are suitable for class as well as for individual self-study and professional development. The series covers core concepts, key methodological approaches, and important issues within key social science disciplines and also across these disciplines. All texts are authored by established experts in their fields and offer a solid methodological background, accompanied by pedagogical materials to serve students, such as practical examples, exercises, case studies etc. Textbooks published in this series are aimed at graduate and advanced graduate students, but are also valuable to early career and established researchers as important resources for their education, knowledge and teaching.

The books in this series may come under, but are not limited to, these fields:

- Sociology
- Anthropology
- Population studies
- Migration studies
- Quality of life and wellbeing research

Ajay Gupta

Qualitative Methods and Data Analysis Using ATLAS.ti

A Comprehensive Researchers' Manual

Ajay Gupta
Human Resource Management
VES Business School
Mumbai, Maharashtra, India

ISSN 2730-6135   ISSN 2730-6143 (electronic)
Springer Texts in Social Sciences
ISBN 978-3-031-49649-3   ISBN 978-3-031-49650-9 (eBook)
https://doi.org/10.1007/978-3-031-49650-9

© The Editor(s) (if applicable) and The Author(s), under exclusive license to Springer Nature Switzerland AG 2024

This work is subject to copyright. All rights are solely and exclusively licensed by the Publisher, whether the whole or part of the material is concerned, specifically the rights of translation, reprinting, reuse of illustrations, recitation, broadcasting, reproduction on microfilms or in any other physical way, and transmission or information storage and retrieval, electronic adaptation, computer software, or by similar or dissimilar methodology now known or hereafter developed.
The use of general descriptive names, registered names, trademarks, service marks, etc. in this publication does not imply, even in the absence of a specific statement, that such names are exempt from the relevant protective laws and regulations and therefore free for general use.
The publisher, the authors, and the editors are safe to assume that the advice and information in this book are believed to be true and accurate at the date of publication. Neither the publisher nor the authors or the editors give a warranty, expressed or implied, with respect to the material contained herein or for any errors or omissions that may have been made. The publisher remains neutral with regard to jurisdictional claims in published maps and institutional affiliations.

This Springer imprint is published by the registered company Springer Nature Switzerland AG
The registered company address is: Gewerbestrasse 11, 6330 Cham, Switzerland

Paper in this product is recyclable.

This manuscript is dedicated to my Parents, Suman, Archana, Abhishek, and Rajeev.

Preface

The book provides an in-depth roadmap for qualitative research methods, data collection challenges, and ATLAS.ti analysis. This self-contained book walks readers through the process of using ATLAS.ti for qualitative research and data processing. Qualitative data analysis is inextricably linked to research methods and data collection. They enable researchers to develop methods and rationales before going into the field to collect data. This occurs before the data analysis stage. As a result, before mastering data analysis techniques, researchers must become acquainted with research methods and field challenges. Learning data analysis without understanding and addressing the nuances of qualitative research methods and field impediments is a flawed approach.

This book is a comprehensive researchers' manual as well as a stand-alone resource for qualitative data analysis using ATLAS.ti across methods and disciplines. It encompasses and rationalizes the key concepts and methods of qualitative research and several innovative frameworks to deepen the understanding of qualitative research methods and data analysis. Several real-world projects are included in the book to help researchers develop conceptual and theoretical knowledge, gain insights from fieldwork, and visualize data collection, analysis, and methodological interpretations.

The backbone of qualitative research is research methodology. To produce credible research results, researchers must understand the philosophical foundations and methodology of qualitative research. Researchers can manually code their data for analysis without using computer assisted/aided qualitative data analysis software (CAQDAS). ATLAS.ti, on the other hand, allows for systematic and complex qualitative data analysis. ATLAS.ti, a computer-aided/assisted qualitative data analysis software, makes data analysis across research methods and disciplines easier. ATLAS.ti is a comprehensive research program and powerful workbench that allows researchers to access a variety of platforms to investigate underlying problems and phenomena in a scientific, creative, and innovative manner.

When compared to manual coding, the introduction of AI coding in ATLAS.ti 23 saves researchers time on data analysis and coding. Using artificial intelligence, ATLAS.ti analyzes data and retrieves the results of the researcher's instruction in a multifaceted format. With the treasure trove of insights and information, researchers can produce focused and customized results. ATLAS.ti's intuitive and automatic research tools, powered by the latest AI Coding and machine learning algorithms, assist researchers in uncovering actionable insights. This book describes manual, automatic, and AI coding, which allows researchers to perform tasks such as filtering codes, linking groups, comparing, relating, interpreting, and building theory from data. Screenshots, tables, diagrams, and networks supplement research findings and interpretations.

Features of the Book

- The book is intended for novice to advanced ATLAS.ti 9, 22, and 23 users on the Windows platform. It enables researchers to conduct independent analyses of their data.
- The book includes practical advice on overcoming field challenges during data collection, ethical considerations, and step-by-step procedures for gaining access to participants and capturing holistic data.
- The book suggests ethical considerations and protocols for various communities, from data collection to data reporting.
- The book covers a pragmatic approach to qualitative methods, proficiency in capturing field observations, researcher-friendly steps to data analysis beginning with transcription, data preparation, identification of significant information, coding, and culminating in numerous insightful thematic reports in a variety of formats.
- The book contains numerous arguments, interpretations, and explanations for concepts, themes, variables, networks, and emerging concepts.
- The book includes a comprehensive list of real-world research projects spanning multiple fields and inquiry methodologies.
- The book includes several research frameworks, data analysis guideposts on real-world research projects, and detailed procedures covering 266 images and 38 tables.
- Although the book is compelling for qualitative research, it can also be applied to quantitative research and a mixed methods approach.
- Qualitative coding can be used for quantitative analysis by creating an SPSS syntax file and analyzing it with R, SAS, or STATA. Though the software is primarily intended to support qualitative reasoning processes, statistical approaches can be useful when dealing with large amounts of data.
- The book demonstrates a variety of research methods, including grounded theory, narrative, thematic, theme network, qualitative content, discourse, case study, sentiment, document, phenomenology,

and narrative, using primary data sources such as interview transcriptions, field notes, observations, and photographs, as well as secondary data sources such as social media data, articles, diaries, records, and reports.
- A self-contained and comprehensive book that teaches to create customized, informative, and insightful reports in Excel, text, and networks using various data analysis techniques. The straightforward and user-friendly format of the book makes it easy to use, comprehend, and interpret.

The book supplements the ATLAS.ti manual in several ways. The book is organized around major themes such as qualitative research methods, field challenges, ethical protocols and guidelines, and the use of ATLAS.ti in real-world research projects. The book provides an appropriate and comprehensive explanation of concepts, their application, and the generation of thematic findings across disciplines and methods. The book contains numerous screenshots, tables with simple language, and explanations based on real-world research projects.

How Should Researchers Use the Book?

This is not a reading book. Researchers should work on their projects while reading the book. They can learn and enhance their research analysis skills using the book as a research manual. The book covers four fundamental aspects: research methods, ethical protocols, field challenges, and conceptual clarity about working knowledge of ATLAS.ti. Advanced applications on several real-world projects have been drawn, and rigorous primary and secondary data analysis has been discussed. The book includes several examples across research methods-narrative, thematic, qualitative contents, discourse, case study, sentiment, document, phenomenology, narrative, etc. using primary data-interview transcription, field notes, observations, images, pictures, etc.-and secondary data-social media data, articles, diaries, records, reports, etc.

The book discusses real-world issues researchers face when seeking respondents' consent, conducting interviews, recording data, informal interaction, nonverbal information, and dealing with unexpected challenges during data collection. It includes the psychological disposition of the researchers in disclosing the identity to respondents, as well as issues associated with it.

Regardless of their research methods, frameworks, networks, diagrams, and screenshots assist researchers in moving forward with their work. Every method has a framework, processes, working guidelines, and themes that are interpreted. From beginning to end, the book demonstrates systematic research roots. Psychologists, sociologists, social workers, management studies, economists, educational scientists, engineers, criminologists, anthropologists, and organizational psychologists use the software. Visualization techniques, networks, excel spreadsheets, and text format were used to present the research findings, each with a significant numerical value. CAQDAS is an incredibly easy application to learn and use. Users initially find the CAQDAS difficult to grasp. However, for those who gain experience, it is an easy-to-learn application.

Mumbai, India   Ajay Gupta

Acknowledgements

Many people blessed this book. My parents taught me the value of courage, hard work, and honesty. Without their divine support, I couldn't have completed the manuscript. A heartfelt thanks to my wife, who granted me space to pursue my goal. Her unwavering support made the adventure inspiring.

Dr. Ezechiel Toppo, my guide and mentor, although no longer with us, deserves special recognition. He continues to be inspired by his life journey of struggle and overcoming against all odds. His personal story is an exceptional example of overcoming adversity.

Dr. Anil Sutar, my mentor, encouraged and supported me during and after my Ph.D. journey. His advice changed how I thought and solved difficulties. I'm grateful for his unrelenting support in finishing my Ph.D.

I thank Mr. Venkatanaryanan Ganapathi, who edited the manuscript and offered remarkable suggestions.

Dr. Anil Khandelwal, former Bank of Baroda CMD, has always guided and encouraged me. Dr. Satish Modh, Director of VESIM Business School, encouraged and supported my endeavor in my academic journey.

I thank Mr. Thomas Muhr, the founder of ATLAS.ti, for allowing me to write the book.

I want to thank Dr. Susanne Friese, a professor and the author of ATLAS.ti Manual and Qualitative Data Analysis, for being an inspiration to me.

I'm grateful to the editors for letting me write this book. I take the risk of bringing up concerns in qualitative methods and data analysis to help researchers across disciplines. Teachers, friends, coworkers, nature, and environment inspired me to write this book. I appreciate God for providing me the ability to think, the courage to face challenges, and the time to do this great work. Time, space, and modesty compel me to stop here.

About This Book

The book covers three critical areas for ATLAS.ti 9, 22, and 23 users:

1. Qualitative methods and design
2. Field challenges and data collection
3. Data analysis and presentation

The book contains real-life examples to immerse readers in the fundamentals of qualitative research, its rationale, and why people conduct qualitative research. It covers qualitative research methods, frameworks, research design, and the use of ATLAS.ti.

The second section delves into the challenges that researchers face while gathering data. It includes, among other things, effective communication, obtaining consent, data collection, ethical considerations, psychological issues, conducting interviews, and writing field notes.

The third section explains how to analyze textual, multimedia, Twitter data, survey, focus group, and geo-data, as well as images, diagrams, and social media comments from Facebook, Twitter, Instagram, YouTube, TikTok, VK, Twitch, and Discord. The findings were explained and interpreted using text, network, and Excel. ATLAS.ti uses machine learning and artificial intelligence to analyze data with great precision. Researchers can contribute methodological and empirical evidence to improve data analysis, reporting, and interpretation.

ATLAS.ti enables researchers, practitioners, and social scientists from a variety of disciplines to analyze qualitative data using qualitative research concepts, formulas, and reporting research results. Researchers using mixed methods can effectively analyze qualitative and quantitative data using the software. The manual's 266 figures and 38 tables on various real-world projects are intended to train users to analyze data independently.

Contents

1	**Introduction**	1
	Meaning and Importance of Qualitative Research	1
	Characteristics of Qualitative Research	3
	The Philosophical Foundations of Qualitative Research	4
	Difference Between Qualitative and Quantitative Research	5
	Research Paradigm	7
	Research Perspectives	7
	Symbolic Interaction and Phenomenology	8
	Ethnomethodology and Constructivism	9
	Structural and Psychoanalytical	10
	Thick and Thin Description	11
	Approaches to Qualitative Research	12
	Meaning of Exploration	16
	Steps in Qualitative Research	17
	The Problem to Be Investigated	18
	Development of Conceptual Framework	19
	Formulating Aims and Objectives	22
	Research Design	22
	Sampling Strategy	23
	Data Collection	23
	Data Analysis	24
	Interpretation and Discussion of Results	25
	Presenting the Report	26
	Recommended Readings	27
2	**Types of Data, Data Collection, and Storage Methods**	31
	Importance of Qualitative Data	31
	Methods of Collecting Qualitative Data	32
	One-on-One Interviews	34
	Observation	35

	Focus Groups	36
	Storing Data	36
	Approaches to Qualitative Inquiry	37
	Narrative Study	37
	Phenomenological Research	37
	Types of Phenomenological Data Analysis Methods	40
	Grounded Theory	41
	Challenges in Using Grounded Theory	42
	Prerequisite for Grounded Theory Analysis	43
	Ethnographic Research	45
	Case Studies	47
	Research Diary and Field Notes	48
	Transcription of Data	49
	Contents of Transcripts	51
	Challenges in Transcription	52
	Framework for Qualitative Research	54
	Recommended Readings	57
3	**Challenges in Data Collection**	61
	Becoming Familiar with the Settings of the Study	61
	Challenges in Data Collection	63
	Ethical Concerns in Qualitative Research	65
	Ethical Challenges and Practices	66
	Ethical Issues in Studying Internet Communities	68
	A Caution About Google Forms	70
	Information from Informal Discussions	70
	Disclosing Researcher's Identity	72
	Two Incidents	73
	The First Incident	73
	The Second Incident	75
	Contents of Letter of Introduction	76
	Challenges in Getting Consent	78
	When and Where to Conduct Interviews	78
	Types of Interviews	82
	How to Begin an Interview	83
	Types of Interview Questions	84
	Probing Techniques During Interviews	84
	Finishing an Interview	87
	What to Write in Field Notes	87
	Describing Field Experiences	88
	Before the Interview	90
	During the Interview	90
	After the Interview	91
	While Waiting for the Interview	92
	Recommended Readings	93

4	**Codes and Coding**	99
	What is a Code	99
	Code Types	101
	Open Coding	102
	Inductive Coding	102
	Deductive Coding	103
	Abductive Coding	104
	Hybrid Coding	104
	In Vivo Coding	104
	Axial Coding	106
	Selective Coding	107
	Approaches to Coding in Qualitative Research	108
	Major Approaches to Coding in Grounded Theory	109
	Coding Process in Grounded Theory	110
	Codes, Categories, and Patterns	113
	Methods in Coding Cycle	117
	First Cycle Coding	117
	Second Cycle Coding	119
	Computer Assisted/Aided Qualitative Data Analysis (CAQDAS)	120
	Advantages and Limitations of Computer-Assisted Analysis of Qualitative Data	121
	Advantages	121
	Limitations	121
	What the Researcher Should Know About CAQDAS	122
	What the Software Does	122
	What the Software Does not Do	122
	Recommended Readings	123
5	**Data Analysis Methods**	127
	An Overview of ATLAS.ti	127
	About the Examples Used in the Discussions in This Book	128
	A Comprehensive Framework for Data Analysis Using ATLAS.ti	128
	Analysis of Qualitative Data	130
	Codes in ATLAS.ti	132
	Free Codes	134
	Open Codes	135
	Quick Coding	136
	Auto-coding	137
	Splitting Codes	138
	When and How to Split Codes	140
	Steps for Splitting Codes in ATLAS.ti	141
	Redundant Coding	141
	When and How to Merge Codes	143
	Auto-Coding in ATLAS.ti	143
	The Auto-Coding Window	146

	Creating a Research Project in ATLAS.ti	148
	Understanding Documents in Qualitative Research	149
	Meaning of Quotations	149
	Identifying Quotations in ATLAS.ti	150
	Identifying Quotations from Responses	153
	Coding Window in ATLAS.ti	154
	Generating Codes from Quotations	155
	Steps for Generating a Code	156
	Memos	157
	Writing a Memo	159
	Steps for Writing a Memo in ATLAS.ti	160
	Steps in Writing Memo Types	160
	Coding Field Notes	161
	Coding Field Observations	162
	Comments	163
	Writing Comments	163
	Steps for Writing Comments	165
	Writing Comments for Documents	165
	Writing Comments for Quotations	165
	Writing Comments for Codes	167
	Writing Comments for Memos	169
	Understanding Comments and Memos	170
	Themes	171
	Code-Theme Relationships	172
	Language of Qualitative Research	176
	Presenting Results and Findings	176
	Recommended Readings	178
6	**Using Networks and Hyperlinks**	181
	Overview	181
	Nodes and Links	183
	Presenting Networks	184
	Using Networks	185
	Relationships and Hyperlinks	187
	Difference Between Code–Code Relationships and Hyperlinks	190
	Creating Hyperlinks	190
	Creating Hyperlinks with Multiple Documents	193
	Tabs and Tab Groups	193
	Properties of Relationships	195
	Neighbors and Co-occurring Codes	196
	Case Analysis Using Networks	196
	Usefulness of Local and Global Filters	197
	Setting up Global Filters	199
	Recommended Readings	201

7	**Discovering Themes in ATLAS.ti**	203
	Data Analysis	203
	Data Analysis Using Query Tool	204
	Code Co-occurrence Analysis	206
	C-coefficient	208
	Sankey Diagram of Code Co-occurrence	208
	Applications of C-coefficient	209
	Frequency Count and Listed Quotations	212
	Operators and Code Co-occurrence	214
	Code-Document Table	214
	Binarized Data	218
	Normalized Frequencies	219
	Sankey Diagram for a Code-Document Table	220
	Showing Themes in One Report	222
	Recommended Readings	223
8	**Generating Themes from Smart Codes and Smart Groups**	225
	Smart Codes and Normal Codes	225
	Smart Groups and Normal Groups	228
	Forming Smart Codes and Smart Groups	230
	Set Operators	231
	Functions of Proximity Operators	233
	Steps for Generating Reports Using Set Operators	236
	Generating Reports with Proximity Operators	237
	Smart Codes and Their Interpretation	239
	Interpreting Smart Groups	240
	Thematic Visualization of Networks	242
	Recommended Readings	243
9	**Group Creation for Entities and Report Generation**	245
	Document Groups	245
	Steps for Creating a Document Group	247
	Creating a Network of Document Groups	248
	Steps for Creating a Network	248
	Reports	249
	Single or Group Document	250
	Reports for Document Groups	250
	Reports for Quotations	250
	Code Groups	251
	Code Groups and Code Categories	252
	Steps for Forming Code Groups	252
	Creating a Network of Code Groups	255
	Making an Informative Network	256
	Reports on Codes	257
	Memo Groups	258
	Creating a Memo Group	259

	Creating a Network of Memo Groups	261
	Creating a Network	261
	Reports from Memos	261
	Single Memo	261
	Reports for Memo Groups	262
	Network Groups	264
	Creating a Network Group	264
	Networks of Network Groups	265
	Network of a Network Groups	266
	Reports for Networks	267
	Report for Single Network	267
	Network Groups	267
	Data Classification Window	268
	Recommended Readings	268
10	**Survey Data Analysis**	**271**
	An Overview of Survey Data	271
	Types of Data that Can Be Imported	273
	Coding Options for Survey Data	276
	Using Stop and Go Lists	279
	Steps in Data Analysis	280
	Analysis of Survey Data	280
	Sample Research Project	281
	The Objective of the Study	281
	Data Collection Process	282
	Code Document Tables	287
	Code Co-occurrence	288
	Analysis of Survey Data Using AI Coding	289
	Generating a Text Report	291
	Recommended Readings	297
11	**Social Network Comments and Twitter Data Analysis**	**299**
	Preparation of Twitter Data	299
	Importing Twitter Data	300
	Importing Data into ATLAS.ti	301
	Coding Tweets	303
	Analysis of Twitter Data	305
	Interpretation of Themes from Twitter Data	308
	Analysis of Social Network Comments	308
	Analysis of Viewer Comments on YouTube	309
	Analysis of YouTube Comments Using AI Coding	314
	Recommended Readings	319
12	**Analysis of Focus Group Data**	**321**
	Focus Group Discussion	321
	Nature and Types of Focus Group Data	324

	Types of Focus Group Discussion	327
	Piggyback Focus Groups	327
	Frameworks for Focus Group Data Analysis	328
	Flowchart of Focus Group Discussion Analysis	329
	Computer-Aided Analysis of Focus Group Data	330
	Coding Options for Focus Group Data	330
	Coding Focus Group Discussion Data	331
	Two-Level Analysis of Focus Group Data	332
	Focus Group Data Analysis	333
	Krueger Focus Group Analysis Tips	334
	Analysis of FGD Data in ATLAS.ti	334
	Steps	335
	Analysis of Focus of Group Discussion Data in ATLAS.ti	337
	Recommended Readings	341
13	**Multimedia and Geo Data Analysis**	**345**
	Multimedia Data	345
	Preparing Transcripts	346
	Five Approaches to Multimedia Data Analysis	346
	Transcription Software	348
	Importing Multimedia Data	349
	Creating Quotations and Codes from Multimedia Data	351
	Working with Multimedia Data and Ready Transcripts	353
	Working with Video Projects	354
	Collecting Multimedia Data	354
	Working with Video Data	356
	Findings and Interpretations	358
	Working with Geodata	359
	Geo Document in ATLAS.ti 23	361
	Creating Quotations and Codes in Geo Documents	361
	Example of a Research Project in Geo Document	362
	Recommended Readings	365
14	**Team Projects in ATLAS.ti**	**367**
	Chapter Overview	367
	Reliability and Validity in Qualitative Research	368
	Intercoder Agreement in Teamwork	371
	Challenges in Interpretation of Intercoder Agreement	373
	Prerequisite Knowledge for Working in an ATLAS.ti Team Project	374
	Team Project Scenarios	374
	Setting up a Team Project	375
	Setting a up a Master Project	376
	A Project with 10 Interviews	377
	Creating Codes in a Team Project	378
	Coding Method in a Team Project	379

	Intercoder Agreement in ATLAS.ti	381
	Preparing Data for Inter-coder Agreement Check	381
	The Lead Researcher's Role	382
	Percentage Agreement	387
	Establishing Intercoder Percentage Agreement	388
	Calculation of Intercoder Agreement	389
	Interpretation of Results of Intercoder Analysis	390
	Recommended Readings	390
15	**Literature Review and Referencing**	395
	An Overview of the Literature Review	395
	The Literature Review Process	397
	Export Literature from Reference Manager Tool to ATLAS.ti	398
	Importing a Saved Folder into ATLAS.ti	400
	Auto-Coding Options for Literature Review	402
	Text Search and Regular Expression Search	402
	Named Entity Recognition	404
	Concepts	406
	Coding Preference	409
	Literature Review in ATLAS.ti	410
	Reporting In-Text Citations and References	416
	Literature Reviews in ATLAS.ti—Good Practice	417
	An Innovative Method for Retrieving Literature	418
	Using Search Project	420
	Reporting and Interpretation	421
	Recommended Readings	422

Index 423

About the Author

Dr. Ajay Gupta has 29 years of diverse experience, including Indian Air Force, Banking, and Management Education. Dr. Gupta is an Erasmus Mundus fellow from the University of Milan, Italy. He holds a Ph.D. in organizational psychology from the Tata Institute of Social Sciences and a Master in Management from the Asian Institute of Management in Manila, Philippines.

Dr. Gupta has taught at T. A. Pai Management Institute, Manipal, an AACSB Management Institute in India, and currently working as a professor and associate dean of Human Resource Management with Vivekananda Business School, Mumbai, and Chairperson of "The Centre for Qualitative Research." Dr. Gupta is Professional ATLAS.ti, Senior Trainer and Consultant (https://atlasti.com/trainers), in Asia-Pacific Region.

Dr. Gupta has conducted several workshops and Faculty Development Programs on Qualitative Research and Data Analysis using ATLAS.ti in many prestigious institutes like Tata Institute of Social Sciences (TISS), International Institute for Population Sciences (IIPS), National Institute of Industrial Engineering (NITIE), MNNIT, Allahabad, IIM Shillong, National Institute for Research in Tuberculosis (NIRT), Sydenham Institute of Management Studies and Research, Indian Council of Medical Research (ICMR), AURO University, University of Rajasthan, Amity University, Banasthali Vidyapith, Mumbai University, Savitribai Phule Pune University, MET Institute of Management, Bank of Baroda, Apex Academy, National Law University, Delhi, and Child Rights and You, etc. He has conducted a webinar invited by the University of British Columbia, Canada, and delivered sessions on Qualitative Research and Data Analysis at the University of Cartagena, Spain.

Dr. Gupta is an invited faculty to teach for the Executive Post Graduate Diploma in Organization Development and Change Management, Executive Postgraduate Diploma in Human Resources Management, Master of Arts in Human Resource Management and Labour Relations, Ph.D., and M.Phil.

program, at Tata Institute of Social Sciences, and Business Ethics at NITIE, Mumbai.

Dr. Gupta has developed a course called "Organizational Psychology" and teaches it along with human resource management, business ethics, leadership and change management, reward management, people dynamics in organizations, people management, organization theory and design, performance management, industrial relations, applied economics, qualitative research, and business research methods.

Corporate Demons-Creating Catalyst Leaders, a book on organizational psychology, and *The Handbook on Management Cases: A Practitioner's View, Vol I* are published books. *Cultures-The Hidden Truth* and *The Handbook on Management Cases: A Practitioner's View, Vol II* are other forthcoming books.

LIST OF FIGURES

Fig. 1.1	Steps in inductive and deductive methods	17
Fig. 1.2	Steps in qualitative research	18
Fig. 2.1	Narrative study framework	38
Fig. 2.2	Types of phenomenological research	39
Fig. 2.3	Systematic approach to grounded theory research	45
Fig. 2.4	Constructivist approach to grounded theory	46
Fig. 2.5	Ethnographic data coding framework	48
Fig. 2.6	Coding framework for case study research	48
Fig. 3.1	How to begin interview and capture field-notes	91
Fig. 4.1	Inductive, deductive, and abductive codes	103
Fig. 4.2	Axial codes network	108
Fig. 4.3	Coding process in grounded theory research	112
Fig. 4.4	Code types in code manager	112
Fig. 4.5	A network of selective code	114
Fig. 4.6	Codes and categories	115
Fig. 5.1	Hyperlinks for quotations	130
Fig. 5.2	Data analysis steps	131
Fig. 5.3	Coding windows In ATLAS.ti 22/23	133
Fig. 5.4	Auto-coding windows In ATLAS.ti 23	133
Fig. 5.5	Free coding option windows in ATLAS.ti 22	134
Fig. 5.6	Open, free, in vivo, and axial codes	137
Fig. 5.7	Code manager, grounded and density	138
Fig. 5.8	Split Code options in ATLAS.ti 22	139
Fig. 5.9	Split Code window in ATLAS.ti 22	139
Fig. 5.10	Splitting codes	140
Fig. 5.11	Redundant coding	142
Fig. 5.12	Merging codes in ATLAS.ti 22	144
Fig. 5.13	Word cloud and word list	145
Fig. 5.14	Auto-coding text search option in ATLAS.ti 23	146
Fig. 5.15	Auto-coding sentiment analysis option in ATLAS.ti 23	147
Fig. 5.16	Auto-coding named entity option in ATLAS.ti 23	147
Fig. 5.17	Creating a project	148

Fig. 5.18	Adding files in ATLAS.ti	148
Fig. 5.19	Identifying quotations and coding them	151
Fig. 5.20	Quotations and codes	151
Fig. 5.21	Quotations, codes and memos	153
Fig. 5.22	Window for projects in ATLAS.ti 22/23	154
Fig. 5.23	Coding window in ATLAS.ti	155
Fig. 5.24	Generating codes	157
Fig. 5.25	Writing a memo	159
Fig. 5.26	How to set a memo type	161
Fig. 5.27	Visualization of themes from field notes and transcripts	163
Fig. 5.28	Comments for entities	166
Fig. 5.29	Writing comments	167
Fig. 5.30	Writing a comment for document	168
Fig. 5.31	Comment for quotation	168
Fig. 5.32	Comment for code	169
Fig. 5.33	Comment for memo	169
Fig. 5.34	Memo and comment network	171
Fig. 5.35	Codes-theme relationships framework	176
Fig. 5.36	Codes-theme relationship in ATLAS.ti	177
Fig. 6.1	Codes, links, relations and hyperlinks	182
Fig. 6.2	Network group and degree	183
Fig. 6.3	Second class links	185
Fig. 6.4	A complex arrangement of networks, nodes, and links	186
Fig. 6.5	Dummy code in the network	187
Fig. 6.6	Adding nodes to a network	188
Fig. 6.7	Code to code relationships	188
Fig. 6.8	Hyperlink relations	189
Fig. 6.9	Link source and target quotations	192
Fig. 6.10	A network showing hyperlinks relations	192
Fig. 6.11	Document view in a new tab group	194
Fig. 6.12	Neighbors and co-occurring codes	194
Fig. 6.13	Setting a global filter for memo group	198
Fig. 6.14	Global filters for document, code, network and memo group	199
Fig. 6.15	Entity managers	200
Fig. 6.16	Global filter in document group	201
Fig. 7.1	Query writing and analysis tools in ATLAS.ti 22 & 23	204
Fig. 7.2	Query tool and options in ATLAS.ti 22 & 23	205
Fig. 7.3	Global filter in ATLAS.ti 22 & 23	206
Fig. 7.4	Code co-occurrence in the query tool	207
Fig. 7.5	Code co-occurrence table and Sankey diagram	210
Fig. 7.6	Research questions and Sankey diagram	211
Fig. 7.7	Code co-occurrence table and query tool	212
Fig. 7.8	Code co-occurrence table with c-value	214
Fig. 7.9	Set operators and co-occurs	215
Fig. 7.10	Results using or and co-occurs	215
Fig. 7.11	Absolute code-document table	216
Fig. 7.12	Normalized table and row-relation frequencies	216
Fig. 7.13	Row-relation and table relation frequencies	218
Fig. 7.14	Binarized code-document table in ATLAS.ti 23	219

Fig. 7.15	Absolute and normalized code-document table	221
Fig. 7.16	Code document table and Sankey diagram	222
Fig. 7.17	Absolute and normalized frequencies	222
Fig. 7.18	Set and proximity operators	223
Fig. 8.1	Smart code in explorer and code manager	226
Fig. 8.2	Options for creating smart codes	227
Fig. 8.3	Smart codes and edit scope	228
Fig. 8.4	Smart group and edit scope in query tool	229
Fig. 8.5	Smart group in code manager	229
Fig. 8.6	A smart group network	230
Fig. 8.7	Operators in ATLAS.ti 22	231
Fig. 8.8	'Or' and 'and' operators	232
Fig. 8.9	'One of' and 'not' operators	233
Fig. 8.10	Set operators	234
Fig. 8.11	Query tool and proximity operators	235
Fig. 8.12	A network of query tool and proximity operators	235
Fig. 8.13	Report using set operators	237
Fig. 8.14	Text report using set operator in code manager	238
Fig. 8.15	A text report using proximity operator	239
Fig. 8.16	Smart codes using edit scope	240
Fig. 8.17	Smart group using edit scope	241
Fig. 8.18	A smart group and themes network	241
Fig. 8.19	Thematic visualization of a network	242
Fig. 8.20	Add neighbors option	243
Fig. 9.1	Creating document groups	246
Fig. 9.2	Comment for document group	247
Fig. 9.3	Document group and network	248
Fig. 9.4	Report for a single document	249
Fig. 9.5	Report for quotations and rename	250
Fig. 9.6	A network of codes, categories and code group	253
Fig. 9.7	Code group manager in ATLAS.ti	254
Fig. 9.8	Forming code groups	254
Fig. 9.9	Creating code groups	255
Fig. 9.10	Importing function	256
Fig. 9.11	A network of code groups	257
Fig. 9.12	A report for a code group report	258
Fig. 9.13	Creating a memo group in memo manager	259
Fig. 9.14	Creating memo group in memo group manager	260
Fig. 9.15	Creating memo group in memo manager	260
Fig. 9.16	A memo writing option	262
Fig. 9.17	A report for memo group	263
Fig. 9.18	Memo group and interpretation of network	263
Fig. 9.19	Network group and comment	264
Fig. 9.20	Creating a network group	265
Fig. 9.21	Adding neighbors to a network	266
Fig. 9.22	Excel report for network group	267
Fig. 9.23	Data classification window	268
Fig. 10.1	Importing survey data into ATLAS.ti 22 & 23	273
Fig. 10.2	Importing survey data into ATLAS.ti versions 22 & 23	274

Fig. 10.3	Codes and comments in code manager in ATLAS.ti 22 & 23	275
Fig. 10.4	Coding options In ATLAS.ti 23	276
Fig. 10.5	Search text 'or' option	277
Fig. 10.6	Treemap in ATLAS.ti 23	278
Fig. 10.7	Stop and go lists	279
Fig. 10.8	Steps in survey data analysis	281
Fig. 10.9	Survey data in ATLAS.ti 23	283
Fig. 10.10	Interpreting survey data	283
Fig. 10.11	Quotation manager and word frequencies	284
Fig. 10.12	Code group, code categories and sub-codes	285
Fig. 10.13	A network of a theme	287
Fig. 10.14	Themes and their relative significance for male and female respondents	288
Fig. 10.15	Themes and their relative significance for male and female respondents in a Sankey diagram	288
Fig. 10.16	Sankey diagram based on C values for themes	289
Fig. 10.17	Project window in ATLAS.ti 23	290
Fig. 10.18	Ai codes and code categories in ATLAS.ti 23	290
Fig. 10.19	Code co-occurrence and code manager	291
Fig. 10.20	A network of organisational practices	292
Fig. 10.21	Sankey diagram of themes	293
Fig. 10.22	Code co-occurrence and code-document analysis	293
Fig. 10.23	Ai summaries and memos	294
Fig. 10.24	Option to convert memos in documents	294
Fig. 10.25	Text report using the edit scope tool	295
Fig. 10.26	Textual report using set and proximity operator	295
Fig. 10.27	A network of a theme and sub-themes	296
Fig. 10.28	Bart chart of theme	297
Fig. 11.1	Import tweets window in ATLAS.ti 22	300
Fig. 11.2	Import and export window of ATLAS.ti 23	301
Fig. 11.3	Tweets, codes, and code groups	302
Fig. 11.4	Coding tweets in ATLAS.ti	302
Fig. 11.5	Word tree, and concepts in ATLAS.ti 23	304
Fig. 11.6	The concepts window in ATLAS.ti 23	304
Fig. 11.7	Collection and analysis of twitter data	305
Fig. 11.8	Code co-occurrence table, bar chart and quotations	306
Fig. 11.9	Code co-occurrence table, and smart code	307
Fig. 11.10	Theme generation and Sankey diagram	307
Fig. 11.11	A thematic network	308
Fig. 11.12	Social network comments window	309
Fig. 11.13	Supported social networks	310
Fig. 11.14	Comments from Youtube	310
Fig. 11.15	Social comments in quotation manager	311
Fig. 11.16	Code groups and comments	311
Fig. 11.17	Youtube comments, codes and code groups	312
Fig. 11.18	Code groups and codes	312
Fig. 11.19	Concepts and quotations	313
Fig. 11.20	Quotation manager for positive sentiment	313
Fig. 11.21	Network for positive sentiment	314

Fig. 11.22	Text report for positive sentiment	314
Fig. 11.23	Youtube comments in ATLAS.ti 23	315
Fig. 11.24	YouTube comments with AI coding	316
Fig. 11.25	Ai codes, code categories and group	317
Fig. 11.26	Code category and codes	317
Fig. 11.27	Themes and report option	318
Fig. 11.28	A theme network	318
Fig. 11.29	Sentiment analysis	319
Fig. 11.30	A sentiment network	319
Fig. 12.1	Focus group data analysis framework	325
Fig. 12.2	Flowchart for focus group discussion analysis	325
Fig. 12.3	Focus group coding pattern	326
Fig. 12.4	Adding codes in focus group discussion coding	326
Fig. 12.5	Framework for data analysis (Ritchie & Spencer, 1994)	326
Fig. 12.6	Coding of focus group data	336
Fig. 12.7	Focus group coding with Pattern 1	337
Fig. 12.8	Focus group coding with Pattern 2	337
Fig. 12.9	Focus group coding-custom pattern	338
Fig. 12.10	Focus group coding network for custom coding	338
Fig. 12.11	Network of focus group codes	339
Fig. 12.12	Word tree view	340
Fig. 12.13	Focus group code co-occurrence	340
Fig. 13.1	Textual report of video data	347
Fig. 13.2	Text report and quotation manager	348
Fig. 13.3	Coding video data and transcript	350
Fig. 13.4	Video document view in ATLAS.ti	351
Fig. 13.5	Using manual transcripts in multimedia documents	352
Fig. 13.6	Multimedia and automated transcripts	352
Fig. 13.7	Snapshot of a video frame	354
Fig. 13.8	Screenshot of the video project	355
Fig. 13.9	Code groups and codes list	355
Fig. 13.10	Sankey diagram of code document table	356
Fig. 13.11	Video coding and quotations	356
Fig. 13.12	Quotation manager and comment	357
Fig. 13.13	Emerging trends, Sankey diagram and quotation	357
Fig. 13.14	Theme and sub-themes for changing trends	358
Fig. 13.15	Theme and sub-themes for endorsers' personalities	358
Fig. 13.16	Network of a theme	359
Fig. 13.17	Geo document in ATLAS.ti 23	360
Fig. 13.18	Document and quotation manager	360
Fig. 13.19	Geo documents and snap shots	361
Fig. 13.20	Edit comment window	362
Fig. 13.21	Quotation manager and report options	363
Fig. 13.22	A network of my professional journey	364
Fig. 13.23	A network of my academic journey	364
Fig. 13.24	A network of academic and professional journey	365
Fig. 14.1	Intercoder agreement window	372
Fig. 14.2	Flowchart in team project	376
Fig. 14.3	Codes, definitions and code groups in excel format	379

Fig. 14.4	Importing codebook into ATLAS.ti	380
Fig. 14.5	Codes imported into ATLAS.ti	380
Fig. 14.6	Snapshot of a project	382
Fig. 14.7	Snapshot of projects in ATLAS.ti 22	383
Fig. 14.8	Pre-merge summary for inter-coder agreement	384
Fig. 14.9	Project merge option	384
Fig. 14.10	Merge report for two coders	385
Fig. 14.11	Adding coders, documents and codes for Ica	385
Fig. 14.12	Inter-coder analysis results	386
Fig. 14.13	Excel report for inter-coder analysis	386
Fig. 15.1	Flow diagram for referencing	396
Fig. 15.2	Literature review in ATLAS.ti	397
Fig. 15.3	Steps for conducting a literature review in ATLAS.ti	398
Fig. 15.4	Reference manager option	399
Fig. 15.5	Reference manager import selection list	401
Fig. 15.6	Imported research project from Zotero	401
Fig. 15.7	Text search options	404
Fig. 15.8	Regular expression and text search result	404
Fig. 15.9	Regular expression and text search window	405
Fig. 15.10	Named entity recognition option	405
Fig. 15.11	Named entity recognition window	406
Fig. 15.12	Named entity recognition result window	407
Fig. 15.13	Named entity recognition coding window	407
Fig. 15.14	Concepts window	408
Fig. 15.15	Word cloud and word list window	408
Fig. 15.16	Concept and apply proposed code window	409
Fig. 15.17	Treemap for selected documents	411
Fig. 15.18	Regular expression search	411
Fig. 15.19	Code lists for documents	412
Fig. 15.20	Code co-occurrence analysis	412
Fig. 15.21	Sankey diagram for code co-occurrence analysis	412
Fig. 15.22	Code document analysis	413
Fig. 15.23	Sankey diagram for code document analysis	413
Fig. 15.24	Query tool	414
Fig. 15.25	Query tool and report for query	414
Fig. 15.26	Code co-occurrence and code-document analysis	415
Fig. 15.27	Code co-occurrence table	416
Fig. 15.28	Concepts and apply proposed code window	417
Fig. 15.29	Project search and regular expression search view	420
Fig. 15.30	Concept and articles association window	421
Fig. 15.31	Concepts, query tool, scope and text report	422

List of Tables

Table 1.1	A comparison of qualitative and quantitative research (Firestone, 1987; Merriam, 1998; Potter, 1996)	6
Table 1.2	Perspectives in qualitative research	7
Table 1.3	Approaches to qualitative research (Creswell & Poth, 2016)	13
Table 1.4	Types of sampling method in qualitative research	24
Table 1.5	A framework for presenting and interpreting findings	27
Table 2.1	Observation types	35
Table 2.2	A comparison of hermeneutic and transcendental phenomenology	40
Table 2.3	Framework for qualitative research	55
Table 3.1	Steps in collection of field data	79
Table 3.2	Types of interview questions	85
Table 4.1	In vivo coding	106
Table 4.2	Axial codes	107
Table 4.3	Major approaches to coding in grounded theory	110
Table 4.4	Approaches to coding in qualitative research	111
Table 4.5	Codes, code categories, axial codes and selective codes	113
Table 4.6	Code and categories	115
Table 5.1	A framework for data analysis framework using ATLAS.ti	129
Table 5.2	Data analysis process	132
Table 5.3	Guidelines for making field notes	162
Table 5.4	Coding field observations	164
Table 5.5	From codes to smart groups	174
Table 5.6	Stages of coding and description	175
Table 6.1	Symbols and their meaning	191
Table 7.1	Colors and descriptions	213
Table 7.2	Excel report of absolute and relative values gender-wise	218
Table 7.3	Binarized excel report	219
Table 7.4	Codes and theme relations	223
Table 9.1	Sub-codes, code categories and code groups	252
Table 10.1	Survey concepts mapped in ATLAS.ti and their meanings	274
Table 10.2	The survey questions	282

Table 10.3	Themes, code categories, sub-codes and groundedness	286
Table 12.1	Conducting a focus group discussion	324
Table 12.2	Focus group coding pattern	331
Table 12.3	Focus group coding patterns	332
Table 13.1	Transcription software compatible with ATLAS.ti	349
Table 14.1	Percentage method of inter-coder agreement	388
Table 15.1	Auto-coding options	403
Table 15.2	References for documents	415

CHAPTER 1

Introduction

Not everything that counts can be counted; not everything that can be counted counts
(Albert Einstein)

Abstract This chapter provides an overview of qualitative research. It discusses the characteristics, and philosophical underpinnings of qualitative research and its characteristics, comparing it, where necessary, with quantitative research. Five critical approaches to qualitative research are explained with the help of suitable examples: grounded theory, case study, phenomenology, narrative, and ethnography. Research paradigm and various perspectives (symbolic, ethnomethodological, and psychoanalytic) of qualitative research perspectives, thick and thin description facilitate researchers in determining the direction of their research. Approaches to qualitative research, meaning of exploration and steps in qualitative data analysis have been explained.

MEANING AND IMPORTANCE OF QUALITATIVE RESEARCH

Qualitative research is a mode of naturalistic inquiry that seeks an in-depth understanding of individual experiences, events, and social phenomena within their natural settings (Denzin, 1971). The focus of qualitative studies is on the "why" of an experience, or a social phenomenon, rather than the "what". Thus, the emphasis of qualitative research is on gaining a comprehensive or holistic understanding of the social setting in which the research is conducted, and how it shapes meaning-making.

Our social world is a complex one. Different people view the same thing differently which depends on their socioeconomic backgrounds and individual experiences, awareness and educational levels, ideological and philosophical orientation, and several other influences. Qualitative Research is important

© The Author(s), under exclusive license to Springer Nature Switzerland AG 2024
A. Gupta, *Qualitative Methods and Data Analysis Using ATLAS.ti*,
Springer Texts in Social Sciences,
https://doi.org/10.1007/978-3-031-49650-9_1

because it enables a deeper understanding of the phenomena, experiences, and contexts, as well as their interactions. Qualitative research allows the researcher to raise and investigate questions that are complex and cannot be easily studied in terms of numbers.

The test of the researcher's abilities is in how the experiences and feelings of the subjects are conveyed to the reader, which must be as close to the way the researcher experienced them during interactions with her/his subjects. Therefore, the language of qualitative research findings should appear as natural as possible. This is possible when the researcher captures the participants' shared experiences in the way they described them. People like to use narratives and metaphors while talking about their experiences (Polkinghorne, 2005) and thus, researchers need to grasp the meanings of words and expressions in the context of their study, and then document them. Metaphors vary from culture to culture and are specific to the language they are used in Lakoff and Johnson (1980).

Qualitative research seeks to study meanings in subjective experiences. The relation between subjective experience and language is a two-way one: language is used to express meaning; but it also influences how meaning is constructed. But giving words to experiences can be a challenge for both researcher and respondent. The right words and expressions can help build a better understanding between the researcher and the respondent. Properly worded feelings, and the way they are explained, are necessary for quality of research data. Some linguists even state that social reality, as experienced by a person, is unique to that person's language; those who speak different languages would perceive the world differently (Ahrens & Chapman 2006).

The authenticity of the findings is enhanced when there is a minimum gap between captured and shared experiences about the phenomenon. The researcher captures the information shared by the respondent and conveys this information to audiences in such a way to appear natural, with minimum/no gap between what the respondent shared and what the researcher conveyed.

Qualitative Research Example In the present context of the debate over the treatment of transpersons, a researcher may consider it as important to study how transpersons perceive the world and the factors that influence their perception. The researcher will use a qualitative research approach to gain a comprehensive understanding. This would include interacting with transpersons, as well as the people who interact with them in various ways. The interactions may be done in various ways: interviews, conversations, observing their daily lives, and even non-verbal communication with the participants. The researcher will then try to discover patterns of information emerging from these interactions that will help in gaining insights into the various dimensions of the lives of the transpersons: sociocultural, experiential, socioeconomic, and psychological. The findings and interpretations are shared with an interested audience.

Characteristics of Qualitative Research

Qualitative research takes place in a social context and has a few distinct characteristics:

1. *It is conducted in a natural setting* (Creswell, 2009; Denzin & Lincoln, 2005). Social reality emerges from an interaction between the researcher and respondent(s) in a given context. Social reality exists in all cultures and concerns the conditions influencing human experience and social practice. All lived experiences occur in social reality. In Qualitative studies, data is collected in the field from the subjects who experienced the issues or problems under study. The researchers in the study talk to people, observe their behaviours, and interact with them over time to obtain data for their study.
2. *The researcher is a key instrument in the study* (Lofland et al., 2006) and in fact, is inseparable from it. Qualitative researchers collect data by themselves using various means: examining documents, observing behaviours, and interviewing the subjects of their study. The study 'instruments', which may consist of open-ended questions, are designed by the researchers themselves. Key responses must be consistent with research inquiry methods. They should be appropriately classified according to demographic, contextual, and environmental factors. The patterns in meaning-making, social relationships, and related life circumstances may be associated with individual biographies and life histories.
3. *Qualitative Research Uses Multiple methods* (Creswell & Poth, 2016). As a rule, qualitative researchers use multiple forms of data, such as observations, interviews, open-ended questionnaires, and so on. The data is reviewed and organized into categories or themes from which researchers make meaning. The reason for using more than one method is to capture deeper and holistic information which may not be possible with a single approach.
4. *Qualitative research calls for more complex reasoning* (Thorne, 2000). Patterns, categories, and themes (recurring features in the data that are relevant to the research question) are built from the "bottom-up" by organizing the data inductively. Inductive processes involve researchers working back and forth between the themes and the database until comprehensive insights are obtained. Sometimes, deductive thinking is also applied to build the themes that are constantly being checked against the data. The inductive-deductive process requires the researcher to use complex reasoning skills in the study.
5. *The participants make meaning* (Hancock et al., 2001). Throughout the qualitative research process, the researcher's focus is on learning the meaning that the participants (not the researchers) make about the problem or issue. These meanings will vary from participant to participant, suggesting multiple perspectives and diverse views on a topic. Thus,

the themes identified by the researcher should reflect these views and perspectives.
6. *Qualitative Research Design is emergent* (Williams, 2008). This means that it is not possible for the researcher to strictly adhere to the initial research plan. The researcher must adopt a flexible approach and be ready to modify their methods without losing sight of the aim of the study, which is to learn about the problem or issue from participants.
7. *Qualitative research Calls for Reflexivity* (Palaganas et al., 2017). In qualitative studies, researchers "position themselves", meaning that they explain the reasons and rationale for their interest in the subject of their study, as well as their perspective and how it informs the interpretation of their data, and what they expect to gain from it. Wolcott (1994) explains reflexivity thus: "Our readers have the right to know about us. And they do not want to know whether we played in the high school band. They want to know what prompts our interest in the topics we investigate, to whom we are reporting, and what we personally stand to gain from our study".
8. *The reflexivity of accounts,* as defined by Garfinkel, stresses how ethnographic findings are inextricably linked to the members' ways of making sense of and shaping social activities (Weider, 1974). Reflexivity requires an awareness of the researcher's contribution to constructing meanings attached to social interactions and acknowledging the possibility of the investigator's influence on the research. Several practices are associated with the "doing" of self-reflexivity. Self-reflexive researchers make notes about others' reactions to them. They also include themselves in writing reflexive notes. Use of the first-person (e.g., "I said," or "They reacted to my questions by...") is not only allowed but even encouraged. Using the first person, "I," reminds the reader of the researcher's presence and influence.
9. *Qualitative research aims at a holistic understanding of the subject of study* (Ormston et al., 2014). This involves reporting multiple perspectives and identifying the many factors that affect a situation or people's experiences. The researcher avoids being bound to study cause-and-effect relationships; instead, they try to study the complex interactions of the factors.

The Philosophical Foundations of Qualitative Research

The philosophical foundations of a qualitative study reveal the researcher's commitment and implications for research design and methodology. There are several schools of thought in qualitative research: positivism, post-positivism, constructivism, advocacy, and pragmatism.

- *A positivist researcher* takes a scientific approach to their study, with emphasis on the collection of empirical data, which is based on prior hypotheses, through a series of logically related procedures. The researcher is convinced that there is one objective reality. The primary objective is to generate functional relationships between causal and explanatory variables to derive logical truths.
- *A post-positivist researcher* also believes in a single reality but draws from social constructionism to understand reality. Researchers use multiple methods to identify a valid reality because all methods are imperfect. The ontological perspective of qualitative research embraces multiple truths and realities (Erlingsson & Brysiewicz, 2013). Individuals understand reality in different ways that reflect their perspectives. Qualitative research is based on the subjective and looks at human realities instead of the concrete realities of objects.
- *Constructivist researchers* strive to comprehend the phenomenon's subjective meaning. It employs more open-ended inquiries to elucidate the process of interpersonal interactions. The researcher interprets the phenomenon via the cultural, socio-economic, and demographic lenses of their subjects, as well as through the researcher's own cultural and social experiences. It is also referred to as interpretive research.
- *The advocacy strategy* of researchers covers the issues confronting marginalised groups in society, organisations, educational, and media environments, among others. By providing action plans, researchers hope to improve the lives of individuals and the organisations in which they live and work.
- *Pragmatist researchers* are concerned with the outcomes of their work—the actions, circumstances, and repercussions of their research. Researchers are free to employ whatever methodologies, strategies, and processes they see are appropriate to meet the needs of their subjects. A pragmatic researcher frequently employs a hybrid strategy because their concern is with what works and how it may be addressed. Interpretation of data and results of analyses necessitate the use of social, political, cultural, and historical lenses to comprehend the issues and propose answers.

Difference Between Qualitative and Quantitative Research

The key objective of qualitative research is to explore ideas to generate new theories or concepts. It necessitates a debate on—and understanding of—the topic of study to generalize the concepts that emerge from the analysis. While quantitative research is oriented toward the generalization of the findings, qualitative research emphasizes understanding. Generalization is possible only after the researcher has arrived at an understanding of the topic (or issue) of the study, Therefore, the question of using the results of a qualitative study to generalize the outcome is fundamentally a misleading one. The nature of qualitative research is not to generalize but to understand. This is why qualitative studies are exploratory—they seek to understand concepts in their specific contexts. However, not all qualitative studies are exploratory. For example, case study research is not a form of exploratory research. Similarly, research conducted on large samples in an exploratory manner can be quantitative. Exploratory research is often qualitative and primary in nature. It is used to investigate a problem that is not clearly understood or defined.

On the other hand, Quantitative research is usually deductive. A classic example of a deductive approach is the research hypothesis. A hypothesis is a statement that links variables already known to the researcher (from prior research or extant literature on the subject). The relationship between variables should be presumed to be known. Thus, if the research involves the study of motivation and high performance, the variables are already known. In such cases, the researcher's coding preference will be deductive for seeking information from the data related to motivation and high performance. Accordingly, codes are assigned to the selected segments.

By using a deductive approach, the researcher can prove a point by explaining it concerning existing theories or concepts. Here, the explanations provided by the researcher revolve around the hypothesis for either supporting or rejecting it.

A summary of the characteristics of qualitative and quantitative research, highlighting their differences, is given in Table 1.1 which provides an overview of the characteristics of qualitative and quantitative research, showing the differences between the two types, including their research designs and the methodological approaches, as well as their philosophical underpinnings.

Table 1.1 A comparison of qualitative and quantitative research (Firestone, 1987; Merriam, 1998; Potter, 1996)

	Qualitative	*Quantitative*
Focus	Quality (features)	Quantity (how much, numbers)
Philosophy	Phenomenology	Positivism
Method	Ethnography/observation	Experiments/Correlation
Goal	Understand, find meaning	Prediction, test hypothesis
Design	Flexible, emerging	Structured, predetermined
Sample	Small, purposeful	Large, random representation
Data Collection Methods	Interviews, observation, documents, and artifacts	Questionnaires, scales, tests, inventories
Analysis	Inductive (by the researcher)	Deductive (using statistical methods)
Findings	Comprehensive, description, detailed, holistic	Precise, numerical
Researcher	Immersed	Detached

Research Paradigm

A research paradigm is variously defined as:

- "The set of common beliefs and agreements shared between scientist about how problems should be understood and addressed" (Kuhn, 1970)
- "Research paradigms can be characterized by the way Scientists respond to three basic questions: ontological, epistemological and methodological questions" (Guba, 1990)
- "Social scientists can ground their inquires in any number of paradigms. None is right or wrong, merely more or less useful in a particular situation. They each shape the kind of theory created for general understanding" (Babbie, 2020). According to Kuhn (1970), paradigm contains "universally recognized scientific achievements that, for a time, provide model problems and solutions for a community of researchers" i.e.,
  - what is to be observed and scrutinized.
  - the kind of questions that are supposed to be asked and probed for answers in relation to this subject.
  - how these questions are to be structured.
  - how the results of scientific investigations should be interpreted.
  - how is an experiment to be conducted, and what equipment is available to conduct the experiment.

Research Perspectives

Approaches to Qualitative Research are governed by various perspectives. Each perspective is underpinned by a different theoretical assumption and how the object of inquiry is understood, as well as methodological focus. Thus, while choosing the perspective, researchers should examine the concepts of their research framework and the goal they want to achieve.

The three main perspectives are presented in Table 1.2 which is followed by a short description of each.

Table 1.2 Perspectives in qualitative research

Symbolic Interaction and Phenomenology	*Tries to understand subjective meanings and attributions by individual sense*	1st perspective
Ethnomethodology and Constructivism	Is interested in the everyday routine of the subjects and their construction of social reality	2nd perspective
Structuralist or Psychoanalytical	Proceeds from an assumption of latent social configurations and unconscious psychic structures and mechanisms	3rd perspective

Symbolic Interaction and Phenomenology

Symbolic interactionism is the process of communicating based on individual experiences about the phenomenon and its meanings. People communicate their subjective viewpoints and social worlds using language and symbols. Individuals imbue objects, events, and behaviors with subjective significance. They act in accordance with their beliefs and not merely what is objectively true. The individual interacts with society and other individuals and creates meaning through interpretations and viewpoints (Mead, 1934). Symbolic interactionism is a dynamic theoretical perspective that views human actions as constructing self, situation, and society. It assumes that language and symbols play a crucial role in forming and sharing our meanings and actions. It views interpretation and action as reciprocal processes, each affecting the other. Individual and collective actions and meanings are consequential. Symbolic interactionism inspires theory-driven research and leads to new theoretical implications. It opens our views of the meanings, actions, and events in the worlds we study.

Both ethnomethodology and symbolic interactionism employ a *verstehen* strategy for comprehending social reality. Verstehen is a German word that means empathic understanding of human behaviour. It is a methodological framework founded upon the principles of Verstehen. However, because each perspective asks different questions, they arrive at various responses and, consequently, "understandings" of varying types. Both are influenced by pragmatic philosophy, which encourages the articulation of truth claims regarding the practical organization of daily life. Ethnomethodologists are concerned with everyone's interpretation strategies to realize (observable) reality. However, their emphasis on the member's interpretative techniques detracts from symbolic interactionists' concentration on negotiated (social) activity.

Perspectives on ethnomethodology and symbolic interaction diverge (Wilson, 1970). While symbolic interactionists question the definition of situations, how do rules of interaction emerge? How is societal stability possible? (Becker, 2017; Denzin, 2017). Ethnomethodologists inquire how individuals perceive, describe, and propose a definition of the situation. How do individuals make a rule emergent, and how do they employ it? And how do individuals view and describe stable social action? (Zimmerman & Wieder, 2017). Three examples of symbolic interaction are discussed below.

Example of Symbolic Interaction Two employees may have radically different perspectives on their organization's promotion processes, which depends on their understanding, interactions, and experiences with the context, people, and culture. This understanding determines their worldview. One employee may view the promotion process positively and express happiness with it, while the other may view the phenomenon negatively and express dissatisfaction with it. Both employees have distinct perspectives on the same phenomenon.

This is an example of two employees' subjective perspectives of the same phenomenon based on their experience and interactions with contexts, people, and cultures. In business organizations, the perception of unfairness in the promotion process might create a subjective perspective for everyone. Individuals receiving promotions will demonstrate their happiness and appreciate the people, culture, and environment.

Conversely, an individual missing promotion will demonstrate unhappiness about the process, people, and environment. People form mental models of phenomena, develop their perspectives, and accordingly communicate with others and their natural environments. Symbolic interactionism is concerned with the subjective perspectives of individuals as formed by their social and cultural environments.

Example The general manager might evoke a heart-warming emotion in someone who has had mostly exciting experiences with general managers. But another person, who general managers may have reproved, may feel fear and disgust. Here the general manager is a symbol, and the meaning of a symbol is interpreted through social interaction depending on individual context.

Example Why do some people engage in unethical conduct when they are aware that it will lead to their downfall? The definition of the working environment that individuals create constructs their beliefs. They feel content and secure and communicate their positive outlook to coworkers and superiors. Thus, the symbolic significance of unethical behavior supersedes the facts regarding unethical behavior and the associated risk.

ETHNOMETHODOLOGY AND CONSTRUCTIVISM

Ethnomethodology provides a phenomenological perspective on the real world. It focuses on how people comprehend or perceive the real world (Sudnow,). Symbolic interactionism views the individual as "making sense" of intransigent reality within the interacting matrix that constrains the entire spectrum of experience. The factual reality significant to the symbolic interactionist is comprised of the actors' interpretations of things, events, and people, which are components of the social structures they construct. The word Ethnomethodology has Latin roots: *Ethno* (meaning people), *methodus* (way of teaching), and *logos* (to study). It is the study of ordinary members of society in everyday situations and how they use common sense. Researchers study social interaction to understand the local culture. It requires the researcher to use common sense in local context. In other words, researchers should seek meaning for local words used by respondents in their contexts. However, in unfamiliar situations, understanding may require gaining experience of doing things the local way.

Ethnomethodologists are interested in uncovering taken-for-granted rules. They collect data using open-ended, in-depth interviews, participant observations, documentaries, as well as the personal experiences of their subjects.

The focus of Ethnomethodology is on social interactions, particularly on the methods of sense-making and reality-constituting processes. Methods of interpretation in ethnomethodology are usually discourse, document, conversation and genre analysis.

On the other hand, phenomenology focuses on the cognitive aspects of sense-making in the natural world and of abstract mental phenomena. Phenomenology is the study of the lived experience of human beings. It is the ways people perceive and understand phenomena. Phenomenology, within psychology (phenomenological psychology), is the psychological study of subjective experience.

Example In one study, the researcher is interested in understanding the 'look-busy' culture in organizations, and how it is perceived in the workplace. Often, the 'look-busy' culture is regarded positively by management and seen to have significant effect on employees' professional development. For the study, the researcher may employ a variety of data collection techniques, including in-depth interviews, focus group discussions, informal interaction with respondents, and the researcher's own experience with non-verbal communication, to unravel the phenomenon and management's rationale for the 'look busy' phenomenon. The researcher will aim at elucidating management views of the phenomenon and the implications of this viewpoint for individuals and organizations.

STRUCTURAL AND PSYCHOANALYTICAL

Research perspective followed under psychoanalytical and genetic structuralism is hermeneutic analysis of underlying structure. Researchers record interactions, collect photos and films for the subject under investigation. The fields of application are usually gender, generation, biographical and family research.

There are distinct reconstruction perspectives in qualitative research methodology:

- The attempt to describe general fundamental mechanisms that actors use in their daily life to "create" social reality as is assumed, for instance, in ethnomethodology.
- Investigations of the first type provide information about the methods used by everyday actors to conduct conversations, overcome situations, structure biographies, etc.
- Investigations of the second type provide object-related knowledge about subjectively significant connections between experience and action, views on such themes as health, education, politics, social relationships, responsibility, destiny, life plans, inner experiences, and feelings.

Thick and Thin Description

Thick description is the technique in the ethnography of providing the contextual meaning of people's behavior by emphasizing their words, and actions and interpreting the social meaning of it (Geertz, 1973). An ethnographer must present not only facts but also a commentary and their interpretations of the meanings. On the contrary, a thin description is a factual account without any interpretation of contextual factors. Thick description is the expression of culture. Thick description seeks to present and explore the multifaceted complexities of the situation under investigation, the intentions and motivations of the actors involved, and the context of the situation (Given, 2008). Investigations of the thick description provide information about the methods used by everyday actors to conduct conversations, overcome situations, structure biographies, etc.

Thick description is an approach in ethnographic research that attempts to understand the phenomena under investigation by searching out and analysing symbolic forms, such as words, images, and people's behaviours. Using thick descriptions, the researchers describe the context of the research. As a result, readers gain an understanding of the situation and context under the study. Researchers should use multiple data sources, interviews, and repeated observations to gain rich data so that interpretations offer enough description. In other words, researchers should provide descriptions of narratives and various examples between the researcher, respondents, and their contexts. A prolonged engagement with respondents, the researcher might develop intimate engagement with respondents, and personal biases may shadow the description (Given, 2008, p. 692). The goal is to offer a cultural description of the phenomenon so that readers understand the situation. Thick description includes unusual, deviant, or unexpected information to gain insights into the phenomenon and explore the path of future research.

According to Ponterotto (2006), the thick description describes and interprets observed social action within the study context (p. 543).

According to Denzin (1989), a thick description is a detailed and contextual description of the phenomenon, including emotion and webs of social relationships that connects people in that context (p. 83).

According to Schwandt (2001), interpretation of the phenomenon makes it thick (p. 225).

Opposite to thick description is the thin description, which is a superficial account that does not explore cultural members' underlying meanings (Holloway, 1997, p. 154).

Investigations of the thin description provide object-related knowledge about subjectively significant connections between experience and action, views on such themes as health, education, politics, social relationships, responsibility, destiny, life plans, inner experiences, and feelings.

Example From a structural standpoint, the researcher can explain the research findings with a thick description. Thick description is the process of describing and interpreting observed phenomena about their natural surroundings. A thick description is a record of individuals' subjective explanations and meanings, which may contain voices, feelings, actions, and meanings.

In the example of the 'look busy' culture in business organisations, where people appear busy without doing any work, the researcher can capture their words, actions, and justifications for their actions in their contexts. The phenomenon can be described by providing a variety of situations including deviant cases.

- After working hours, some people gaze at their computer screens and appear busier than normal working hours.
- Some people appear uncomfortable while leaving the office. They look at people gazing at computer screens.
- Some people stand outside the manager's cabin keeping several papers in their hands.
- Few people don't bother about people sitting after working hours. They leave the office with confidence on their faces.
- During working hours, people spend much time gossiping and spending time on mundane work.
- All these happen after working hours.

Accordingly, the researcher can justify the emergent themes by categorizing selected responses on contextual, sociocultural, and other parameters. The interpretations will include the researcher's comments about the phenomenon, which can further be categorized according to the hierarchy. In the example, the researcher provides the finer details of the phenomenon and contextual factors. This helps readers to visualize the context where the phenomenon takes place.

Approaches to Qualitative Research

The main approaches to qualitative research are summarized in Table 1.3. A brief description of each approach is provided after the table.

The selection of the research method depends on the type of problem being studied and its context, the objectives of the study, and the conceptual framework developed by the researcher.

- *Narrative Research*. Narrative Research involves analysis of the research subjects' accounts of their perceptions or experiences to offer an interpretation. Narrative research aims at an in-depth study of the meanings that people make of their experiences. Data for narrative research is obtained from a small sample of participants, usually through open-ended interviews and questions. Sometimes, narrative research is also done through the analysis of written documents.

- *Phenomenology*. A phenomenological study is qualitative research in which the researcher describes the lived experiences of individuals. The focus of this method is on understanding the phenomenon or event that has impacted the subject(s) of the study.
- *Grounded Theory*. Grounded theory "discovers theory" in systematically obtained data. Although it is primarily a tool used in social science research, it is widely applied in other disciplines as well. The term 'grounded' means that the theory is grounded in field data. Data collection is generally field based, where interviews play a significant role in the data collection process. However, other techniques like observations, multimedia resources, and documents are also useful. Grounded theory is not about describing the phenomenon. The researcher analyses data to determine if they fit in with existing theories or generate concepts to explain their findings.
- *Ethnography*. Ethnography is a qualitative research design where the researcher describes and interprets a culture, the study group's shared and learned patterns of values, behaviors, beliefs, and language (Harris, 1968). An ethnographic study is a systematic description of people and cultures, including habits, customs, beliefs, and even differences among them. An ethnographer observes and studies through immersion in the settings of the subject people to study their day-to-day lives. The researcher lives with (or among) the subjects, interviewing them, observing their lives, and interacting with them. Data collected in an ethnographic study will include, besides the researcher's notes and participant responses, pictures and images, and artifacts that are critical to the understanding of the topic of study.

Table 1.3 Approaches to qualitative research (Creswell & Poth, 2016)

Characteristics of each research type	Narrative research	Phenomenology	Grounded theory	Ethnography	Case study
Focus	Explores the life of an individual	Understanding the essence of the experience	Developing theory or theories that are grounded in empirical data	Describing and interpreting the characteristics of a culture-sharing group	Developing an in-depth description and analysis of a case or multiple cases

(continued)

Table 1.3 (continued)

Characteristics of each research type	Narrative research	Phenomenology	Grounded theory	Ethnography	Case study
Type of problem best suited for design	The need to tell stories of individual experiences	The needing to describe the essence of a lived phenomenon	Grounded theory in the views of participants The theory is grounded in the data	Describing and interpreting the shared patterns of the culture of a group	To gain an in-depth understanding of a case or cases
Discipline/ Background	Drawn from the humanities, including anthropology, literature, history, psychology, and sociology	Drawn from philosophy, psychology, and education	Drawn from sociological studies	Drawn from anthropology and sociology	Drawing from psychology, law, political science, medicine
Unit of analysis	One or more individuals	Several individuals with shared experiences	Study of a process, action, or interaction that involves many individuals	Study of a group that shares the same culture	Study of an event, program, or activity, more than one individual
Data collection methods	Primarily interviews and documents	Primarily interviews with individuals, but documents, observations, and art may also be considered data	Primarily interviews with 20–60 individuals	Observations and interviews, but other data sources can also be accessed in extended field studies	Multiple sources, such as interviews, observations, documents, artifacts
Approach to Data analysis	Analyzing data for stories, "restoring" the stories, and developing themes, often chronologically	Analyzing data for significant statements, meaning units, textual and structural description, and description of the "essence."	Analyzing data through open coding, axial coding, selective coding	Analyzing data through a description of the culture-sharing groups; themes that define the group's characteristics	Analyzing data through the description of the case and themes of the case as well as cross-case themes

(continued)

Table 1.3 (continued)

Characteristics of each research type	Narrative research	Phenomenology	Grounded theory	Ethnography	Case study
Written reports	Develop narratives about an individual's life	Describe the "essence" of the experience	Generate a theory illustrated in a figure	Describe how a culture-sharing group works	Detailed analysis of one or more cases
General structure	Introduction (problem, questions) Research procedures (narrative, significance of individual, data collection, analysis outcomes) Report of stories Individuals theorize about their lives Narrative segments identified Patterns of meaning identified (events, processes, epiphanies, themes) Summary (Adapted from Denzin, 1989, 2001)	Introduction (problem, questions) Research procedures (phenomenology and philosophical assumptions, data collection, analysis, outcomes) Significant statements Meanings of statements Themes of meanings An exhaustive description of the phenomenon (Adapted from Moustakas, 1994)	Introduction (problem, questions) Research procedures (grounded theory, data collection, analysis, outcomes) Open coding Axial coding Selective coding and theoretical propositions and models Discussion of theory and contrasts with extant literature (Adapted from Strauss & Corbin, 1990)	Introduction (problem, questions) Research procedures (ethnography, data collection, analysis, outcomes) Description of culture Analysis of cultural themes Interpretation, lessons learned, questions raised (Adapted from Wolcott, 1994)	Entry vignette Introduction (problem, questions, case study, data collection, analysis, outcomes) Description of the case/cases and their context Development of issues Description of the selected issues Assertions Closing vignette (Adapted from Stake, 1995)

- *Case Study*. According to Yin (2003), the case study approach is used to answer the how and why questions about the phenomenon of study. In contexts requiring a case study approach, the conditions influence the phenomenon under investigation, and boundaries between phenomenon and context are unclear. With this approach, the researcher can test theoretical models in real-world situations. A case study can include individuals (single or several), cases or cases, events, or events. The selection of a case (or cases) depends on the purpose and objectives of the study. Many researchers (Denzin & Lincoln, 2011; Merriam, 1998; Yin, 2003) describe case studies as an inquiry strategy. Case studies are widely used in various disciplines, particularly in the social sciences. They entail an in-depth investigation of an individual, group, or event. Depending on the research question being addressed, case studies can be descriptive or explanatory. They are used to gain an in-depth, multi-faceted understanding of a complex issue in a real-life context.

Meaning of Exploration

The investigation can be qualitative or quantitative (Glaser & Strauss, 1967). Exploration and exploratory research include both types of data, regardless of their proportion and significance in any one study or the entire chain of studies.

When researchers have little or no scientific knowledge about the group, process, activity, or situation they want to investigate but have reason to believe it contains elements worth discovering, they explore. To effectively investigate a given phenomenon, they must approach it with two distinct mindsets: flexibility in the search for data and openness to what they will find and where they can be found. The main goal of exploratory research is to produce inductively derived generalizations about the group, process, activity, or situation under study as a result of these procedures. The researcher then weaves these generalizations into a grounded theory that explains the object of study, the construction of which is best described in a series of publications by Glaser and Strauss (1967) and Glaser (1978). During exploration, both quantitative and qualitative data may be collected. But qualitative data predominates in most exploratory studies.

It is more accurate to classify exploration as inductive and confirmation as deductive. In other words, researchers do think deductively during exploratory studies, but they do so primarily within their emerging theoretical framework rather than within established theory and a set of hypotheses deduced from it. They engage in confirmation, but it is their emergent generalizations that are confirmed rather than the predictions they may have made earlier.

Confirmatory researchers, on the other hand, sometimes observe regularities that lead to generalizations about the group, process, activity, or situation they are investigating, despite research design constraints. Some of these chance discoveries may have been reached through inductive reasoning, but such induction is not systematic in confirmatory work. Thus, social science researchers who want to be absolutely clear about the nature and scope of their research should describe it as qualitative-exploratory, quantitative-exploratory, qualitative-confirmatory, or quantitative-confirmatory. Exploration and inductive reasoning are important in science because deductive logic alone will never yield new ideas or observations (Stebbins, 2001).

Steps in Qualitative Research

A qualitative study begins with an exploration of the subject of the researcher's interest, usually a search for concepts and theories in extant literature, which may help the researcher find the answers to the questions raised by them. The absence of satisfactory answers to the research questions motivates the researcher to conduct a deeper study of the topic of interest. The process followed provides a logical path for understanding the concept, resulting in the creation of new knowledge about the subject of study.

Regardless of the approach and methods used, a qualitative study broadly follows the steps shown in Fig. 1.2. However, depending on the research objectives, variations can and do occur. In descriptive research, the goal is to describe, highlight, and comprehend the phenomenon, while explanatory research seeks to analyze, explain, and examine the issues under investigation. A third type, exploratory research, seeks to discover, understand, and explore the issues under investigation. Depending on the purpose of the research, researchers may use inductive, deductive, or both methods.

Figure 1.1 shows the steps followed in data analysis using inductive and deductive methods.

An exploratory study employs an inductive approach that begins with research questions and progressing to data collection, analysis, findings, and the development of themes. The study ends with construction of a theory. Explanatory and descriptive research employs the deductive method, which is theory driven. It begins with research questions/hypotheses, then moves

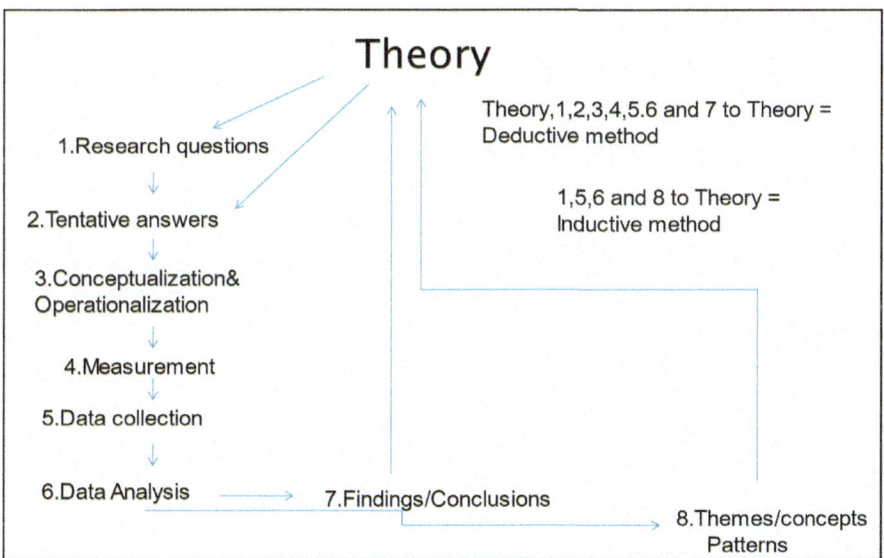

Fig. 1.1 Steps in inductive and deductive methods

Fig. 1.2 Steps in qualitative research

on to developing concepts and operational definitions of the concepts under investigation.

The steps in a qualitative study, using appropriate measurement techniques, data collection tools, sampling, fieldwork and data collection, data analysis method, findings, and confirming/rejecting theory, are described in Figs. 1.1 and 1.2.

The Problem to Be Investigated

This is determined by the researcher's interest and, based on their understanding, the need for an in-depth study of the lived experiences of an event or phenomenon, which cannot be explained with existing knowledge. The absence of satisfactory answers to the research questions is the primary motivation for a deeper study of the topic of interest.

> *Important* The author of this book conducted an extensive literature on employee morale for his Ph.D. He found that most studies on morale were based on military contexts in post-World War II European settings. These studies were mostly survey-based and in their conclusions, the authors recommended a qualitative approach for future research on the subject. This author, who is based in India, reviewed studies on employee morale in business settings and found significant research gaps, indicating both the need for qualitative research on the topic, as well as its potential for contributing to existing knowledge. A pilot study was first conducted to understand the contexts and components of employee morale in businesses. Based on the findings of previous studies and the pilot study, the author developed the methodology for qualitative research of employee morale, which included an open-ended questionnaire.

Justifying the Need for a Study

Justification of the need for research is based on a comprehensive literature review of the topic. Drawing on previous studies on the subject, the researcher highlights the gaps in existing knowledge, and then tries to explain how the study will address these gaps and improve understanding, both at a conceptual level as well as, if the context requires, to inform policy, advocacy, and actions aimed at improving the situation.

DEVELOPMENT OF CONCEPTUAL FRAMEWORK

Qualitative researchers are often called upon to explain their motives for their study. On the face of it, the question may seem a simple one. While responses may vary, they must convince the audience of the researcher's understanding of the topic, as well as the rationale of their approach. We will try to examine this further.

Existing knowledge might not be sufficient to satisfy the researcher's need to know more about the characteristics of the phenomenon of interest. Another reason is that no prior knowledge may be available on the subject, which then becomes the motivation for a study. Moreover, extant literature may suggest a qualitative approach for addressing the issue of lack of enough knowledge about the subject.

Qualitative research always takes place in a social context in which the researcher aims to understand their subjects' interactions and experiences with an event or phenomenon. Since the aim of the study would be to understand, the researcher will not use well-defined concepts or hypotheses. Unlike in quantitative studies, hypotheses may emerge only from the themes identified in the study. These hypotheses are then tested for generalizability in a larger social context.

A qualitative study begins with the researcher exploring the subject of researcher's interest, usually with an assessment of concepts and theories in existing literature. At this stage, the researcher is looking for answers or explanations for the questions raised by them. The absence of satisfactory answers to the research questions provides the motivation and rationale for a deeper study of the topic of interest. The results and findings of the study are expected to contribute new insights and knowledge.

In qualitative research, the researcher is inseparable from the study and therefore, must document their observations and experiences to understand the meaning the study's participants make of their contexts. The observations made by the researcher are the data that is analyzed to produce a set of results and findings with which the researcher will present their insights into the topic.

Since qualitative research is context-specific, data is collected in natural settings. The field experiences of the researcher are a crucial component of research. They are an important part of the data and hence must be documented by the researcher.

Researchers should understand and apply three important concepts that improve rigor and quality of the research. They are "Bracketing, Sensitizing Concepts, and Reflexivity".

Edmund Husserl developed bracketing, also known as epoche, as a phenomenological reduction process that could lead to the ideal description and understanding of the universal essence of the investigated phenomenon. It is a fundamental methodological concept and scientific process in which a researcher suspends their assumptions, biases, theories, or previous experiences in order to see and describe the essence of a specific phenomenon. Although researchers from various philosophical, epistemological, or theoretical traditions may use bracketing based on their own perspectives and divergent meanings, the fundamental components of bracketing remain consistent.

Creswell and Creswell (2003) defined bracketing as a method by which researchers separate their personal experiences from the study. Researchers refrain from passing judgment on the ordinary, everyday way of seeing things (Moustakas, 1994). Researchers use the bracketing method to reduce the potentially negative effects of unacknowledged preconceptions about the research and, as a result, to increase the rigor of the research.

Researchers should recognize they are contributing to knowledge. Hence, their prior experiences, assumptions, and views can influence the study process and outcomes. Reflexivity ensures the integrity of reporting and the accuracy of research. It involves analysis of the researcher's judgments, as well as the data gathering and reporting processes.

Writing memos throughout data collection and analysis as a means of examining and reflecting on the researcher's engagement with the data is one method of bracketing (Cutcliffe, 2003).

Herbert G. Blumer coined the term "sensitizing concept," which allows researchers to focus on developing empirically grounded concepts. It is derived from the participants' point of view, using language or expressions that sensitize the researcher to potential lines of inquiry. They are social constructs that help in the investigation of other social settings. Sensitizing concepts are words, phrases, or symbols used in context to explore fruitful lines of inquiry and analysis. It enables researchers to enter the realm of the studied population, shedding light on context-specific meanings.

Sensitizing concepts are derived from symbolic interactionism on an epistemic level. In his book, 'Symbolic Interactionism: Perspective and Method', Herbert G. Blumer focuses on notions that stimulate the senses. Based on the works of pragmatist philosophers like James Dewey, symbolic interactionism emphasizes experience and interaction about the subject under investigation. To comprehend this experience, the researcher also needs to comprehend the significance that social actors imbue their beliefs and actions. Sensitizing concepts are a necessary and logical methodological consequence of this premise. Sensitizing concepts are distinctive, natural phrases used within an investigated community that the researcher can use to produce more generic social constructs that are beneficial for investigating different social environments. Sensitizing concepts are concepts that sensitize the researcher to capture the meaning of an emergent concept/word/thought, facilitating the discovery of the concept's meaning in the current situation. It aids researchers in developing social constructs that may be used to investigate other social environments.

In qualitative research, three key benefits are associated with using sensitizing concepts. First, they are a crucial methodological tool for entering the world of meanings of a studied group. Second, they provide the ability to transcend the inherent difficulty of collecting unique case-specific data. Thirdly, they allow the researcher to focus on generating empirically grounded notions.

According to this methodology, the ideal method for capturing meanings is to employ concepts that contain the words and ideas that study subjects use to explain their reality. Sensitizing concepts are constructs derived from the study participants' perspective, utilizing their language or expressions, making the researcher aware of potential avenues of investigation. They are distinct, natural concepts used by an examined population, which the researcher might utilize to build more general social constructs useful for analyzing various social environments.

Reflexivity is a foundational element of bracketing. Reflexivity is necessary for identifying and acknowledging one's own biases and assumptions before approaching a study (Ahern, 1999). The major purpose of reflexivity is for the researcher to recognize the influence of their biases on the outcomes of the investigation. Ruby (2016) defines reflexivity as "the capacity of any system of meaning to refer back to itself, so becoming its object." It entails recognizing that researchers are a part of the social context they examine (Frank, 1997). This understanding is the product of a genuine evaluation of the values and interests that may influence research (Porter, 1993). Researchers should not permit their presumptions to influence the data collection process or impose their interpretations on the data. This method is called bracketing (Crotty, 1996). Bracketing is a method for establishing the validity of the data gathering and analysis methods, removing researcher bias from the study (Porter, 1993). Bracketing is the first step of "phenomenological reduction," in which the researcher sets aside as many prior experiences as humanly possible to comprehend better the study participants' experiences (Moustakas, 1994).

Collection of field data is followed by its transcription. The transcription must include observation notes, conversion of audio and audio-visual information, and data from other modes into a textual form. The transcription must include both data types: the participants' (or subjects') responses and the researcher's field notes. Data transcription is the first step to its systematic arrangement.

Formulating Aims and Objectives

The researcher must adhere to the framework of the research's design. The strategy for data collection and analysis is determined by the type and nature of problems the researcher wants to understand. The research design describes data needs and how the data will be collected. The data collection methods and instruments are decided accordingly. Research design includes a statement of purpose, the context of the study, methodology and data collection methods, measurement methods, and data analysis techniques.

Data collection methods in qualitative research include open-ended questions for eliciting descriptive responses, interviews (audio, video), observations, and focus groups. Data analysis can be performed manually or using computer-assisted qualitative data analysis software. Researchers must know qualitative research methods, concepts, rationale, skills, and working knowledge of the software used for analysis.

This book explains the use of ATLAS.ti for analyzing qualitative data. ATLAS.ti uses textual reports, networks, bar charts, Sankey diagrams, code co-occurrence, and code-document tables to help qualitative researchers to analyze data and present the results of the analysis.

Research Design

Research design is an essential component of qualitative research. It does not just describe the methods for data collection, analysis, and report writing. Research design covers the entire research process, from conceptualizing a problem to writing the narrative, (Bogdan & Taylor, 1975). According to Yin (2003), the research design for a qualitative study is the logical sequence that connects the empirical data to a research question and its conclusions. It addresses the need for philosophical and theoretical perspectives on the quality and validation of a study. The essential components of the research design for any study are (i) the introduction to the study, (ii) methods of data collection, analysis, and representation, (iii) report writing, and (iv) validation of results. The research design process must also include the researcher's worldview, paradigms, sets of beliefs, and assumptions made by the researcher while deciding on the study.

Research design is the process of transforming a research idea into a research project. Three essential components of research design are theoretical, methodological, and ethical considerations.

Sampling Strategy

Qualitative research requires a non-probability sampling method that is used in exploratory research, i.e., convenience, quota, voluntary response, purposive, and snowball sampling types. In convenience sampling, the researcher gathers data from whatever cases can be conveniently accessed. In the quota sampling method, the researcher selects cases from within different subgroups. For snowball sampling, the researcher relies on participant referrals to recruit new participants.

The purposive sampling method is used by the researcher to select only those respondents who can provide the data and information required for the understanding of central phenomenon (Creswell). A study of senior managers' motivation in the pharmaceutical industry may place senior managers, chief managers, AGM, DGM, and GM in separate subcategories to ensure that the researcher includes subcategories in the sample, with adequate representation for gender, age groups, and seniority.

Table 1.4 shows the type of sampling in qualitative research and basic description.

Data Collection

There are four approaches to data collection: observations, interviews, documents, and audio-visual materials (Creswell & Poth, 2016). The researcher can play the role of observer, participant, insider, or outsider. Data is collected and documented in the field notes diary. The time spent before and after the data collection with the respondent is the period of observation. The researcher can also get access to the settings which they can observe as an insider. Even when located outside the context, the researcher can act as the observer. The researcher is also an observer during interviews and while engaging in informal interactions.

In a qualitative study, the researcher collects data from unstructured and semi-structured, focus groups, email, face-to-face, online focus groups, telephonic, open-ended, audiotape/videotape interviews, which are then transcribed verbatim.

The researcher should maintain a journal throughout the course of the study. The journal should include, among others, letters from participants and

Table 1.4 Types of sampling method in qualitative research

Sl. No	Type of sampling	Purpose
1	Maximum variation	Considers the wide variations in population characteristics, and identifies significant common patterns
2	Homogeneous	Reduce variation, simplify analysis, and facilitating group interviews
3	Critical case	Permits logical generalization and maximum application of information to other cases (One case/small number of cases can be chosen to gain insight about similar other cases)
4	Theory-based	Finds examples of a theoretical construct and thereby elaborate on and examine it
5	Confirming and disconfirming cases	Confirms the importance and meaning of possible patterns, seek exceptions, look for variation
6	Snowball or chain	Identifies cases of interest from people who know people who know what cases are information-rich
7	Extreme or deviant case	Learn from highly unusual manifestations of the phenomenon of interest
8	Typical case	Highlights what is normal
9	Intensity	Information-rich cases that manifest the phenomenon intensely but not extremely
10	Politically important	Attracts desired attention or avoids attracting undesired attention
11	Random purposeful	Adds credibility to the sample when the potential purposeful sample is too large
12	Stratified purposeful	Illustrates subgroups and facilitates comparisons
13	Criterion	All cases that meet some criterion; useful for quality assurance
14	Opportunistic	Follow new leads, taking advantage of the unexpected
15	Combination or mixed	Triangulation, flexibility; meets multiple interests and needs
16	Convenience	Saves time, money, and effort but at the expense of information and credibility

Source Miles and Huberman (1994, p. 28). Qualitative data analysis: A sourcebook of new methods, (2nd, d.). Thousand Oaks, CA: Sage

public documents such as official memos, minutes, archival materials, management notes, autobiographies, biographies, photographs, videotapes, and other documents relevant to the research.

DATA ANALYSIS

Data analysis for textual, video, audio, photo, and images are driven by the research questions or objectives of the study. The researcher reads and identifies the segments of information in the data that are of relevance or interest, highlights them, and codes them appropriately. ATLAS.ti permits inductive, deductive, In-vivo, quick, and axial coding to be used (Refer to Saldana's book to know more about coding).

After coding, the next task of the researcher is to look for patterns in the information. Homogeneous and related codes can be merged, linked, or grouped. If a code can hold more than meaning, the researcher may consider splitting the code into more codes. Then the codes may be categorized and grouped depending on the way they are related to each other. While creating code categories or groups, it is preferable to give them thematic names so that the meanings of the codes, or code categories, for the group are clear. Depending on the data size and information emerging from codes, the researcher can show relations between code groups and code categories for creating themes. In ATLAS.ti, themes are identified from smart codes, code categories, code groups, and smart groups.

Document categories can be formed anytime during the data analysis process. The classification of documents should be done so that themes can be analyzed across contextual, demographic, cultural, and social parameters.

ATLAS.ti facilitates creating a textual report, networks of themes, excel reports, Sankey diagrams, code co-occurrence, and code-document tables for data analysis. Data analysis using ATLAS.ti is explained in Chaps. 5–8.

INTERPRETATION AND DISCUSSION OF RESULTS

Themes should be interpreted based on their significance to the study. Themes emerge from codes, field notes, researchers' understanding of the data, and associated challenges. Themes are supported by key quotations (significant information) and memos based on the researcher's field observations about the research questions and observations of the context. The themes identified by the researcher, and their interpretation, should reflect the nature of the subject or phenomenon of study.

In addition, themes should also be linked with the research questions. The researcher supports the themes significant quotations and memos. In ATLAS.ti, the significance of themes is "measured" by their groundedness and c-values (Chap. 4). The researcher must also show the association between theme/s and research questions.

Theme-to-theme relations are shown by their association with the research question/s, which are then supported by selected quotations and memos. A theme-to-theme relation helps researchers to understand the significance of themes.

The interpretation of themes may be based on the context of the study, considering various influencing factors: personal, social, and professional demographic, socio-cultural, socio-economic and others. Interpretation of themes helps readers, researchers, policymakers, managements gain insights into an issue.

PRESENTING THE REPORT

The themes identified are linked to the research's objectives and their significance explained with the help of with supporting quotations, memos, networks, tables, and appropriate diagrams. Contextual factors like socio-economic and socio-demographic can be used to analyze the emerging themes. While presenting the report, researchers should explain how their findings will help respondents, organisations, stakeholders, policymakers, and management. They should also explain how issues are interrelated, organized in order of significance and how relevant steps can address the prevailing issues in the organization. Appropriate suggestions and recommendations may also be made.

The report (or thesis) should also address limitations of study. Deviant cases should be highlighted, with the researcher explaining how deeper understanding can be gained by examining such cases. While presenting the report, protocols to address ethical concerns should be followed and researchers should ensure that no personal, professional, and social harm is caused to the respondents and the organisations (or institutions or communities) they belong to. They should take adequate care to protect any identifiable information about respondents and organisations.

Table 1.5 shows the framework for reporting and interpreting qualitative research findings. Themes should be reported and interpreted using research questions. Short narrations or significant pieces of information that support themes should be reported and interpreted. The researcher should create contextual interpretations of findings and their interconnections. By doing this, the reader can visualize the phenomenon, their contexts, and cultural settings. While making interpretations, researchers should use field experiences (memos) to strengthen the findings. Since themes are based on pattern emerging from data, researchers should also report critical issues emerging from the data analysis that don't find place in pattern. They should show the importance of such issues and suggest the direction of future research.

It is vital to present pictures, networks, diagrams, screenshots, tables, and visualization techniques to support the significance of themes. Criteria such as credibility, validity, ethical challenges, vulnerability, bias should be adequately addressed while reporting and interpreting qualitative research findings.

Table 1.5 A framework for presenting and interpreting findings

	Data Types	Significant Segments	Codes	Code-categories (Sub-themes)	Themes	Future direction
Research Questions	Responses	Observations	Examples	Types		
	Text	Field-notes	Sentence/Word	Inductive	Attributes	Constructs	Deviant cases
	Audio	Shyness/avoidance	Paragraph	Deductive	Relationships	Code categories relationship
	Images	Non-verbal clues	Contextual information	In-vivo	Sentiments	Core concept	Newly emerged issues
	Geodata Social media Video	Excitements Anxieties	Clipping/image Location Field-observations	Abductive Hybrid		

Recommended Readings

Ahern, K. J. (1999). Ten tips for reflexive bracketing. *Qualitative Health Research, 9*(3), 407–411.

Ahrens, T., & Chapman, C. S. (2006). Doing qualitative field research in management accounting: Positioning data to contribute to theory. *Handbooks of Management Accounting Research, 1*, 299–318.

Babbie, E. R. (2020). *The practice of social research*. Cengage AU.

Becker, H. S. (2017). Practitioners of vice and crime. In *Pathways to data* (pp. 30–49). Routledge.

Bogdan, R. C., & Taylor, S. J. (1975). Introduction to qualitative research methods: A phenomenological approach to the social sciences. *No title*.

Creswell, J. W., & Creswell, J. (2003). *Research design* (pp. 155–179). Sage Publications.

Creswell, J. W., & Poth, C. N. (2016). *Qualitative inquiry and research design: Choosing among five approaches*. Sage Publications.

Creswell, J. W. (2009). *Research design. Qualitative, quantitative, and mixed method approaches* (3rd ed.). SAGE Publications.

Crotty, M. (1996). *Phenomenology and nursing research*. Churchill Livingston.

Cutcliffe, J. R. (2003). Reconsidering reflexivity: Introducing the case for intellectual entrepreneurship. *Qualitative Health Research, 13*(1), 136–148.

Denzin, N. K. (1971). The logic of naturalistic inquiry. *Social Forces, 50*(2), 166–182.

Denzin, N. K. (1989). *Interpretive biography* (Vol. 17). Sage.

Denzin, N. K. (2001). *Interpretive interactionism* (Vol. 16). Sage.

Denzin, N. K., & Lincoln, Y. S. (Eds.). (2011). *The Sage handbook of qualitative research. sage*.

Denzin, N. K. (2017). Symbolic interactionism and ethnomethodology. In *Everyday life* (pp. 258–284). Routledge.

Denzin, N. K., & Yvonna S. L. (2005). Introduction. The discipline and practice of qualitative research. In N.K. Denzin & Y.S. Lincoln (Eds.), *The Sage handbook of qualitative research* (pp. 1–32). SAGE Publications.

Erlingsson, C., & Brysiewicz, P. (2013). Orientation among multiple truths: An introduction to qualitative research. *African Journal of Emergency Medicine, 3*(2), 92–99.

Frank, G. (1997). Are there life-after categories? Reflexivity in qualitative research. *The Occupational Therapy Journal of Research, 17*(2), 84–98.

Firestone, W. A. (1987). Meaning in method: The rhetoric of quantitative and qualitative research. *Educational researcher, 16*(7), 16–21.

Geertz, C. (1973). *The interpretation of cultures: Selected essays*. Basic Books.

Given, L. M. (Ed.). (2008). *The Sage encyclopedia of qualitative research methods*. Sage Publications.

Glaser, B. G., & Strauss, A. L. (1967). *The discovery of grounded theory*. Aldine.

Glaser, B. G. (1978). *Theoretical sensitivity*. University of California.

Guba, E. (1990). *The paradigm dialog*. Sage.

Hancock, B., Ockleford, E., & Windridge, K. (2001). *An introduction to qualitative research*. Trent Focus Group.

Harris, M. (1968). Emics, etics, and the new ethnography. In *The rise of anthropological theory: A history of theories of culture* (pp. 568–604)

Holloway, I. (1997). *Basic concepts for qualitative research*. Basic Books.

Kuhn, T. S. 1970a. Logic of discovery or psychology of research. I. Lakatos & A. Musgrave (Eds.), *Criticism and the growth of knowledge* (pp. 1–23). Cambridge University Press.

Lakoff, G., & Johnson, M. (1980). The metaphorical structure of the human conceptual system. *Cognitive Science, 4*(2), 195–208.

Lofland, J., Snow, D., Anderson, L., & Lofland, L. (2006). *Analyzing social settings: A guide to qualitative observation and analysis* (4th ed.). Wadsworth.

Mead, G. H. (1934). *Mind, self, and society* (Vol. 111). University of Chicago press.

Merriam, S. B. (1998). *Qualitative Research and Case Study Applications in Education. Revised and Expanded from "Case Study Research in Education."*. Jossey-Bass Publishers, 350 Sansome St, San Francisco, CA 94104.

Miles, M. B., & Huberman, A. M. (1984). Qualitative data analysis: A sourcebook of new methods. In *Qualitative data analysis: a sourcebook of new methods* (pp. 263–263).

Moustakas, C. (1994). *Phenomenological research methods*. Sage Publications.

Ormston, R., Spencer, L., Barnard, M., & Snape, D. (2014). The foundations of qualitative research. *Qualitative Research Practice: A Guide for Social Science Students and Researchers, 2*(7), 52–55.

Palaganas, E. C., Sanchez, M. C., Molintas, V. P., & Caricativo, R. D. (2017). Reflexivity in qualitative research: A journey of learning. *Qualitative Report, 22*(2).

Polkinghorne, D. E. (2005). Narrative psychology and historical consciousness. *Narrative, identity, and historical consciousness* (pp. 3–22).

Ponterotto, J. G. (2006). Brief note on the origins, evolution, and meaning of the qualitative research concept thick description. *The Qualitative Report, 11*(3), 538–549.

Porter, S. (1993). Nursing research conventions: Objectivity or obfuscation? *Journal of Advanced Nursing, 18*(1), 137–143.

Potter, J. (1996). Discourse analysis and constructionist approaches: Theoretical background. British Psychological Society.
Ruby, J. (Ed.). (2016). *A crack in the mirror: Reflexive perspectives in anthropology.* The University of Pennsylvania Press.
Schwandt, T. A. (2001). *Dictionary of qualitative inquiry* (2nd ed.). Sage.
Stebbins, R. A. (2001). What is exploration. *Exploratory Research in the Social Sciences, 48*, 2–17.
Strauss, A., & Corbin, J. (1990). *Basics of qualitative research.* Sage Publications.
Sudnow, D. (1967). *Passing on; The social organization of dying.* Prentice-Hall.
Sundow, D. (1978). *Ways of the hand.* Harvard University Press.
Thorne, S. (2000). Data analysis in qualitative research. *Evidence-Based Nursing, 3*(3), 68–70.
Weider, D. L. (1974). *Language and social reality.* Mouton.
Williams, J. P. (2008). Emergent themes. *The Sage Encyclopedia of Qualitative Research Methods, 1*, 248–249.
Wilson, T. P. (1970). Conceptions of interaction and forms of sociological explanation. *American Sociological Review*, 697–710.
Wolcott, H. F. (1994). *Transforming qualitative data: Description, analysis, and interpretation.* Sage.
Yin, R. K. (2003). Design and methods. *Case Study Research, 3*(9.2), 84.
Zimmerman, D. H., & Wieder, D. L. (2017). Ethnomethodology and the problem of order: Comment on Denzin. In *Everyday life* (pp. 285–298). Routledge.

CHAPTER 2

Types of Data, Data Collection, and Storage Methods

Abstract The chapter discusses the importance of qualitative data, the various types of qualitative data, and data collection and storage methods. Qualitative inquiry methods include grounded theory, phenomenology-hermeneutic and transcendental, narrative, case study, and ethnography. Data collection techniques in qualitative studies include one-on-one interviews, focus groups, records, documents, archives, and observations. Research diaries and field notes are critical to the data collection process, which are discussed in detail with relevant examples. The chapter also includes a brief discussion of the ethical concerns that researchers may face and suggestions on how they can be addressed. In addition, techniques for the transcription of data, and the information that researchers should keep in mind while transcribing, are also described. Both manual and automated methods of transcription of data are described in this chapter. The chapter also examines the benefits and drawbacks of transcribing processes. Frameworks for qualitative research—from conceptualization of the study to reporting of results and findings—are also explained.

IMPORTANCE OF QUALITATIVE DATA

Qualitative data is non-numerical in nature. It can only be observed and recorded. Qualitative data is important for determining traits and characteristics of the subjects of the study, whether people or phenomena, to understand their experiences from which the researcher can make interpretations in the study's context.

Qualitative Data is collected using various methods: observations, one-to-one interviews, focus group discussions, and others. Researchers collect qualitative data from the field in two ways: i) by recording interviews and discussions, from documents, diaries, and photographs; and ii) by observing nonverbal communication, such as clues from sighs, expressions of surprise, laughter, and anxiety, as well as from the information shared during informal interactions. The information in (i) is explicit and can be retained in its original form, whereas in (ii), the data is implicit, and its value depends on how quickly and completely researchers document their observations. Both forms of data must be collected and analyzed by the researchers.

Qualitative data is mostly exploratory in character and thus, data collection methods are mainly aimed at gaining insights into people's emotions or perceptions, or their motivations for behaving in a certain manner. Since this data cannot be measured, researchers use data collection methods that are largely unstructured and flexible. However, a study with large samples can also be used as a quantitative exploratory approach.

Methods of Collecting Qualitative Data

Obtaining the respondent's consent to share information is essential to the data collection process. The common practice is to contact potential participants (through email, letters, or other methods), explaining the subject of study, its relevance, and expected benefits for the respondents and their organizations (or community). Email correspondence leaves a credible communication trail between the researcher and their correspondents. However, researchers can use other means to obtain respondents' consent for participation in the study. The email should contain information attached as a separate letter, which addresses ethical concerns and is signed by a person in authority. The letter should promise that the respondent's name would not be disclosed; nor would the information, participating individuals' and organization's name be shared, as also any sensitive information that might directly or indirectly cause injury to individual or organizational reputation and image.

After the consent of participants is obtained, the researcher should develop a protocol for data collection and recording, which will be shared with the respondents. As far as possible, observation protocols should not be shared as the participants can become self-conscious and modify their behaviors. In other words, respondent should avoid writing that they will observe the respondent's words, expression, shared information before, during and after the interview.

Data shared during informal interactions can reveal much insightful information that helps the researcher find the truth. The researcher must also be aware of the debate over the capture and use of data obtained during informal discussions. It is the opinion of this author that the duty of the researcher is to capture holistic information. The objective is to help the respondents and organizations and, therefore, capturing information from the informal discussion, and using it in analysis, will help the researcher to study an issue from multiple perspectives. Therefore, all types of information captured during the data collection process should be regarded can be used as long as there is no breach of ethical protocol.

Ethical concerns remain paramount throughout study. While presenting the outcomes of the study, researchers should not reveal respondent-specific information unless they have written permission to do so.

Although information can be shared if individuals and organizations waive confidentiality conditions, it is the researcher's moral and ethical responsibility to ensure that no personal and professional harm results from the disclosure of information.

Data collection involves obtaining permissions and consent, developing a sampling strategy, recording information manually and digitally, capturing field notes, and storing the data (Creswell & Poth, 2016). Researchers should collect data in multiple phases, and from more than one source. They should decide which type of purposeful sampling is the most suitable for their research.

Researchers conducting studies within their organizations must be aware that they may be creating 'dangerous' knowledge (Glesne, 2016). It is research that examines 'your own backyard'. Therefore, Glesne recommends the use of multiple validation strategies to ensure that the findings are accurate and insightful.

Respondents can give their informed consent over email or by signing the consent form. Many organizations have difficult and time-consuming processes in getting management's consent to allow their employees to be interviewed. This approach may not always be feasible. In such cases, the researchers may approach respondents for their consent, informing them that their participation is entirely voluntary and that they are at liberty to withdraw from the study at any time.

Bogdan and Biklen (1997) suggest that site selection should address the following considerations:

- Why has the site has been chosen for study?
- How much time will the researcher spend at the site?
- How will the result be reported?
- What are the expected benefits of the study to the participants (and, where relevant, the organization)?

Collection of primary qualitative data is primarily done with the following methods:

One-on-One Interviews

One-on-one interviews are one of the most commonly used data collection instruments. The researcher (or interviewer) collects data directly from the interviewee. Interviews in qualitative research are informal and unstructured. In fact, they are mostly conversational in nature. The questions are open-ended, and the flow of the interview decides what questions must be asked and when.

Kvale and Brinkmann (2009) suggest seven logical steps for conducting an interview, starting from thematizing the inquiry, then designing the study, interviewing the participants, transcribing the interviews, analyzing interview data, verifying the validity, reliability, and generalizability of the findings, and reporting the results of study. Rubin and Rubin (2012) are of the opinion that the sequence need not be fixed. Rather, they must allow the researchers to change the question, the sites chosen, and the situation to study.

Creswell recommends the following steps which are embedded in the larger research process.

- The first question should be general, open-ended, and focus on understanding the key phenomenon in the study.
- Carefully identify respondents who can best answer the research questions.
- Decide the types of interviews needed to capture the most useful information. The interview could be telephonic, focus group, or one-on-one.
- Every type of interview has its pros and cons. A telephonic interview is suitable when the researcher does not have direct access to respondents. The researcher cannot have informal communications in this type of interview. A one-on-one interview is preferable, where respondents are not hesitant to speak and share ideas. The researcher should determine a setting suitable for one-on-one interviews.
- Focus group interviews are suitable for respondents to cooperate when the time to conduct a one-on-one interview is limited, especially when respondents are hesitant to provide information (Krueger & Casey, 2009). In a focus group, the researcher should encourage all participants to talk and monitor participants who tend to dominate the conversation. However, many respondents can speak at length without providing much useful information. Similarly, many shy respondents may hesitate to speak. This can provide a challenge to the researcher.
- Ensure the quality of recording equipment to record interviews.
- The researcher must refine the interview questions, develop research instruments, assess the degree of observer bias, frame questions, collect background information, and adapt the research procedure (Sampson, 2004).

- The respondent's consent agreement in physical form must always be available with the researcher. The respondent often asks for a physical copy even though the consent has been received through email.
- During the interview, be polite, respectful, and observant, and learn to probe to seek deeper information. Simultaneously, learn to intervene when the respondent loves their voices so much that further questioning becomes difficult.

Observation

In this method of data collection, the researcher is immersed in the settings of respondents, becoming a part of the community, and taking notes of their lives, behaviors, and interactions. Besides taking notes, other documentation methods, such as video and audio recording, photography, and similar methods, can be used.

Four types of observation methods are available for the researcher to choose from for data collection (Table 2.1). The researcher may also change from one type to another as if the situation requires it.

The Observation method is one of the key tools of data collection in qualitative research. The method and type of observation are determined by the objectives of the study and the research questions. While observing their subjects, researchers use all their senses (sight, hearing, touch, smell, and taste) to study the physical settings, participants, interactions among them, their activities, conversations, and behaviors (Angrosino, 2007).

John Creswell suggests the following series of steps for observation.

- Select the site and obtain permission to gain access to the site.
- Identify the components of observations and the time required for them.
- Decide your role as an observer. It is better to practice remaining as an outsider and then as an insider observer.

Table 2.1 Observation types

Observer as an Insider	*Observer as an Outsider*
Complete participant The researcher is fully engaged with the people they are observing (Angrosino, 2007)	Nonparticipant or observer as participant The researcher is outside the study group, watches the subjects, and takes field notes from a distance. People in the study group are aware of the presence of the researcher
Participant as observer The researcher participates in the activities at the study site. The researcher gains insider views but is hindered from recording data while the activity is in progress	Complete observer The researcher is neither seen nor noticed by the people being studied

- Prepare documents for recording notes in the field. This includes descriptive and reflective notes (Angrosino, 2007).
- Describe events and activities, as well as your reflections, including insights, ideas, confusions, hunches, and breakthroughs.
- Thank the participants after the observations are over.
- Prepare your full notes immediately after the observation. Offer a thick and rich narrative description of the people and events under observation.

Focus Groups

The Focus group discussion (FGD) is a frequently used approach to collect qualitative data for an in-depth understanding of social issues. The participants in an FGD are purposefully selected. Group size is usually limited to 6–10 people and discussions are moderated by the researcher.

Focus Group Discussions require considerable researcher skills to control the flow and direction of the conversations, ensuring that they do not stray from the topic, and that disagreements among the participants do not turn into arguments. Hence the researcher should be aware that the moderation challenges increase with group size.

> **Important** There is a crucial difference between Interviews and Focus Group Discussions. In interviews, the researcher's principal role is that of an "investigator" in which the researcher asks questions and engages with one respondent at a time. On the other hand, in an FGD, the researcher's role is that of a "facilitator" or "moderator". Here, the researcher is peripheral to the process, with the role of a facilitator or moderator of the discussions among participants.

Storing Data

Researchers should know how data are organized and stored. Davidson (1996) suggested that backing up information collected and noting changes made to the database is sound advice for all research studies. Researchers are advised to develop a master list of the types of information gathered. While storing data, they must ensure that the anonymity of participants is protected by disguising their names and other information that can, potentially, identify them.

Approaches to Qualitative Inquiry

John Creswell and Poth (2016) describe five approaches to qualitative inquiry and research design. They are grounded theory, narrative study, phenomenology, ethnography, and case study methods. In addition, there are qualitative research analytical methods—Qualitative content analysis, thematic analysis, thematic network analysis, document analysis, sentiment analysis and data mining, discourse analysis, and conversation analysis. These methods are often used in combination, but they can also be used as stand-alone methods.

Narrative Study

"Narrative inquiry is stories lived and told" (Clandinin & Connelly, 2004). In a narrative study, every interview (or observation) tells a story that the researcher reflects upon and presents to the audience with insights. Generally, data for a narrative study is collected through interviews with the participants around the topic of interest. In some studies, data collection might also involve the analysis of documents. The contents of the narrative are presented in story form, which is then interpreted through content analysis.

After collecting data, the researcher creates a demographic profile of the respondents and then codes the data. The data is then analyzed to show, among other things, the demographic characteristics of the subjects of the study: at individual or group levels, and other features. The researcher creates a 'structure of stories about the experiences and perceptions of the respondents in the study while presenting the findings and insights.

Researchers should present stories using plot, chronology, three-dimensional space, and themes (Clandinin & Connelly, 2004) as shown in Fig. 2.1. They should use variables such as setting, characters, situation, context, and epiphany (a sudden revelation emerging from narrative analysis).

Phenomenological Research

Nathanson (1973) believed that experience is the source of all knowledge and thus stressed that researchers must experience "natural objects" before they theorize about them. The role of the researcher is to identify this phenomenon, i.e., an object of human experience (Van Manen, 2016), and collect data from respondents who have experienced the phenomenon and made meaning out of it. Then, the researcher describes the collective experiences of the respondents to explain "what" they experienced and "how" they experienced it (Moustakas, 1994).

In Nathanson's view, collecting data in phenomenological research can be difficult. In a phenomenological study, information is gathered (through interviews, journals, and/or observations) from individuals who have experienced or lived in a situation (or phenomenon). Unstructured interviews are preferred because they allow participants to provide explanations and opinions

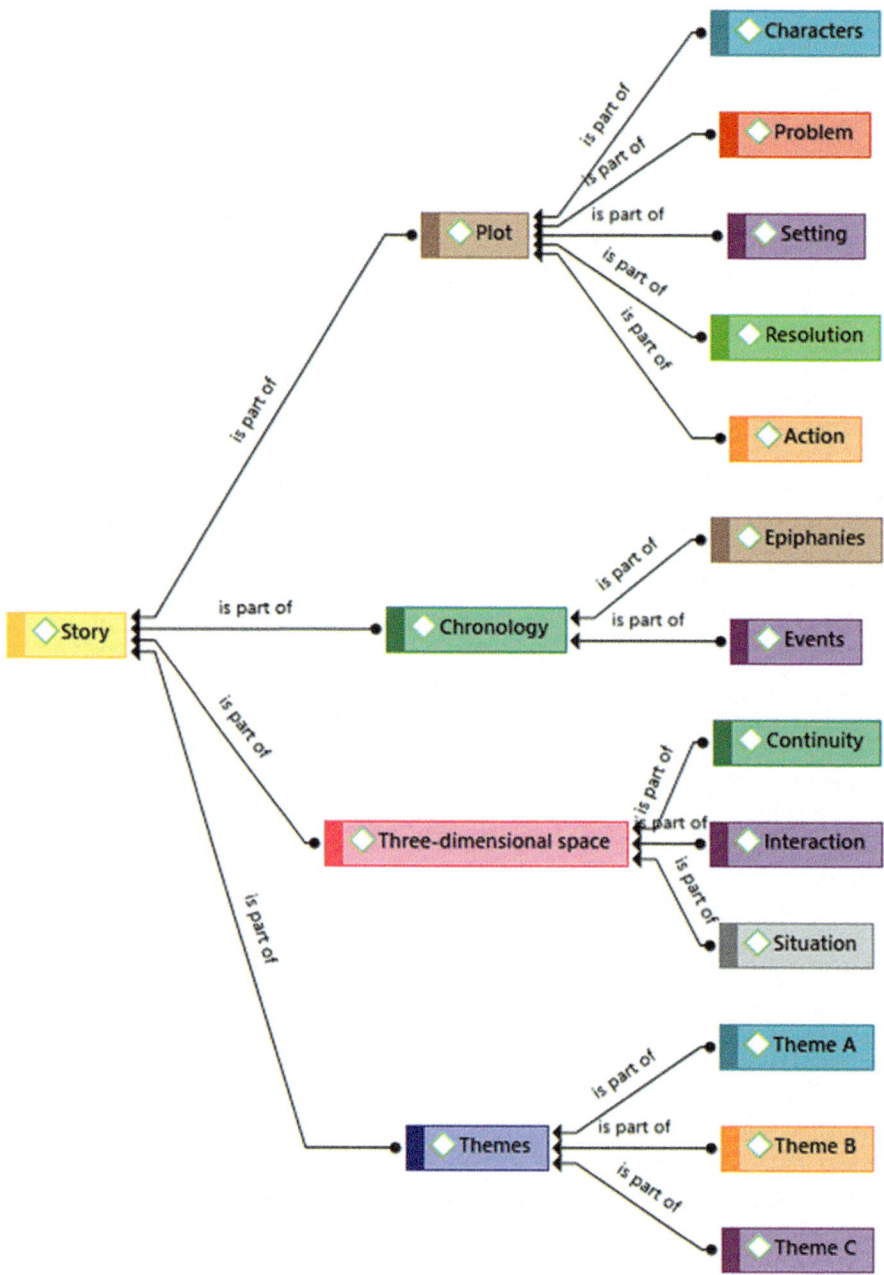

Fig. 2.1 Narrative study framework

in their own words (Tracy & Robles, 2013). Interviews also allow researchers to explore complexities surrounding the phenomenon that may not always be visible. The interviews are recorded and transcribed.

The Two Approaches to a Phenomenological Study

There are two approaches: hermeneutic and transcendental or psychological (Fig. 2.2). Hermeneutic phenomenological studies interpret respondents' lived experiences in life texts (hermeneutics). According to Van Manen, personal experiences, conversational interviews, and close observation are described. Researchers investigate empirical (experiences) and reflective (analysis of experiences) activities
Psychological phenomenology is more about describing participants' experiences (Table 2.2). It is the psychological study of the subjective experiences of respondents with a focus on making sense of the structures of the meaning of the lived experiences of the participants in the study. Moustakas (1994) suggested the following procedure for transcendental phenomenology should follow this sequence: identifying the phenomenon, bracketing out one's experiences, collecting data, analysing data, and developing themes followed by textural and structural descriptions

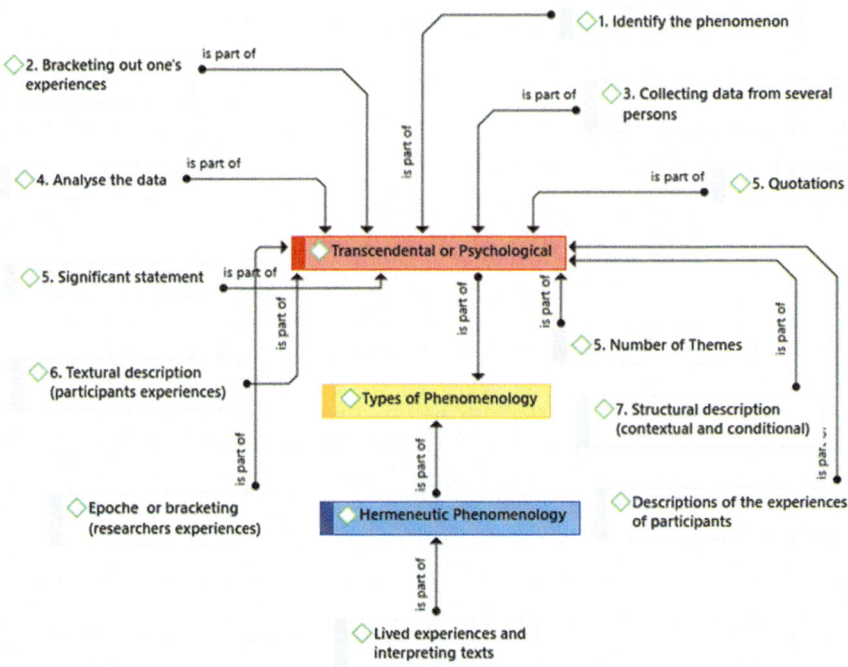

Fig. 2.2 Types of phenomenological research

Table 2.2 A comparison of hermeneutic and transcendental phenomenology

	Hermeneutic phenomenology	Transcendental phenomenology
Phenomenon	To discern the meaning of texts, gestures, and lived experiences of the participants about the phenomenon	To discern the meaning of texts, gestures, and lived experiences of the participants about the phenomenon using the conditional and contextual lens
Researcher	The researcher reflects on field experiences and is part of the study	The researcher reflects on the subjectivity involved during the data collection and analysis outside the study
Data collection and participants	In-depth interviews from respondents between 5 and 25 in number	In-depth interviews from respondents between 5 and 25 in number
Developing themes	Develop themes using key quotations and memos based on the researcher's field experiences	Develop themes using key quotations and the researcher's subjective experience
Theme interpretations	The researcher interprets empirical and reflective accounts of the phenomenon and the participants' lived experiences	The researcher describes the phenomenon, including subjective experiences in data collection and analysis. It brackets the phenomenon and offers textural and structural description using the contextual and conditional lens

Types of Phenomenological Data Analysis Methods

Researchers have described mainly the following types of phenomenology.

Descriptive phenomenology, which is also known as Husserlian or transcendental phenomenology, describes the essential meanings and structures of the phenomenon without preconceptions. The data analysis involves bracketing, horizontalization, clustering, and textualization. Bracketing is keeping aside the researcher's assumptions; horizontalization refers to equal treatment of every statement or expression; clustering refers to grouping/categorizing statements into themes; and textualization refers to writing a comprehensive description of the phenomena and their essences.

Interpretive phenomenology, which is also known as existential or psychological phenomenology, assumes complete bracketing is impossible. Therefore, data analysis involves constructing meaning between the researcher and the respondents. The data analysis involves reading, reflecting, and writing. Reading is identifying significant information, reflecting means analyzing significant information concerning the study's context, objective, and literature, and writing means creating narratives and their interpretations to understand the phenomenon.

Hermeneutic phenomenology, which is also known as Heideggerian or philosophical phenomenology, emphasizes that data analysis is circular and dynamic. It emphasizes constant dialogue between the researchers, respondents, data, and literature. The data analysis involves naïve reading, structural analysis, comprehensive understanding, and critical reflection. Naïve reading is getting a sense of preliminary themes; structural analysis identifies sub-themes; comprehensive understanding synthesizes themes and sub-themes and holistic interpretation of the phenomenon. Critical reflection evaluates interpretations concerning research questions, literature, and implication of the study.

Narrative phenomenology, or Ricoeurian or linguistic phenomenology, emphasizes data analysis creatively and expressively. The researcher uses narratives and language to construct the meaning of the phenomenon. The data analysis involves emplotment, configuration, and refiguration. Emplotment organizes data into a chronological sequence of events or actions; configuration means shaping data into a narrative form in phases: beginning, middle, and end. Refiguration means inviting readers to engage with narratives and interpret them in their ways.

Grounded Theory

Data collection for grounded theory research is done in the field. Interviews are important to the data collection process in this approach. Other techniques like observations, and focus group discussions are also used. In certain situations, sources of data like multimedia and documents are also useful. In grounded theory studies, data collection and data analysis are interactive processes that take place simultaneously. This is because emergent knowledge may call for more data for the validation of theory or theories.

Grounded theory is not about describing (or explaining) the phenomenon. Researchers using this approach should aim to 'discover' the knowledge, concept, or theory that is present in the data (Strauss & Corbin, 1998) or, if it is not available, they should generate one. In addition, researchers should also develop an abstract analytical schema of the process. A schema is an internal representation of the world, an organization of concepts and actions that can revise knowledge about the world. The theory emerges from the field data (Strauss & Corbin, 1998), in which the researcher proposes to explain a process, action, or interaction shaped by a large number of participants (Strauss & Corbin, 1998). Thus, grounded theory generates a theory of actions, interactions, or processes through inter-relating categories of information collected from individuals.

Overview of the Approaches to Grounded Theory Research

There are two widely used approaches to grounded theory research: the systematic procedures of Strauss and Corbin (1990, 1998) and the constructivist approach of Charmaz (2005, 2006). In the first, the researcher conducts, typically, 20 to 30 interviews, collecting data till saturation is reached in each category. Data is also gathered through observations and from available documentation, but such data is not often used by researchers (Creswell & Poth, 2016). However, in the opinion of this author, these kinds of data may be crucial to the study as they contain much valuable information and hence, should be used. The researcher may categorize the data based on distinct features like events, happenings, and instances (Strauss & Corbin, 1990) and compare them with emerging categories. Thus, grounded theory research is a constant comparative method of data analysis. While selecting respondents (or participants) for grounded theory research, the researcher should be careful to select only those who can provide rich, holistic data. This process is known as theoretical sampling

The categories formed by the researcher should be derived from core phenomenon (Strauss & Corbin, 1990). Researchers should develop a conditional matrix that connects conditions influencing the phenomenon under the study. Causal conditions are related to the factors that cause the core phenomenon. Contextual and intervening conditions describe the situational factors that influenced the strategies. Finally, researchers provide contexts to the core phenomenon linked to structure and process

Researchers should create a visual model of categories that are directly associated with the core phenomenon. While coding, the researcher should use the axial coding (discussed in Chap. 4) paradigm. Then, they can use selective coding which helps to develop a model, proposition, or hypothesis. Researchers should support their models or hypotheses through narrative statements of respondents (Strauss & Corbin, 1990), a visual picture (Morrow & Smith, 1995), or a series of hypotheses or propositions (Creswell & Brown, 1992)

The constructivist approach is interpretive in nature. It places a greater emphasis on individuals' perspectives, values, and assumptions. It encourages the development of emergent critical questions. The approach encourages questioning the accepted methodology and takes a reflexive stance. The approach promotes the use of gerund-based expressions

CHALLENGES IN USING GROUNDED THEORY

In a grounded theory study, the investigator must, as far as possible, set aside theoretical ideas or existing notions for the analytic or substantive theory to emerge. Despite the evolving, inductive nature of the qualitative inquiry, the researcher must recognize that grounded theory is a systematic approach to research that follows specific steps for data analysis (Strauss & Corbin, 1990). The researcher must determine when categories have reached saturation and when the theory is sufficiently detailed. One strategy for approaching saturation is to employ discriminant sampling, in which the researchers gather additional data from participants similar to those initially interviewed to determine whether the theory holds for these additional participants. The researcher must understand that the primary outcome of grounded theory is a theory composed of specific components: a central phenomenon, causal conditions, strategies, conditions and context, and consequences. Strauss and

Corbin's (1990, 1998) approach may lack the adaptability that some qualitative researchers desire. In this instance, the Charmaz (2006) approach may be used, which is less structured and more adaptable.

Prerequisite for Grounded Theory Analysis

When there is no theory to explain a process or phenomenon, the grounded theory approach may be the most appropriate. While studies may indicate the suitability of various models, it is possible that they were developed and tested on samples and populations in settings and contexts that are comparable with those being studied. Additionally, theories may exist, but these may be incomplete because they may have overlooked the important variables that may be of interest to the researcher. At a practical level, a theory may be required to explain how people experience a phenomenon. A grounded theory can provide such a framework.

The research questions in a grounded theory study will aim at elucidating how individuals experience the process (What was the process? How did it transpire?). After an initial examination of the responses, the researcher returns to the participants with more specific questions, such as.

- What was central to the process (The central phenomenon)?
- What influenced or precipitated the event or phenomenon (Cause-and-effect relationships)?
- What strategies for research questions were used during the process (strategies)?
- What were the results (consequences)?

Such questions are usually asked in the interviews, but other types of data, such as observations, documents, and audiovisual materials, can also be collected. The researcher's objective is to collect sufficient data to develop (or saturate) the model.

Strauss and Corbin (1990, 1998) advocate a systematic approach in which the researcher develops a theory to explain a process, action, or interaction. It entails collecting data from 20 to 30 interviews until the categories and the field notes are saturated. A category is a collection of events, occurrences, and instances (Strauss & Corbin, 1990). The researcher compares emerging categories across data collection phases, a technique referred to as constant comparative data analysis.

Inductive and axial coding methods are used to create categories that are relevant to the subject under investigation. Then, the code categories are grouped and link to form groups. Each group reflects a broad concept. Finally, the researcher interprets the emergent theory through the use of narrative statements (Strauss & Corbin, 1990), as well as a visual representation (Morrow & Smith, 1995). Narrative statements are key quotations that describe the phenomenon.

Strauss and Corbin (1998) also recommend that the researcher should develop a conditional matrix that connects the micro and macro conditions influencing the study phenomenon. Such a matrix will encompass the community, region, and demographic situation. By incorporating a conditional matrix into the grounded theory, the systematic approach to grounded theory is completed. In other words, researchers provide contexts to the central phenomenon. It includes creating code categories linked to structure with process.

Charmaz (2005, 2006) proposed a constructivist approach that advocates a social constructivist perspective in which the researcher highlights the diverse local worlds, multiple realities, and complexities of specific worlds, perspectives, and actions with this approach.

The constructivist approach of grounded theory is interpretive in nature. It places greater emphasis on individuals' perspectives, values, beliefs, feelings, assumptions, and ideologies. As with the systematic approach, field notes, memo writing, and theoretical sampling are important. Since complex terms, diagrams, and conceptual maps obstruct the development of grounded theory in a systematic approach, Charmaz, (2005) advocated for the use of active codes (Source: Charmaz Approach in John Creswell book), such as gerund-based expressions. Next, any conclusions reached by grounded theorists are suggestive, incomplete, and inconclusive. Strauss and Corbin (1990, 1998) serve as illustrations of grounded theory procedures because their systematic approach (Fig. 2.3) is instructive for individuals learning about and applying grounded theory research.

In a grounded theory study, the researcher uses an open (inductive) coding approach. After coding, relationships between codes are established. This process is known as axial coding. A logic diagram is prepared, which helps to investigate the causal conditions that influence the phenomenon, identify the context and intervening conditions, and their effects on the phenomenon.

The last step is selective coding. Researchers link code categories to one core category that represents the central thesis of the research. In selective coding, the researcher may write a 'story line' that connects the categories. Finally, the researcher visualizes a conditional matrix to show the social, historical, and economic conditions that influence the main phenomenon. Although not mandatory, this step helps the qualitative researcher to consider the model from the narrowest to broadest perspective. The results of data analysis are presented by the researcher as a substantive theory.

The theory emerges from the memos written by the investigator who notes down ideas about the evolving theory, which emerges from open, axial, and selective coding. The substantive theory can then be empirically verified with quantitative data to determine whether generalization is possible and valid.

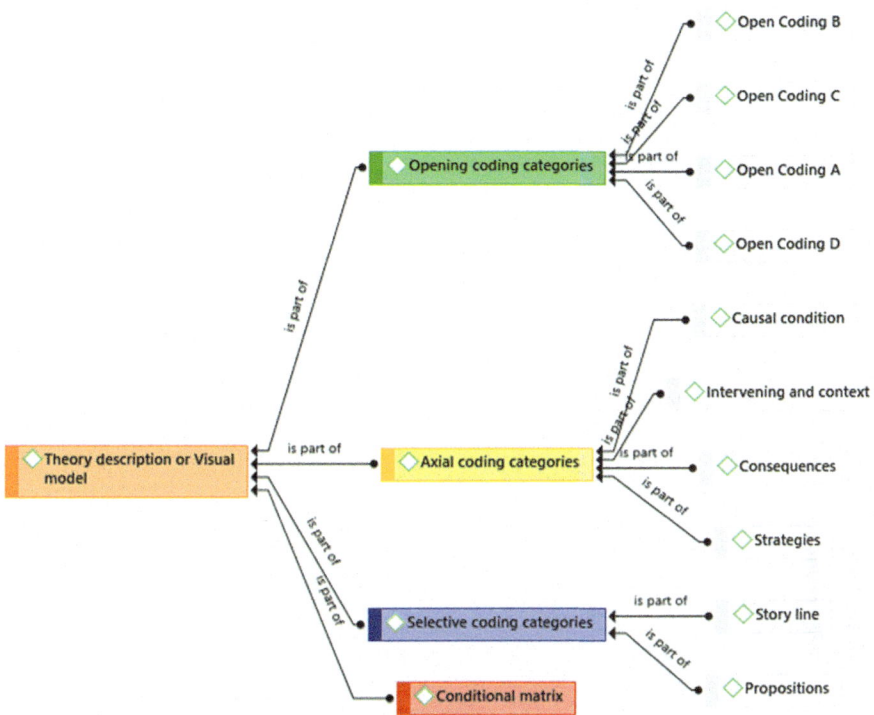

Fig. 2.3 Systematic approach to grounded theory research

Figure 2.4 shows the data analysis process for the constructivist approach in grounded theory research. Two types of data can be used: primary data from interviews, photos, documents, diaries, etc., and the data in personal observations made in field notes. The researcher codes the data and shows relationships between codes. While coding, the researcher also mentions significant and vital information captured in the field notes (Writing field observation is known as memo writing). After coding, the codes are placed in categories based on characteristics, attributes, and meanings emerging from codes. In Fig. 2.4, the codes have been categorized as causal conditions, strategies, intervening and contextual conditions, and consequences.

In the next step, the researcher does selective coding by linking code categories with the appropriate relationships. If the axial coding process is used, the researcher creates relationships between code categories. Finally, themes are generated from the relationships between codes.

Ethnographic Research

In ethnographic studies, the researcher collects 'naturalistic' data (Fig. 2.5). This type of research requires the researcher to be immersed in the study

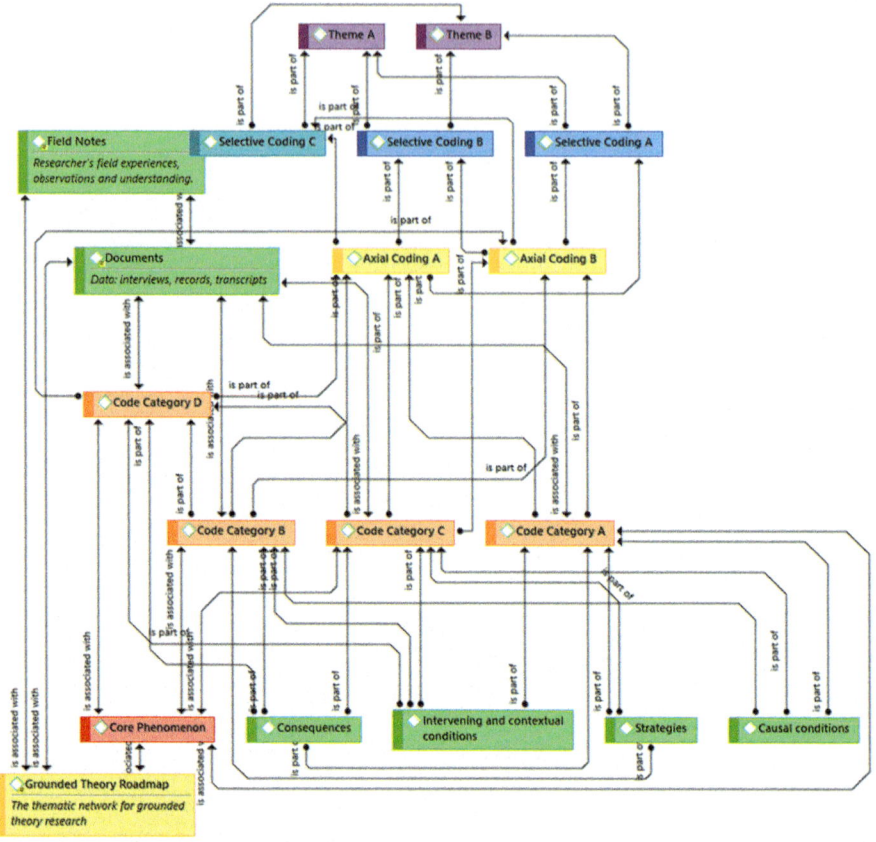

Fig. 2.4 Constructivist approach to grounded theory

setting. This means that the researcher becomes an 'insider' and a part of the community (or social group) that is the subject of the study. The ethnographic method requires the researcher to commit substantial time to fieldwork (observing the subjects and taking field notes).

Collecting Ethnographic Data

An ethnographic study should be conducted in natural settings in which the researcher should aim at discerning patterns of events, cultural themes, and other features that characterize the group. The researcher should make detailed observations of shared behaviors, language, and artifacts of individuals or groups
Neuman and Wiegand (2000) proposed following the rules for taking field notes:
- Take down notes immediately or as soon as possible after the observation
- Keep count of the number of phrases used by subjects
- Never neglect anything as insignificant

(continued)

(continued)

Collecting Ethnographic Data

- Record the sequence of events chronologically, as well as the period in which the events occur
- Field notes should only contain the researcher's observations. They should not reflect judgment or bias. Avoid evaluative judgments or summarizing of retrieved facts and respondents

Case Studies

Data for qualitative research through case study approach is collected from multiple sources (Fig. 2.6). They include interviews, observations (direct and participant), questionnaires, and documents. The use of multiple data collection techniques and sources in case study research strengthens the credibility of the findings. Case study analysis can also be used for theory building.

Case studies offer rich perspectives and insights that can lead to an in-depth understanding of behaviors, issues, and problems. This method can be used to study both simple and complex subjects. The strength of this method is that it uses a combination of one or more qualitative data collection methods to gain insights and draw inferences.

Yin (2009) recommends collecting six types of information in ethnographic case study research: documents, archival records, interviews, direct observations, participant observation, and physical artifacts.

Depending on the research question, objectives of the study, and the characteristics of the study population, these methods can be used individually or in a combination of two or more. However, the approach to data collection is different for each research type, which is discussed in the following section of this chapter.

> **Important** The first decision that the researcher must make is to identify the entities or persons for the study. The cases selected must best address the research questions. The case may be a single individual, a family, a social community, work group, organization, or institution (Flick, 2006). Case study research may involve a single case, multiple cases, or a series of layered or nested cases (Patton, 2002). The sampling process is purposive, with the researcher selecting cases that offer variety, typicality, and opportunity for gaining deeper understanding of the issue. Accessibility to data is also important because it determines the quality of research outcomes.

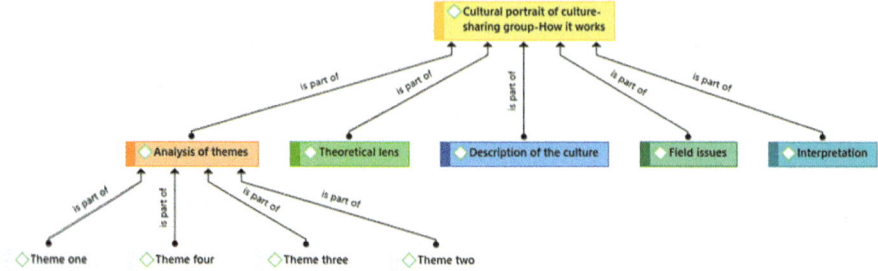

Fig. 2.5 Ethnographic data coding framework

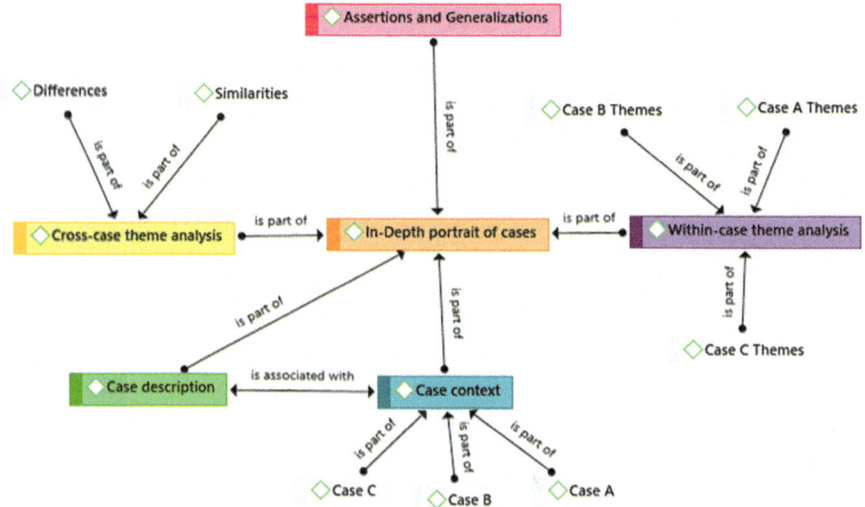

Fig. 2.6 Coding framework for case study research

RESEARCH DIARY AND FIELD NOTES

The research diary is a logbook of the researcher's reflections. In it, the researcher records their ideas, discussions with fellow researchers, feelings about the research process, and how the researcher's feelings evolved as the research progressed. The researcher must make a note of anything and everything pertinent to the study. The diary is a very personal document and reflects the researcher's journey through the research. Sometimes, it is called a fieldwork journal or a research journal and contains, among other things, a day-to-day commentary about the study, including data collection, thoughts, ideas, and inspirations for analysis. In addition, the researcher should also mention challenges, difficulties, achievements, serendipitous findings, literature reviews, surprises, influences, people's contact information, you, surprises, contact information about people, etc.

Field notes are taken contemporaneously with data collection in the research setting (Angrosino, 2007). They are particularly useful in ethnographic studies. They are also known as mental notes, captured on paper as they are made in the researcher's mind. The researcher notes the information in her/his journal after leaving the place of the interview.

Field notes are essential in an ethnographic study where the researcher observes the phenomenon using all their senses. They are unplanned and unstructured; they are usually open-ended, loose—and often unruly and messy. When researchers observe anything of significance to their study, they must first make a mental note of it and then document it with minimum delay.

Field notes may include what the respondents said or did, whether the person was uncomfortable or happy while talking, the facial expressions of the participants, and so on. As Emerson et al. (2001) explain, descriptive writing, such as field notes, embodies and reflects purpose and commitments. It also involves active processes of interpretation and sense-making.

Field notes are sources of rich data and are valuable for getting deep insights into the subject of the study. They help in evolving theory, and methodologies, writing memos (explained in Chap. 3), and coding data (explained in Chap. 4). Most researchers describe their experiences in the field notes, including their observations and insights about a phenomenon.

A field journal will contain, typically, the researcher's observations (and reflections) on

- What are people doing? What are they trying to accomplish?
- How, exactly, do they do this? What specific means and strategies do they use?
- How do the participants talk about, characterize, and understand what is going on?
- What assumptions are they making?
- What do I see going on here? What did I learn from these notes?
- Why did I include them?
- How did my observations and experiences challenge (or strengthen) my assumptions, position?

Transcription of Data

Data in field studies is collected through audio and audio-visual means. The data collected must then be converted to a text form. The process of converting audio or audio-visual data to text is called transcription. While transcribing, researchers must preserve the original data (recordings) for proving, whenever called upon to do so, the authenticity of the transcripts. The examiner or audience has the right to—and can—question the data sources of the transcripts. There are two points that researchers should be mindful about:

- avoid the use of software to convert audio or video data to text because there are high chances of critical information being lost.

- avoid outsourcing transcription of data to a third party because of 1) the possibility of violation of confidentiality conditions, and 2) loss of information due to poor quality of the transcription.

Transcription of data takes time and effort. It is often repetitive and there are no easy alternatives or shortcuts. The researcher needs to rewind, replay, pause and play the recording as many times as necessary to ensure that everything that was recorded has been transcribed. The time required for the transcription of field data depends on the nature of the study and the researcher's skills at transcription. A 30-min interview can take about 3 to 5 h for transcription, which should give an idea of the rigor involved in such work. Therefore, the researcher would be well-advised to transcribe and update his field notes every day. The researcher must write the timing of his transcription.

Many times, researchers hire help to transcribe their data. There could be various reasons for doing this, but such practices should be avoided as far as possible. If transcription by a third party is unavoidable, it must be done only in the researcher's presence to make sure that 1) all the data is transcribed and 2) that the data is secure and remains in the custody of the researcher.

Transcription requires skill and is not just a mechanical reproduction of data from one form to another. It is not enough to transcribe what is spoken; the researcher must also include information about the settings, context, body language, and her/his observations of the interview session. Many studies recommend taking photos to help in capturing reality, but there may be ethical issues depending on the situation. However, if the photos and images do not reveal the identity of the participants, the researcher can use them.

Data is of two types: explicit and implicit. Explicit data is collected through interviews, questionnaires, and other survey tools. They can be retrieved easily Implicit (tacit) data is only sensed and is hence, more difficult to capture.

Most researchers rely only on explicit data for their studies. They do not assume anything beyond the descriptive responses they have recorded. For this reason, they miss the "hidden" information, that which is expressed by the respondent through signals and body language. Implicit data can only be observed, but it is of crucial importance for the researcher to not miss and document it because it is lost in a short time. The researcher must make field notes on such information and then transcribe them into text.

To capture implicit data, the researcher must document information in their field diary as soon as possible after the interview. They should find a quiet place to write up the observations. This should be done after every interview because the writing of observations and experiences is a significant component of data collection in qualitative research. The researcher's notes must include the knowledge gained in the data collection process because they may provide substantial evidence to support her/his claims.

While transcribing, the researcher should note every pause and change in the respondents' tone and expressions, including those indicating discomfort, unhappiness, or lack of self-confidence while speaking. This can be a challenge while transcribing audio recordings; however, video recordings can minimize these difficulties to a large extent. Noting pauses may not be necessary for all types of qualitative research. However, researchers should offer a rationale for not showing pauses in their transcription.

The goals for transcription can differ with the approach to a qualitative study (Sandelowski, 1994). In narrative analysis, it is important to capture language patterns and nonverbal communication, such as pauses and false starts. However, in constant comparative analysis or grounded theory, the transcription process needs to focus on the accuracy of the information content. Poland and Pederson (1998) reasoned that what is not said is just as important as what is said. Hence, transcripts may require that researchers include contextual information regarding silences or pauses in conversation. For ease of transcription, Silverman (1998: 264) has provided a set of transcription symbols to denote pauses, verbal stresses, and overlapping talk. Seale and Silverman (1997) have recommended using analytic ideas and transcription symbols of conversation analysis in addition to a verbatim transcription of the text. This includes capturing even the "uh-ms" and pauses in speech.

Contents of Transcripts

Transcription is time-consuming, costly, and prone to errors. Even the most experienced transcriber can make mistakes, or cause discrepancies during the transcription process (Poland, 1995). Hence, transcription necessitates proofreading to ensure that the transcripts are accurate. Many experts recommend having two people who will separately transcribe the data, and then compare the transcripts for agreement (Kvale, 1996:163). This procedure is also prone to errors.

Transcription is not just a technical detail; it is an integral component of research activity (Atkinson & Heritage, 1984). As a result, the researcher must ensure that transcription includes all relevant data on the topic of research.

The theoretical position and application of notation symbols are reflected in the transcribing process (Davidson, 2009; Green et al., 1997; Lapadat & Lindsay, 1999; Ochs, 1979). As a result, subjectivity in the transcription process will lead to content disagreement (Bucholtz, 2000), defeating the purpose and objectives of the study. Transcription errors can significantly impact the data analysis process, leading to inaccurate findings by researchers (Easton et al., 2000; Poland, 1995).

Researchers can transcribe data manually or with the help of software. But first, it is necessary to consider whether data must be captured in audio or video format.

Researchers gather field data in various ways: from telephonic interviews, administering questionnaires (in person or digitally), conducting online or face-to-face audio interviews which are audio or video-recorded, or through observations. Each method comes with a set of advantages and disadvantages. For example, the respondents may be hesitant to express themselves fully in a digital questionnaire; or alter their reactions to indicate an opinion they think the researcher expects of them.

Respondent reluctance can also happen during face-to-face interactions. But in this case, the researcher has the advantage of observing the respondents' behaviors, expressions, and the nonverbal cues being given. It is also possible for the researcher to repeat and reframe the question, probe further, and ask more questions based on the participants' responses.

This flexibility is absent while administering questionnaires. Whether audio or video-recorded, telephonic interviews provide researchers with more room for cross-questioning, probing, and observing the participants' replies, pauses, or avoidance behaviours. In a face-to-face interview (audio or video), the researcher can observe a range of behaviours from the physical gestures, expressions, and tones of the respondents, which can indicate what they are feeling at that time—happy or sad, open or suspicious, comfortable or uneasy, and so on. As a result, the researcher can obtain comprehensive data information during a face-to-face interview.

With audio data, the researcher can listen to the responses and by playing and replaying the recording, can understand what the respondent conveyed. However, capturing expressions and recalling them later to obtain clues is difficult. For this reason, the researcher should make field notes that can be referred to for assisting recall. It is crucial to record the tacit/implicit information that is conveyed through the respondents' tone, expressions, and body language as soon as possible.

Video recordings capture significantly more information than audio interviews. Video transcripts help to catch the settings, gestures, body movements, glances, and facial expressions that audio transcripts cannot (Pomerantz & Fehr, 1997). With audio data, it is virtually impossible for the researcher to reproduce expressions and cues that were observed during and after the interviews. It is for this reason that researchers should document their observations as soon as possible, even immediately after the interview if feasible. Such observations and field notes, as well as information shared by respondents after the interview and during informal discussions should be included in memos.

Challenges in Transcription

Interview data must be transcribed as early as possible. A delay in transcription could result in the loss of key information because recall diminishes with time. Even though the researcher's field notes may be comprehensive, a significant amount of information is retained in the mind, which cannot be put on paper. Therefore, it is imperative that transcription takes place with minimum delay. Video data generally do not suffer from such issues.

Crichton and Kinash (2013) mention that since researchers spend a lot of their research budget on transcription, this often influences the number of interviews that will be conducted. Further, according to Tilley (2003), issues in the process of transcription mean that "researchers must consider the extent of detail necessary in transcripts when making informed decisions about appropriate transcription procedures".

Generally, it takes 6–7 h to transcribe one hour of taped interview (Britten, 1995). The interval between the interview and its transcription can impact the quality of research (Tilley, 2003). Making comprehensive field notes can help address many of the challenges to accurate transcription.

Professionals can be engaged for transcription; however, there is no guarantee of accuracy and completeness. Vital information may be overlooked, resulting in errors (Poland, 1995). However, according to Easton et al. (2000), such errors are common whether researchers do their own transcribing or engage a transcriber.

The main advantage with a researcher transcribing their own data is that they are aware of the context and settings of the study. Moreover, there is also better recall of the "noteworthy moments", which makes it easier to take a holistic approach to transcription and ensure that no crucial information is lost.

Another challenge is in denoting emotions in the transcripts. Emotions, changes in tone, and periods of silence can be captured on audio or video; however, they are difficult to represent in transcripts (Poland, 1995). Here, notation systems come in handy (Poland, 1995).

Interviews are usually conducted in the respondents' language. If the interview language is different from the language used in the study, at least two transcripts will be required: one in the respondents' language and the other, a translated version for the researcher's use. Many scholars transcribe after translating the responses. In this approach, there is the risk of loss of important information in the translation process. According to Twinn (1998), transcripts may benefit from being studied in their original language rather than being translated first. Information in the original language, the language of the respondents, is more accurate than a translated version. It is often difficult to find exact equivalents of the expressions and idioms used in the language of the interviews. According to Yelland and Gifford (1995), conducting interviews in respondents' second languages has a negative impact on data quality. Thus, it is always preferable to design the questionnaire (or interview schedule) in the language respondents are familiar with.

There is also the challenge of transcribing long-duration audio or video recordings which consume large amounts of the researcher's time. To address this issue, one needs to understand it from the perspective of data analysis in ATLAS.ti. In the conventional (manual) approach to data analysis, the researcher reads the data and highlights the information that addresses the research objective. Coding and memo writing follow. The researcher then tries to discover patterns in the code lists, group them, and form smart codes or themes. Then, with the help of a cluster of quotations (key information from responses relevant to the objectives of the study) and memos (researcher's field observations and experiences which are usually not captured in texts) the emerging themes are explained and justified.

A similar approach is used for analysis of audio or video data. One can listen to—or visualize—the information that captures the researcher's interest because they may be crucial to finding the answers to the research questions. Other information, which is not directly or indirectly connected with the research question or unlikely to be important for analysis, can be discarded. Using the approach to text analysis of quotations, the researcher need only clip the important segments of audio or video data and transcribe them for coding. Full transcription of the audio or video data may not be necessary.

Transcribing video clips and coding the important segments only provides more authentic data than the complete transcripts (Crichton & Child, 2005). It allows researchers to selectively focus on what is crucial without missing changes in tone, expressions of strong feelings, pauses, and inflections throughout the analysis process. It also reduces time and cost associated with data transcription and management while simultaneously enhancing data accuracy. As a result, researchers can study interview data in greater depth without consuming much time on full transcriptions. Selective playback also saves the time of audiences and improves their engagement with the topic.

But working with the recordings alone without a transcription may fail to convey crucial points (Heritage, 1984). Transcripts help to make sense of data and convey important information to the audiences (Lapadat & Lindsay, 1999) and hence transcriptions are an important component of the analysis process. Pomerantz and Fehr (1997) argue that working only with recordings makes it difficult to study and isolate phenomena. However, it is easier to evaluate the quality of the transcript when one can compare it with the recording (Pomerantz & Fehr, 1997) because recordings provide the evidence on which transcripts are based (Mondada, 2007).

ATLAS.ti allows the use of both options—directly working on recordings, and with transcripts and recordings simultaneously. The researcher can support the findings of their research with video clips and associated transcripts. This option provides researchers with the flexibility required to explain themes to audiences. While presenting the themes in a text format, the researcher can adjust the pictures from video clips to make them informative and insightful to the audience.

The researcher can also present themes in a network form, which is supported by supporting video clips, images, transcripts, and memos to show their association. Such representations will not only arouse audience interest, but they also allow the researcher to interpret the theme systematically, thereby increasing the chances of the audience's confidence in the research outcomes.

Framework for Qualitative Research

Table 2.3 presents an overview of the qualitative research framework. It can help the researcher develop frameworks for their studies from conceptualization to completion.

Table 2.3 Framework for qualitative research

S. No.	Phase	Steps		Description	
1	Conceptualization	Research objective/research questions		Based on experience, literature, and the researcher's inclination and interest. It begins with a research idea that the researcher wishes to explore in depth	
2	Design	Methodology		GT/phenomenology/narrative/case study/narrative Context People/phenomenon/culture Methods of data collection	Research inquiry methods, context, and data collection techniques besides the rationale behind it. What is to studying people, culture, or behavior
3	Mapping terrain	Context Pilot study Reformulate research questions		Examine the research objective and literature in the present context	Analyze research objectives with existing knowledge. Conduct a pilot study if needed and reformulate research questions
		Research question		Research objective Research questions Research goal	The research question should be aligned with the research objectives and intended research outcomes
		Sampling design		Purposive People Research objective	Carefully selecting respondents to address research objectives

(continued)

Table 2.3 (continued)

S. No.	Phase	Steps	Description	
4	Data collection	Methods	In-depth interview, observations, FGD, survey	Emphasis on multiple data collection methods to ensure triangulation
5	Transcription	Words to words, pause to pause (verbatim) same day with the timing of interview and transcription	Primary data and field notes need transcription on the same day verbatim	
6	Analysis	Reading transcripts identifying key segments, coding, memoing, commenting, code grouping, networks, developing themes	Reading texts through the lens of the research question and identifying informative segments and levels with the initial coding process. Simultaneously writing a memo, comments follow with code groups, creating significant value and creating themes	
7	Interpretation	Code patterns, c-coefficient, concept-to-concept relations, making networks, generating textual and excel report in a simple to complex manner	Significant value emerging from data analysis triggers interpretation. It is based on code pattern, c-coefficient, and degree of concept-to-concept relations	
8	Reporting	Writing and producing in the analysis section	Thematic reporting is based on significant values of emerging concepts supported by essential quotations, memos, and theoretical models/knowledge and creating a new body of knowledge	

Researchers often find it difficult to explain the theoretical underpinnings of their studies and define the key concepts in them. However, concepts are the very foundations of their work and hence, they must make every effort to present them in a logical manner.

Frameworks include the definitions and approaches that will be used in a qualitative study. They cover grounded theory, methodology, methods, and the approach to qualitative analysis. They help researchers across disciplines of study. The framework for a study should include the philosophical assumptions made and point researchers in the appropriate direction. The framework of a study can be divided into three phases—indoor, outdoor, and reporting.

The indoor phase consists of conceptualization, design and mapping terrain where the researcher develops the strategy and structure of research. It includes the pilot study which is also part of the outdoor phase. Mapping terrain covers both indoor and outdoor phases. The outdoor phase includes mapping terrain for data analysis. The last phase is the reporting phase in which the researcher presents and interprets research outcomes concerning research objectives and questions. Reporting and interpretation should be supported by scientific research methods, data analysis, sampling, validity, reliability, saturation, and triangulation. Finally, the researchers should provide evidence of the emergent themes using quotations (informative pieces of data) and memos (researcher's field observations).

Recommended Readings

Angrosino, M. V. (2007). *Doing ethnographic and observational research.*

Atkinson, J. M., & Heritage, J. (Eds.). (1984). *Structures of social action.* Cambridge University Press.

Bogdan, R. C., & Biklen, S. K. (1997). *Qualitative research for education: An introduction to theory and methods.* Allyn & Bacon.

Britten, N. (1995). Qualitative research: Qualitative interviews in medical research. *BMJ, 311*(6999), 251–253.

Bucholtz, M. (2000). The politics of transcription. *Journal of pragmatics, 32*(10), 1439–1465.

Charmaz, K. (2005). Grounded theory in the 21st century: A qualitative method for advancing social justice research. *Handbook of Qualitative Research, 3*(7), 507–535.

Charmaz, K. (2006). Constructing grounded theory: A practical guide through qualitative analysis. Sage.

Clandinin, D. J., & Connelly, F. M. (2004). *Narrative inquiry: Experience and story in qualitative research.* Wiley.

Corbin, J. M., & Strauss, A. (1990). Grounded theory research: Procedures, canons, and evaluative criteria. *Qualitative Sociology, 13*(1), 3–21.

Creswell, J. W., & Brown, M. L. (1992). How chairpersons enhance faculty research: A grounded theory study. *The Review of Higher Education, 16*(1), 41–62.

Creswell, J. W., & Poth, C. N. (2016). *Qualitative inquiry and research design: Choosing among five approaches.* Sage Publications.

Crichton, S., & Childs, E. (2005). Clipping and coding audio files: A research method to enable participant voice. *International Journal of Qualitative Methods*, *4*(3), 40–49.

Crichton, S., & Kinash, S. (2013). Enabling learning for disabled students. *Handbook of Distance Education*, *3*, 216–230.

Davidson, C. (2009). Transcription: Imperatives for qualitative research. *International Journal of Qualitative Methods*, *8*(2), 35–52.

Davidson, F. (1996). Principles of statistical data handling. *No title*.

Easton, K. L., McComish, J. F., & Greenberg, R. (2000). Avoiding common pitfalls in qualitative data collection and transcription. *Qualitative Health Research*, *10*(5), 703–707.

Emerson, R. M., Fretz, R. I., & Shaw, L. L. (2001). Participant observation and fieldnotes. *Handbook of ethnography* (pp. 352–368).

Flick, U. (2006). *An introduction to qualitative research* (3rd ed.). Sage.

Green, J., Franquiz, M., & Dixon, C. (1997). The myth of the objective transcript: Transcribing as a situated act. *Tesol Quarterly*, *31*(1), 172–176.

Glesne, C. (2016). *Becoming qualitative researchers: An introduction*. Pearson. One Lake Street, Upper Saddle River, New Jersey 07458.

Heritage, J. (1984). *Garfinkel and ethnomethodology*. John Wiley & Sons.

Krueger, R. A., & Casey, M. A. (2009). *Focus groups: A practical guide for applied research* (4th ed.). Sage Publications.

Kvale, S. (1996). The 1,000-page question. *Qualitative inquiry*, *2*(3), 275–284.

Kvale, S., & Brinkmann, S. (2009). *InterViews: Learning the craft of qualitative research interviewing*. Sage.

Lapadat, J. C., & Lindsay, A. C. (1999). Transcription in research and practice: From standardization of technique to interpretive positionings. *Qualitative Inquiry*, *5*(1), 64–86.

Mondada, L. (2007). Commentary: Transcript variations and the indexicality of transcribing practices. *Discourse Studies*, *9*(6), 809–821.

Morrow, S. L., & Smith, M. L. (1995). Constructions of survival and coping by women who have survived childhood sexual abuse. *Journal of Counseling Psychology*, *42*(1), 24.

Moustakas, C. (1994). *Phenomenological research methods*. Sage Publications.

Nathanson, H. (1973). Understanding everyday reality: the phenomenological model.

Neuman, W. L., & Wiegand, B. (2000). *Criminal justice research methods: Qualitative and quantitative approaches*. Allyn and bacon.

Ochs, E. (1979). Transcription as theory. *Developmental pragmatics*, *10*(1), 43–72.

Patton, M. Q. (2002). Two decades of developments in qualitative inquiry: A personal, experiential perspective. *Qualitative social work*, *1*(3), 261–283.

Poland, B. D. (1995). Transcription quality as an aspect of rigor in qualitative research. *Qualitative Inquiry*, *1*(3), 290–310.

Poland, B., & Pederson, A. (1998). Reading between the lines: Interpreting silences in qualitative research. *Qualitative inquiry*, *4*(2), 293–312.

Pomerantz, A., & Fehr, B. J. (1997). Conversation analysis: An approach to the study of social action as sense making practices. In T. A. van Dijk (Ed.), *Discourse as Social Interaction, Discourse Studies: A Multidisciplinary Introduction* (pp. 64–91). Sage Publications.

Rubin, H. J., & Rubin, I. S. (2012). *Qualitative interviewing: The art of hearing data* (3rd ed.). Sage.

Sandelowski, M. (1994). Focus on qualitative methods. Notes on transcription. *Research in nursing & health, 17*(4), 311–314.

Sampson, H. (2004). Navigating the waves: The usefulness of a pilot in qualitative research. *Qualitative Research, 4*, 383–402.

Seale, C., & Silverman, D. (1997). Ensuring rigor in qualitative research. *The European Journal of Public Health, 7*(4), 379–384.

Silverman, D. (1998). *Harvey Sacks: Social science and conversation analysis.* Oxford University Press, USA.

Strauss, A., & Corbin, J. (1998). *Basics of qualitative research techniques.*

Tilley, S. A. (2003). "Challenging" research practices: Turning a critical lens on the work of transcription. *Qualitative Inquiry, 9*(5), 750–773.

Tracy, K., & Robles, J. S. (2013). *Everyday talk: Building and reflecting identities.* Guilford Press.

Twinn, D. S. (1998). An analysis of the effectiveness of focus groups as a method of qualitative data collection with Chinese populations in nursing research. *Journal of advanced nursing, 28*(3), 654–661.

Van Manen, M. (2016). *Researching lived experience: Human science for an action sensitive pedagogy.* Routledge.

Yelland, J., & Gifford, S. M. (1995). Problems of focus group methods in cross-cultural research: A case study of beliefs about sudden infant death syndrome. *Australian and New Zealand Journal of Public Health, 19*(3), 257–263.

Yin, R. K. (2009). *Case study research: Design and method* (4th ed.). Sage.

CHAPTER 3

Challenges in Data Collection

Abstract Data collection presents various types of challenges, and the researcher must be prepared for them. They begin with obtaining consent from the participants, gaining familiarity with the settings of the study, data collection challenges and field-notes. The author explains the types of data collected in fieldwork, the importance of conducting interviews at a time and place that are convenient for the respondents, and the challenges associated with obtaining informed consent. The author also explained how to initiate an interview and the various kinds of interviews. Informal conversations are important and the various ways of recording information from such interactions are described. A sample letter of introduction containing information on ethical guidelines to which the researcher is committed is also included. In addition, the author also offers suggestions on what the researcher can do while waiting for the respondent to arrive at the interview site, or for the interview to commence. This chapter also includes a framework for data collection, writing field notes, ensuring their quality, and their use in analysis. Two opportunities for field researchers to engage in critical learning experiences have been discussed.

BECOMING FAMILIAR WITH THE SETTINGS OF THE STUDY

Researchers must become familiar with the research settings as quickly as possible. This is necessary for understanding the people and their interactions with each other, as well as the ways they communicate, the social environment, dynamics, practices, and other features that characterize their community. Specifically, they must gain knowledge and experience with the following components of their study.

1. Respondents
2. Gender norms, such as, for example, gender-based segregation.
3. Sensitive issues and matters that are likely to arouse emotions.
4. Languages spoken.
5. Risks and benefits of research to respondents. The research should have the potential to benefit the respondents and not result in harm.
6. Ethnicity, stereotyping, and hierarchy (i.e., presence of marginalized, dominant, and privileged groups in the settings).
7. Power dynamics.

The researcher should become familiar with their respondents, with an understanding of their nature and personal traits, tastes and preferences, and the social group/s they affiliate themselves. Where language is likely to become a barrier to communication, it is advisable that the researcher acquire a working knowledge of the language spoken in the community.

Regardless of the identity and social status of the respondents, researchers should show sensitivity in their communication and interactions with them. There must be awareness of the power structures in and access to persons of influence in the community so that getting permissions and contacting potential respondents will not be a hurdle to data collection.

At the same time, researchers should examine their own attitudes, tastes and preferences, assumptions they make, and the preconceived notions they may have about the research topic. Familiarity with the study settings and subjects, local issues, and dynamics, will help researchers to strategize their fieldwork and establish rapport with the participants, including those from vulnerable groups. Once mutual confidence is established, the next step is to explain to the respondents the code of ethics adopted for the study.

According to Rosenblatt (1999), there is no single "reliable ethical formula" that can be applied to qualitative research interviews; instead, ethical guidelines are co-constructed as the interview progresses. In response to the "emergent" ethical issues confronting the qualitative interviewer, "it is often suggested that researchers engage in ongoing reflection while responding sensitively to the needs of participants" (Cohn & Lyons, 2003; Shaw, 2003).

Richards and Schwartz (2002) recommended supervision for qualitative researchers, whereas Shaw (2003) recommends research training for social workers conducting research with vulnerable populations, such as the young homeless (Ensign, 2003) and the bereaved (Parkes, 1995; Rosenblatt, 1999), to protect participants from inexperienced researchers. Before conducting research, novice researchers should first gain experience working with vulnerable communities (Ensign, 2003).

There is also a risk of voyeurism, as well as the temptation to focus on the most sensational aspects of a phenomena or event (Ensign, 2003; James & Platzer, 1999) and to study phenomena that attract public and media attention (Humphreys, 1970). In such situations, privacy is compromised when the interviewer probes into areas that at least one interviewee would prefer to keep private. When interviews reveal previously confidential details between a couple, confidentiality is compromised (Forbat & Henderson, 2003).

CHALLENGES IN DATA COLLECTION

Data collection in qualitative research is not without challenges. Challenges are a part of research, and the researcher must maintain a positive view while addressing them. Challenges are present throughout the study, beginning with planning where the researcher is seeking consent to participate from the respondents. During data collection, appointments may be cancelled or postponed. A few respondents may even withdraw their consent. The researcher may have to remind respondents about their appointments and address ethical issues that arise.

Most challenges or difficulties can be foreseen, and the researcher must have a plan to address them at the project's conceptualization and planning phase. While selecting the sample, the researcher must also consider various ways of contacting them. The usual approach is to obtain the respondent's email address (or other contact information), and then write a letter of introduction with a description of the research and its objectives. This approach is useful for organizations or institutions with a formal communications structure.

However, this approach does not work with difficult-to-reach or vulnerable populations. For example, in a study of the young homeless, Ensign (2003) said that only oral assent/consent was possible. The researcher must be aware of vulnerability issues and work to address them. When informed consent has the potential to dilute data quality, researchers should explore methods to gain access to respondents and obtain their insights into the topics of their study. In Anderson's (1996) research with homeless women, the procedure for informed consent included informing participants that while they would not benefit from participating in the research project, other women and girls may benefit from the knowledge gained. Several women in the study stated unequivocally that they would participate if it meant even one woman would be helped in the future. As a result, they were more interested in participating in the study to gain more knowledge from the study and less concerned about the possible associated risks in participating in the study.

Recording informal conversations also requires the consent of respondents. However, researchers should determine if obtaining consent while the natural flow of information is active can inhibit the flow of information. Obtaining informed consent prior to, during, or after informal conversations in natural, everyday settings, especially in the form of a written document, may appear intrusive and even impractical at times (Akesson et al., 2018; Swain & Spire, 2020).

Informed consent to participate in the study is essential. The rationale of the study must be clearly explained. Questions and doubts, spoken or unspoken, should be addressed to the respondent's satisfaction. Under no circumstances must the respondent be compelled to participate in the study.

A few data collection experiences are discussed in this chapter. The purpose is to help researchers to understand the challenges and the need to be organized. The examples should also help the researcher in developing an awareness about the information that emerges during the data collection process, and what they can do with the data.

In one qualitative study, a purposive sampling method was used for sample selection. Based on the objectives of the study, the researcher prepared a profile of the respondents. The topic of the research was "A comparative study of middle managers' morale in two public sector banks in India". The objectives of the study were:

- To study the middle managers' morale in two public sector banks.
- To study components that influence middle managers' morale.
- To develop a model and measures to improve middle managers' morale.

In qualitative studies like the one in the example, the researcher does not decide on the sample size at the start of data collection. Sample size is determined by the emerging information and the researcher's understanding of when data saturation has been reached and at which stage data collection is stopped.

The author collected data from top management personnel (serving as well as retired), union leaders, and middle managers from each bank. Among the union directors, the researcher interviewed those who were working in the branches he visited. Middle managers for the sample were drawn in equal numbers from each branch of the bank visited by the researcher.

The challenge in this study was in how to approach the respondents. The author had previously worked in the management cadre of a bank and thus, was familiar with the work culture, employee perceptions and attitudes. But this advantage also posed a dilemma for the author—whether he should disclose his previous identity to the respondents. Despite consulting several people, and reading up on ethical issues, the researcher could not find a satisfactory answer.

Work culture in the banking sector is characterized by mutual suspicion and employees are generally reluctant to share information. The dilemma before the author was whether disclosing his banking experience would be an obstacle to the flow of information. In the end, trusting his experience and intuition, the author decided not to disclose his identity, and not to collect data from respondents he was acquainted with.

Ethical Concerns in Qualitative Research

Since the nature of qualitative research is such that it requires engagement at deeply personal levels, ethical issues are inevitable. Data collection requires the researcher to build rapport with the subjects of the study so that they will trust the researcher enough to share personal, intimate, and controversial details about themselves. In this interaction between the researcher and the respondent, there is the implied expectation of the respondent that the information shared will be treated with respect and sensitivity.

Ethical issues are of vital importance in qualitative research. The researcher must inform the participants about the purpose and aim of the research, how it will be conducted, and under what conditions, as well as the kind of information they are required to share. The researcher must also make sure that participants have the information sought from them. Furthermore, it is also reasonable for participants to expect that the findings of the study will be shared with them. The researcher should obtain the participants' informed consent before commencing data collection.

Ethical concerns must be anticipated by the researcher and addressed as transparently as possible. It is crucial to ensure anonymity, and that the respondents are made aware of how the data provided by them will be used. At the end of the interview, the researcher should thank the respondents for sparing their time. Indeed, the researcher must try to make the participants feel that they are partners in the study. The interviewer must also be prepared for the possibility that, during the interviews, some respondent might avoid or prefer to skip answering, or may want more time for answering some questions for various reasons. This is normal. The researcher must accept such decisions and not force the respondent to answer.

Since a key objective of qualitative research is to obtain insights that could help the subjects, it is also binding on the researcher to not record any information that can harm the participants (and organizations/institutions that may have a stake in the study).

In many interactions, a participant will express personal opinions about the subject s/he is required to share information about. This is understandable, even natural. Should such interactions be recorded? The answer is yes because the perceptions of the participants are essential for the research even though they are not part of the interview schedule. The researcher's focus must always be on the topic and any data or information that is likely to provide insights into the topic must be recorded.

Researchers should adhere to the principles of the ethics of non-maleficence and beneficence-centered studies. Nonmaleficence refers to not causing harm to others or putting them at risk of harm. Beneficence refers to the prevention and removal of harm, as well as the promotion of good. Nonmaleficence and beneficence may conflict in certain circumstances; it is the researcher's responsibility to determine if the risks of the research outweigh its benefits (Silva,

1995). If the risks outweigh the benefits, it would be extremely unethical to conduct the research.

The researcher must follow up with the respondent for a reply using, where necessary and appropriate, mutual, or influential contacts. The objective of this approach is to get the consent of the respondent to participate in the study. The researcher must make it clear from the beginning that all ethical concerns will be addressed.

Qualitative research seeks to explain, with the help of available theories and concepts, how people and situations interact with each other. In this sense, qualitative research is nomothetic and deductive. Besides, although all researchers are sensitive to how their descriptions are interpreted, they must also be realistic enough to appreciate the need to represent the views of participants and respondents as faithfully and accurately as possible. In this sense, qualitative research is inductive and exploratory.

Many times, researchers do not clearly communicate with the respondents about the nature of the interviews they plan to conduct and the mode of recording (audio or audio-visual). On one occasion, the author found that the researchers had not taken the consent of the respondents to record the interview. They had assumed that the respondents would not allow recording and hence, quietly recorded the interview. But after some time, the respondent became aware that their conversation was being recorded, they asked the researcher to stop the interview and to leave the place immediately. Such situations must be avoided.

Communications between researcher and respondent should be clear and cover all aspects of the study, including date, time and place of the interviews, the type of discussions that will be held, and clear understanding between researcher over anonymity, privacy, and confidentiality of data, and how the findings will be shared.

Ethical Challenges and Practices

Sometimes, however, even concealing the information is not enough to ensure anonymity. According to Punch (1994), many institutions and public figures are nearly impossible to obscure, so their participation in research may expose them to some degree. In such cases, ensuring the confidentiality of portions of the data may be important to participants and influence their willingness to participate.

In the case of participants seeking public visibility, researchers do not require consent to conduct research using publicly accessible information like political rallies, demonstrations, and public meetings. On the other hand, even if the researcher is using pseudonyms to mask identities, consent of the respondent is necessary before conducting research with publicly available information. This caution is necessary to prevent the revelation of the identities of people, communities, and participants from vulnerable, marginalized, and socially stigmatized communities.

In some instances, even after anonymization, quotations, speech mannerisms, and comments about the context may provide ample information to identify participants, and it is not always possible for the researcher to predict which data may lead to identification (Richards & Schwartz, 2002). Additionally, changing too many details can render the data meaningless (Richards & Schwartz, 2002). Therefore, it may be possible to negotiate the disclosure of information and identities with those involved in the research. However, despite all precautions, the researcher may still have to make difficult decisions regarding what should not be reported because by doing so, there is a high risk of harm resulting from compromising the anonymity of participants.

According to Baez (2002), absolute confidentiality refers to the convention of confidentiality, which is upheld primarily to safeguard research participants. Even after ensuring anonymity, revealing the identities of vulnerable populations, such as drug users, minors, or stigmatized individuals, would be harmful. Inevitably, the researcher is responsible for determining which aspects of a person's story or life circumstances must be altered to preserve confidentiality (Parry & Mauthner, 2004; Wiles et al., 2008).

Researchers vary in their willingness to change. In some studies, the researchers even altered their respondents' quotations (Kaiser, 2009). When disseminating research results, researchers must consider whether the specific quotations and examples they provide could lead respondents to be identified through deductive disclosure. According to Tolich (2004), the primary concern is whether the respondents' acquaintances will be able to identify them.

Deductive disclosure, also known as internal confidentiality (Tolich, 2004), occurs when the characteristics of individuals or groups identify them in research reports (Sieber, 1998). Given that qualitative studies frequently contain detailed descriptions of study participants, qualitative researchers are especially concerned about confidentiality breaches caused by deductive disclosure. As a result, qualitative researchers face a conflict between conveying detailed, accurate accounts of the social world and protecting the identities of the participants in their research.

There is a risk of losing insight if information is concealed. Making data unidentifiable has the potential to alter or destroy its original meaning. The modifications render the data inapplicable to answering the current research questions (McKee et al., 2000; Parry & Mauthner, 2004). Readers are typically unaware of how data has been altered and, as a result, are unable to consider the significance of changes for their interpretations of the data or for the reliability of the data (Wiles et al., 2008). According to Corden and Sainsbury (2006), respondents have strong opinions about how their words or personal characteristics are changed in research reports. Consequently, the code of ethics requires obtaining additional consent when data cannot be altered.

Some research indicates that respondents may express a desire to publish data that would have been deemed too sensitive for publication by researchers.

Researchers tend to interpret displays of painful emotion by respondents as a sign that sharing their data would be harmful; however, respondents may want their data published if sharing the data makes them feel empowered or if they believe they are helping others (Beck, 2005; Carter et al., 2008; Dyregrov, 2004; Hynson et al., 2006; James & Platzer, 1999; Wiles et al., 2006).

Beauchamp and Childress (1979) delineated four principles for ethical practices in health care that can be applied to the conduct of research across disciplines. The four principles are autonomy, beneficence, nonmaleficence, and justice. Autonomy is the capacity to make independent decisions. The aim of beneficence is to benefit humanity through research. Nonmaleficence refers to refraining from causing harm to participants, their groups, or their communities. Acting in a fair and equitable manner is defined as justice.

This raises the question of whether complete anonymity and confidentiality are appropriate or even possible for all types of research (Goodwin et al., 2020). In addition, Goodwin et al. suggest that before making promises of anonymity to research participants, the level of anonymity that can be achieved should be considered. This involves posing questions regarding the research design, such as: What are the practical, ethical, and interpretive advantages and disadvantages of concealing the research location? If it is to be disguised, how can this be accomplished effectively? Is the location sufficiently "typical" for a pseudonym to be effective, or should the research design be modified to include multiple locations?

While these principles should be viewed as values that guide the ethical conduct of research, in practice, researchers may be required to make trade-offs when pursuing research objectives. Instead of strictly adhering to all four principles, it may be necessary for researchers to adopt a utilitarian stance and consider the effects of their research in terms of maximizing the greatest good for the greatest number.

Ethical Issues in Studying Internet Communities

Brownlow and O'Dell (2002) recommend that researchers must promote a set of generally accepted guidelines for qualitative internet research that respects data sources. There are three commonly used types of internet-based methods. The study of information available on websites or interactions in online discussion groups without the involvement of researchers is known as passive analysis. The involvement of researchers in communities to determine the accuracy of responses using semi-structured interviews, focus groups, internet-based surveys, or the Internet for traditional research is known as active analysis.

The fundamental ethical principles of human scientific research are informed consent, privacy, and confidentiality. To determine whether informed consent is required, first determine whether online community communications are "private" or "public." This distinction is important because informed consent is required "when research participants' behavior occurs in a private context in which an individual can reasonably expect not to be observed or reported."

Researchers, on the other hand, "may conduct research in public places or use publicly available information about individuals (such as naturalistic observations in public places and analysis of public records or archival research) without obtaining consent, and research involving observation of participants in, for example, political rallies, demonstrations, or public meetings should not require Research Ethics Board review because the participant can be expected to comply."

Privacy and confidentiality: Researchers can easily and unintentionally violate people's privacy on the Internet. A researcher may violate a newsgroup participant's confidentiality by quoting the exact words of the participant, even if the participant's personal information is removed. Because powerful search engines like Google can index newsgroups, anyone can use the direct quote as a query to retrieve the original message, including the sender's email address. Requests for participation must be direct and explicit. Permission must also be sought to quote their responses verbatim. In addition, participants should be made aware that their email addresses may be traceable.

Another reason researchers should contact individuals before quoting them is because the author of the posts may not be interested in protecting their privacy. Nonetheless, extensive quotes without attribution may also be considered an infringement of another person's intellectual property. Hence, it is always advisable to seek permission.

Before studying an internet community, Eysenbach and Till (2001) recommend that the following points be considered.

1. Intrusiveness: Discuss the extent to which the research conducted is intrusive ("passive" analysis of internet postings versus active participation in the community's communications).
2. Perceived privacy: Discuss (ideally with community members) the perceived level of privacy in the community (Is it a closed group that requires registration? What is the total number of members? What are the group's expectations?
3. Vulnerability: Talk about how vulnerable the community is; for example, a mailing list for sexual abuse victims or AIDS patients is a highly vulnerable community.
4. Potential harm: Given the preceding considerations, consider whether the researcher's intrusion or publication of results has the potential to harm individuals or the community as a whole. Determine whether informed consent is required or if it can be waived (and, if so, how it will be obtained).
5. Confidentiality—How can participants' anonymity be protected? (Because originators can be easily identified using search engines when using verbatim quotes, informed consent is always required.) Intellectual

property rights—In some cases, participants may seek publicity rather than anonymity; thus, postings without attribution may be inappropriate.

A Caution About Google Forms

The user of google forms has the responsibility to protect the personal, sensitive or any identifying information about the respondents. Google form is General Data Protection Regulation (GDPR) compliant. Google Forms can also capture email data, names, and other identifying information. However, it is the user that is responsible for ensuring the safety of information that can harm the users, organisations, and other stakeholders. If alternative forms and survey software platforms are available, Google forms should be avoided. Researchers should remember that google forms is basically a data processor whose services are available to users. Google Forms is not used for more complex, in-depth research or surveys that may involve sensitive or personal information.

Users should design their Google survey form with data protection and security in their minds for which familiarity with its features and settings is necessary. While sharing the form with colleagues for editing, the form settings should be suitably adjusted to allow collaborators to edit the form. Depending on the settings, the user can restrict usage of a Google forms. The researcher must also ensure that that respondents cannot view the answers, or any information shared by other respondents. Some key considerations for the design and use of Google Forms are given below:

1. No unauthorized sharing of information.
2. Respondents can request users to delete their data anytime. This protects the data.
3. Respondents should have the right to change/modify/correct the information shared by them at any time.

The respondents in the study should also be informed about data protection features of the forms they are using.

Information from Informal Discussions

The objective of any research is to arrive at the truth about the subject of study. The epistemological perspective known as constructivism refers to the idea that knowledge is created through the process of learning. There is no single truth or reality, according to the constructivist theory. In fact, there are various realities that must be understood. The constructivist philosophical paradigm, according to Honebein (1996), is a theory that that people build their own knowledge and understanding of the world from their experiences and reflecting on those experiences. Kim (2005) proposed that people build new knowledge from their experiences through the processes of accommodation and assimilation. Constructivist research methods contend that "reality is socially produced," with the goal of comprehending "the world of human experience" (Cohen & Manion, 1994, p. 36; Mertens, 2005, p. 12). The constructivist researcher acknowledges the influence of the participants' background and experiences on the research and frequently relies on their "views of the situation being researched" (Creswell, 2003, p. 8). Hence, the researcher should make all possible efforts to evoke the 'truth' of their subjects' realities.

Researchers should be aware that the information emerging from informal interactions may be deeper and more meaningful than what is gained in formal communications. A crucial phase in qualitative research is gathering data from informal interactions. Respondents are more likely to unconsciously provide a wealth of informative data in informal conversations, which may have been missed during formal interactions.

Even though the researcher may have obtained the informed consent of the participants, the respondent may not reveal sensitive information to be recorded because they may be apprehensive about the potential consequences. Thus, the respondent may hesitate in sharing their views. However, after the interview, when the recorder is turned off, the respondent may be more than willing to share information. In such circumstances, the researcher documents the information in their field notes, describe these experiences in memos (Chap. 5). Informal interactions may reveal considerably more detailed information which researchers should be careful not to miss. To the extent that confidentiality conditions are respected, noting information from casual conversations does not constitute breach of research ethics.

According to Rutakumwa et al. (2020), in some cases, not recording an interview may be the best option. According to Swain and King (2022), there are two sorts of informal conversations that can be used as data: observed conversations that take place during researchers' field notes and participatory conversations that involve an engaging interaction between the researcher and other participants. Observed conversations are also referred to as 'listened-in' discussions that can be analyzed. Researchers continue to employ informal talks nowadays (Angotti & Sennott, 2014; Berg & Sigona, 2013; Vigo Arrazola & Bozalongo, 2014).

Swain and King (2022) presented an ethical practice perspective that is similar to Whiteman's (2012) embedded approach to research in which the researcher is frequently required to make ethical and methodological decisions in the moment, based on situated and sometimes unpredictable contexts, rather than strictly adhering to a set of predetermined ethical principles. One of the consequences of conducting inductive research in this manner is that it makes gaining prior informed permission more difficult.

Disclosing Researcher's Identity

The decision to not disclose author's banking experience was governed by likely perceptions of the respondents towards the researcher and his motives. There was the possibility that the participants might become more guarded in the responses, thereby affecting the quality of the data. An outsider, meaning someone with no knowledge of banking, may have a better chance of being accepted by the respondents who would, then, be more forthcoming. Because of his experience in the banking sector, the author was aware of the mindset of bank employees, as well as the cultural barriers in the banking sector. Hence, the author decided to not disclose his banking experience.

Since the author's research was a comparative study of middle managers' morale, it was natural that the first choice of the study site was the bank where he had worked. But he chose another bank, which was bigger and was significantly different from the first in terms of various performance parameters. Data collected from this bank helped in a better understanding of the determinants of middle managers' morale in each bank.

The next challenge was communication. As the researcher, the author had to explain the aim of the study to the participants and obtain their consent to participate. But meeting the top management was an even bigger challenge (Coincidentally, the top managements of both banks worked from head offices which were located at the same place).

The author decided not to tell respondents that he had worked in a bank. He also decided to select only those respondents who were not known to him or knew about, directly, or indirectly. Finally, he decided that data collection would begin with the top management—the past and present Chairperson and managing director of each bank.

The author communicated with several people who could help to connect him with top management. He emailed the CMD of one bank, explaining the purpose of the research, sharing the letter from the institute the author was affiliated with to establish his bona fides. After a wait of a few days, when he did not receive any response, he called the secretariat of the CMD and spoke to the personal assistant about his email and the purpose of the study, and also requested a ten-minute interview. After repeated requests and reminder calls, he was asked to visit the head office of the branch at 5 pm one day. The researcher prepared semi-structured in-depth interviews. Permission had also been granted to audio-record the interview.

The interview with the CMD lasted 43 min. The researcher could also conduct another interview that day—with an executive director who the author was referred to. In addition, he was also given the contact information of a retired CMD who was also interviewed on another day.

In parallel, the researcher reached out to the middle managers of both banks. He wrote emails, telephoned them, and even met them for informal discussions. Some declined to participate but many also agreed to speak to the researcher. The interviews were conducted according to their convenience at their workplaces.

> **Important** The researcher should be aware that directly communicating with potential respondents increases the chances of getting more and better-quality responses than if the interactions are mediated. However, people may provide more sensitive information anonymously, such as in an online survey, than in face-to-face interviews (Tourangeau, 2004). Researchers should decide how surveys are administered via the internet if using paper-based methodologies are likely to influence the respondents into providing socially desirable responses (Frick et al., 2001; Joinson, 1999) and self-disclosure (Weisband & Kiesler, 1996). They should also decide how human involvement in question administration affects responses to sensitive personal questions.

Two Incidents

The researcher would like to share two experiences that should underscore the need for patience and persistence.

THE FIRST INCIDENT

Getting an appointment for a meeting with the CMD was difficult. The author did not get any response to his email request for an interview. Then he called the number mentioned on the bank's website. A woman answered his call, and the author introduced himself, explained his research and drew her attention to the mail he had sent the CMD a few days earlier. He was told that the CMD was traveling and expected to resume office after a week. Accordingly, the researcher decided to complete his interviews with the other respondents before meeting the CMD. After a week, he called the CMDs office again. The same woman took his call. Again, he was told that the CMD was traveling. He called again after a few days. This time, his call went unanswered. Over the next four to five weeks, the researcher also tried other numbers mentioned on the bank's website. The calls went unanswered. The author must have made more than 100 calls during that period.

The author spoke about his experiences to a friend and sought his suggestions for breaking this stalemate. The researcher's friend explained why his call was not answered: his number was blocked. The friend then suggested that the author try calling from his phone. The researcher did and the call was answered immediately: by the woman who he had spoken to earlier. Again, she informed the researcher that the CMD was traveling and was expected to join the office in a few days. The friend then suggested that the researcher visit the head office personally and seek an appointment with the CMD.

The researcher visited the head office after a few days. He was directed by the reception desk to the 10th floor where the CMD's secretariat was situated. He met the lady who answered his calls and introduced himself and explained the purpose of his visit. He was given the same reply. The researcher also enquired why his earlier calls were not answered. The lady's casual response was that many board lines were not working and hence they could not receive calls. Despite the author's efforts he could not meet the CMD. However, the researcher could interview a retired CMD of the same bank.

The purpose of this narration is to stress the importance of alternative plans. If one respondent does not give consent, another respondent with a similar profile must be identified. The lessons learned from this can be summarized as follows:

- Collection of qualitative data calls for rigorous effort.
- Getting appointments for interviews, especially with decision makers in an organization (or institution) requires persistence.
- The researcher must be patient and remain respectful.
- It is also important to know when to stop trying. If the researcher is sure that a respondent is not likely to be cooperative, it is better to change respondents.
- The researcher should document the challenges and experiences of data collection. This strengthens the researcher's observations, as well as claims about the expected/research outcomes. Documentation of field experiences will support the emerging themes or outcome of research.
- The researcher can better decide what information they extract to support themes. In other words, researchers should offer rationale for extracted information. Based on challenges in data collection and field experience, the researcher can decide the significance of information that support the research outcomes. Themes and evidence should address to research questions.

The Second Incident

The second experience concerned a general manager working in the head office of the bank. The author wrote him an email, explaining his research objective and requesting an appointment for an interview that would be audio-recorded. He also attached the letter of introduction and ethical approval of his institution. The researcher called the general manager after a few days and got an appointment for a meeting.

After a short wait, he was called into the interviewee's office. After introducing himself, the author thanked the general manager for giving his time for the study. He nodded his head in acknowledgement; but it was clear to the author that he was trying to conceal his feelings. The interviewee avoided eye contact and the tone of his voice suggested a lack of interest and unwillingness to cooperate. Thinking that it might help, the author told his interviewee that it was the previous CMD who had recommended his name as a potential respondent.

At the mention of the ex-CMD's name, the general manager became angry. "Don't try to influence me by giving reference of the CMD", he said. The researcher was surprised by this reaction and was not sure how to respond. After a few seconds, he said, *"Sir, if you are ok, only then I will continue with the interview; there is no compulsion [on you to participate]. I will leave your office willingly"*. The author continued, *"I assure you that this interview is entirely confidential, and no data or name will be disclosed. It is only for academic purposes. If you want, you can read this ethical letter [again]*. After reading the letter again, the general manager asked if he could keep the copy. The author replied, *"Sir, this is your copy. I have already sent this letter in the email as well"*. The interviewee then had the letter photocopied for his records.

After he had put the letter away in his desk, the manager appeared more relaxed and said, *"Now you can start the interview."* He was reserved during the interview, skipping a few questions, and giving short responses to others. All the answers were formal and general and hence, the researcher was unable to gain any useful insights.

The interview lasted about eighteen minutes, the shortest in his research. The author thanked the general manager and shut his laptop. Before leaving, he again thanked the general manager for his knowledge, experiences, and insightful comments even though they had fallen far short of what the author expected.

At this signal of the interview's end, the general manager's demeanor changed. He smiled and began to look more confident. He then spoke to the researcher about his experiences in the banking industry. The general manager had worked in another bank before its merger with the present one. His experience in the present bank was not a happy one.

This perhaps explained his reactions and behaviour during the interview. The man also narrated incidents that he claimed had adverse influence on his banking career. This informal conversation lasted for over half an hour, and it was during this time that the author obtained rich information. The respondent even discussed the points that he had chosen not to talk about earlier, as well as elaborate on others. The researcher thanked the man again before leaving.

What were the lessons from this experience?

- The researcher must show both patience and empathy with the respondent.
- The researcher must show utmost courtesy to build mutual trust and never fail to thank the respondent for participating in the study.
- The researcher should make every effort to establish his (or her) bona fide, with all supporting documents if required.
- The researcher must be aware that deeper and richer information may emerge from informal interactions and thus, should document all meaningful information.
- The researcher must keep the promises of confidentiality and anonymity. Most importantly, it is ethical to document information that emerged during the interaction without disclosing the organizations or the respondents' names.
- Research of any kind aims to discover the realities and truth that can help people and organizations. Since the purpose of the research is to discover concepts and gain insights, the question of disclosing the identity of respondents or participating organisations does not arise. Therefore, even information shared during informal conversations can be captured as data for analyses as there is no breach of the ethical contract between the researcher and respondent.

Contents of Letter of Introduction

Researchers are often not sure about how they should introduce themselves to their respondents. The content of the letters and style in which they are written are important for creating a positive image in respondents' views. A poorly worded request letter to a respondent is likely to create hurdles for the researcher—from a delayed response to even a refusal to participate in the study. Therefore, the request letter should be emotionally appealing to respondents.

Sample request letter

Ajay Gupta
Ph.D. Scholar, School of Management and Labour Studies,
Institute of Social Science, Mumbai
19th December 2020

Dear Sir/Madam,
Greetings!

RE: REQUESTING YOUR PARTICIPATION IN MY RESEARCH

I am Ajay Gupta, a research scholar at ISS, Mumbai, conducting a study of Employees' morale in the Banking Sector. The objective is to help management increase employee morale, an essential prerequisite for outstanding organizational performance. As the banking sector has abundant potential for career growth of talented employees. Hence, the issue of employee morale is of particular relevance.

Your rich knowledge and leadership experience can make my study possible and provide valuable insights. I understand that your time is valuable and, therefore, I humbly request you to spare a few minutes for an interview at your workplace. Any time convenient for you is acceptable. The duration of the interview will be about 10 minutes and will be audio recorded.

We will treat your opinion as confidential. I am also attaching a letter affirming our commitment to adhere to the code of ethics for this study. It is my moral and ethical responsibility to not disclose the respondents' identity and personal information to anyone.

My research report will aim to empower employees like you professionally, as well as personally. The findings of my study will be shared with you, as well as suggestions and recommendations that will help your personal and professional growth.

I sincerely look forward to your confirmation of participation in this research.

Yours Faithfully,

Thanks and Warm Regards

Ajay Gupta

Ph.D. Scholar, ISS, Mumbai

Challenges in Getting Consent

Getting the informed consent from the sample for participation in the study can often be a challenge. The chances of refusal are more than of acceptance. For Author's study, he could manage a reply to his communications from only one or two CMDs of the several public sector banks that he approached. It was only later that he learnt that he must first contact the personal assistant of senior managers to get an appointment.

It was even more difficult to convince his subjects about his research and its objectives.

Table 3.1 suggests a framework for data collection. The framework is based on scholarly literature on data collection and the author's experiences during his doctoral work. This framework can be useful for researchers in collecting data.

When and Where to Conduct Interviews

Since qualitative research is a naturalistic inquiry, the researcher should interview the respondents in their natural settings: homes, workplace, etc. The time and place should be convenient for the respondent. This is because responses in settings of the respondent's choice are more likely to accurately reflect the subject's interactions with the environment, which is essential for understanding the subject under investigation. Interviews should be conducted at a time and location that is convenient for the respondent and the interviewer should not question the selection or disagree with it. Instead, if necessary, the interviewer may request the respondent to provide a rationale for the selection. This approach may provide important information about the respondent's work or living environment, and issues associated with them.

Sometimes, the respondent may request the meeting to be held in their homes. As far as possible, the researcher should agree to such requests. Interviews conducted at places that the respondent is not comfortable in may not be very helpful in capturing crucial insights and hence may defeat their purpose of the interviews. However, the decision about which location is more suitable for interviews should be left to the researcher.

The timing of an interview is also important. The respondent should be given enough time to answer the interviewer's questions. Hence, it may not be advisable to not conduct interviews during business hours (or the time when respondents are busy with their work). Nevertheless, the researcher should also try to observe their respondents while they are at work (or engaged in their daily routines) to be able to understand their work environments and try to identify factors that could influence their perceptions, attitudes, and behaviors. In addition, the researcher should raise questions about their observations even if they are not in the interview schedule or questionnaire.

Table 3.1 Steps in collection of field data

S. No.	Steps	Activities	Description	
1	Making contact with your subjects and communicating with them	Contacting respondents Write emails, make phone calls, references, etc	Explain the objectives of the research and expected outcomes, how it can benefit the respondent (and the organization if relevant) Provide documentary proof of the researcher's bona fides, and ethical approvals	
2	Address Concerns and seek consent	Follow up till an appointment is given	Be humble and respectful Seek less time than may be required Build trust and motivation Convince the respondent by discussing all ethical concerns	
3	Fieldwork (This is the most crucial part)	Before the interview	Study and observe the surroundings, as well as people's behavior and interactions	Make casual conversation Try to understand how the organization functions Talk about the subject and seek their views
		During the interview	Note non-verbal signals, facial expressions, extreme emotions, and avoidance attitude	Make a note of the respondent's discomfort, anxiety, expressions of happiness, evasiveness, using symbols These observations and notings help the researcher to probe further during informal interaction. The researcher must know how to probe and intervene when the respondent tries to divert the subject or sidestep questions

(continued)

Table 3.1 (continued)

S. No.	Steps	Activities	Description	
		After the interview	Engage in informal discussion, show appreciation for the respondent's knowledge Probe further into points that may have not been adequately discussed	Do not leave the place immediately after the interview. Engaging in informal discussion is crucial to data collection Raise questions to gain more insightful and authentic information
4	Thanking the Respondent	Show appreciation for the respondent's experience and knowledge, thereby making the space for further communications Promise to share a summary of the findings of the research	Continue communication to share research summary and information during your data collection. It helps to build mutual relations. Make the meeting memorable, and the respondent should have a feeling of "a man of enriching knowledge and accomplishment"	
5	Fieldnotes (Extremely significant step to capture many insights and apposite information)	Write about observations of the surroundings, make a note of non-verbal communications, covering all stages of the interview	Find a safe place outside the interview settings and note everything that you observed, heard and felt. All information and data that could not be recorded must be captured The researcher must organize their thoughts and observations and describe them exactly as they occurred	

(continued)

Table 3.1 (continued)

S. No.	Steps	Activities	Description
6	Recall, conceptualize critical information and write (Necessary for creating inductive codes without losing the essence of captured information. It is an integral part of field-based qualitative research)	For recalling critical concepts in the form of codes, ideas, Record 'wow' moments, or serendipitous information	Write as many codes as possible, based on the interview, observations, and experiences Recall extreme and unexpected moments that can be explored for further investigation. Use in vivo codes to capture the essence. Essence means vital information of the subject under investigation. In Vivo codes are verbatim responses that become codes
7	Transcription	Primary data and field notes	Transcription of primary data and field notes must be done on the same day with mention of time of interview and transcription

Thus, overall, the site of the interview is important to the data collection process. Interruptions, settings in which respondents feel uncomfortable, power dynamics, vulnerability, and other adverse conditions are very likely to affect the quality of data. Therefore, the researcher should strive to provide a comfortable setting for their interviewees, taking care that the time and place of the interviews are based on respondents' convenience. Although participants can also be interviewed in their homes if they want, care must also be taken to ensure that distractions are minimized or managed (Holloway & Wheeler, 2010). A wealth of information can be gained from observing the participant in their natural environment (their home, classroom, or place of employment), surrounded by the material culture of the space they have created, and perhaps even interacting with others there. This adds an ethnographic dimension to the exchange (Edwards & Holland, 2013).

In some situations, online interviews may be preferred, especially when interviewer and interviewees live in different time zones (Chen & Hinton, 1999). Or, either or both may be based in remote locations which makes physical meeting difficult. In such cases, interviews may be conducted over phone or video conferencing, the latter option being most suited for focus groups.

The researcher's flexibility can help to ensure data quality. For example, they may request respondents to record their views and opinions, and then send them to the researcher. With this method, the respondent can avoid being identified. This method may also capture more details than face-to-face, focus group, telephonic, or other data collection methods. It is better suited for studying sensitive issues in which the respondent is hesitant to share opinions due to fear, stigma, social taboos, or fear of being identified, which could jeopardize the respondent, community, and groups.

Types of Interviews

Self-interviews can be conducted without the researcher being present at the interview site. The respondent participates in the interview in their own way and space, which helps them to think, reflect, pause, and decide what to say and how much. (Keightley et al. 2012). The respondents record interviews, and take photos based on the guidelines they are given by the researcher.

E interviews are conducted using email exchanges between researcher and respondent. It has several advantages, such as saving time, resources, producing texts, and flexibility of responses, triggering the next communications, etc. Lewis agrees that a written email response 'allows participants greater scope to think over any questions asked and, as such, often encourages more descriptive and well thought out replies' (Lewis, 2006: 5).

Walking and talking has its roots in ethnography. Kusenbach (2003: 463) describes the 'go-along' method as both more limited and more focused than the generic ethnographic practice of 'hanging out'. Clark and Emmel (2010) provide useful advice for walking and talking. The narratives can add detail to the researcher's' understanding and insights. The location must be suitable for eliciting discussion, and encouraging questioning that might not occur in a room setting. Anderson (2004) suggests that conversations in place, or 'talking whilst walking', offer the potential to add new layers of understanding for the social scientist.

Researchers can also take advantage of **social networking sites** (i.e., Facebook) to reach out to potential participants and conduct interviews (Snee, 2008). Chat rooms and other virtual meeting places may be employed (Shepherd, 2003). The researcher might undertake an ethnographic study of a virtual community (Beneito-Montagut, 2011; Kozinets, 2010). All these possibilities change the nature, dynamic and space of the qualitative interview and raise further questions about research in virtual encounters (Hooley et al., 2012; Mann & Stewart, 2000).

The telephone interview may be more acceptable to some participants who, when discussing sensitive topics, want to ensure confidentiality/privacy. It may also be a convenient option for certain respondents who are busy and cannot find the time otherwise. The disadvantages and limitations of telephone interviews include the lack of face-to-face contact, thereby missing non-verbal cues.

How to Begin an Interview

The preferred approach is to start with questions of a general nature. These help the respondent to relax and establish openness between researcher and respondent. As far as possible, the questions should in the order of increasing complexity, the simpler and general questions being asked first and then followed by more specific queries and requests for more details and clarifications. It may also be necessary for the researcher to be ready with sub-questions with which to obtain more information on the subject.

For example, if the researcher is conducting a study of middle managers' morale, the first question could be "What is your opinion about employees' morale in public sector banks?", and the sub-question, "Where do you evaluate them and what suggestions can you offer to improve employees' morale?".

Luker (2009) recommends that researchers should use 'the hook' to start conversations about the topic of research. Once the stage for the interview has been set through the hook, interviewers often like to ask if the respondent has any questions about the interview before the researcher begins. The researcher opens the interview 'proper' by asking general, broad questions of the 'grand tour' type, for example: 'Please tell me how you started skydiving.' As the interview progresses, the questions gradually focus on more specific and targeted enquiries. Further, Luker offers the idea of 'turn signals' between different aspects of the research topic that comprise the interview, which alert the interviewee that you are shifting from the issue that you have just asked them about, and they are currently discussing, to another area of the research topic.

Janesick (2000, p. 382) recommends starting with the question: What do I want to learn from this study? On the other hand, Charmaz (2006, p. 20) proposes broad questions like: What are the fundamental social processes? Wolcott raises even more broad questions, such as, "What is going on here?" Maxwell (2005, p. 65) refers to these questions as provisional questions, pointing out that even early questions are indicative of theory and methodological direction.

Researchers' prior knowledge about the issues influences the design of the interview questions. According to Agee (2009), novice researchers' initial questions are likely to be too broad and lack a reference point to a specific context. It is referred to as "background". However, questions posed by experienced researchers could be different and referred to as "foreground" (Richardson & Wilson, 1997).

Agee (2009) proposes a dialogic process to move researchers from an overly broad perspective to more focused questions in qualitative research because it is about specific people, places, or experiences of people and how they may interpret them. As a result, qualitative research questions must be descriptive, and good qualitative questions cannot be answered with a simple "yes" or "no." As the study progresses, qualitative questions will become more focused;

however, as Maxwell warns, too-focused questions can lead to "tunnel vision" (Maxwell, 2005, p. 67).

Types of Interview Questions

There are multiple types of interview questions qualitative researchers should raise during the interview. Each question serves a different purpose that collectively seeks answers to research questions. Kvale's (1996) types and purpose of questions has been summarized with examples in Table 3.2. These questions help interviewers to develop their interviewing skills. The questions are arranged in order of significance to reveal deeper information in a naturalistic environment.

Probing Techniques During Interviews

Researchers should learn to probe during interviews. Probing may be useful, even necessary, when the respondent exhibits signs of excitement, unease, anxiety, avoidance, or a gives a brief response that is perhaps unclear. Sensitive issues that could make the researcher vulnerable for a variety of reasons invite investigation. Even if informed consent has been granted, respondents in organizations may be concerned about the possible disclosure of sensitive or confidential information. Frequently, respondents may choose to avoid answering questions. The researcher must exercise caution in such circumstances. They should take note of such instances and determine whether the questioning appears to invade the respondent's privacy. In addition, if probing is limited during the interview, the researcher should politely bring up the issue during the informal conversation that follows the interview.

Depending on the issues, the researcher may face emotional responses during the interview when respondents tell their stories infused with anger, resentment, grief, loss, death, violence, and so on (Bloor et al. 2008; Dickson-Swift et al., 2006; Hubbard et al. 2001; Watts 2008). It has the potential to elicit both conscious and unconscious emotions in both the researcher and the respondent. Researchers should learn to empathize and listen more to their subjects, use probing techniques, and ask follow-up questions.

Bernard (2011) defines seven methods of probing during qualitative interviews, the majority of which require careful and well-judged application at various points within a single interview:

1. Silence. This investigation entails remaining silent after an interviewee appears to have finished answering a question, perhaps nodding your head, and waiting for an interviewee to continue and add more to the topic they were discussing. It gives interviewees time to reflect. Allowing silence to persist during an interview can be difficult for interviewers, but it can be effective if used sparingly.

Table 3.2 Types of interview questions

Types of questions	Purpose of questions	Examples
Introductory questions	To kick start the conversation before moving to the main interview	Generic questions, such as "can you tell me about how you feel working here?"
Follow-up questions	To direct questioning to what has just been said	Nodding head with a "hmm", and repeating significant words
Probing questions	To draw out more complete narratives	Could you say more about it, and could you offer some examples?
Specifying questions	To develop more precise descriptions from general statements	What did you think about it? What did you do then? How did you respond?
Direct questions	To elicit direct responses	Have you been rewarded for good work? What do you mean for competition—healthy or unhealthy?
Indirect questions	To pose projective questions	"How do you believe your colleagues regard competition?"
Structuring questions	To refer to the use of key questions to finish off one part of the interview and open up another, or to indicate when a theme is exhausted by breaking off long irrelevant answers	"I would like to introduce another topic…"
Silence	To allow pauses, so that the interviewees have ample time to associate and reflect, and break the silence themselves with significant information	
Interpreting questions	Similar to some forms of probing questions, to rephrase an interviewee's answer to clarify and interpret rather than to explore new information	Is it correct to behave the way you did? Is it right to view competition the way you do?
Throw away questions	To serve a variety of purposes, i.e., to relax the subject when sensitive areas have been breached	Ohh. I forgot to ask you about the culture of your organization

Source Adapted from Kvale (1996, pp. 133–135)

2. Echo. The interviewer reiterates the last statement made by the respondent. This encourages them to continue and broaden their perspectives.
3. Uh-huh. Saying 'yes,' 'I see,' 'right,' etc. as an interviewee speaks validates the interviewee's statements. It is like silently nodding your head.

4. Tell-me-more. This encourages respondents to elaborate and go further by asking follow-up questions such as "why do you feel that way about it?"
5. Long question. This can be useful at the start of interviews in the grand tour format.
6. Leading. In an interview, any question can lead to a direct probe. For example, do you think this is a really bad way to act?'
7. Baiting. A phased assertion is a type of probe in which the interviewer acts as if they already know something. It helps the respondents opening up or offering the correct information.

> **Important** 'Verbal' interviewees are more likely to go off on tangents because they tell you far more than you need to know for your research topic. Bernard suggests making a 'graceful' interruption and getting the interview back on track. Interviewees who are 'nonverbal' respond to questions with monosyllabic or 'don't know' responses. Furthermore, terminating an interview may be the best course of action when continuing the interview fails to provide relevant information.

In their 2017 book "Critical approaches to questions in qualitative research," Swaminathan and Mulvihill suggest typical interview questions:

1. Can you tell me more about that?
2. Can you give me an example?
3. Am I understanding you correctly?
4. If I might summarize... Did I understand that properly?
5. Is there another story you recall that might illustrate what you mean?
6. How did that come to take place?
7. What do you wish had happened?
8. What if _____ had changed in the following ways?
9. What do you think is happening here?
10. Is there anything else you would like to share with me about...

Such questions expand the scope for comprehending the phenomenon under investigation. They assist researchers in enhancing their thinking and analysis.

Finishing an Interview

After an interview, researchers frequently turn off their recording devices. According to Luker (2009), the researcher must be prepared to record the information that emerges during informal conversations with respondents. However, such a practice may have ethical implications. (Wiles, 2012).

Frequently, researchers attempt to foster an environment in which respondents feel comfortable expressing their opinions. Researchers attempt, to a certain extent, to generate an environment in which respondents experience rapport-building emotions that are conducive to the interview. These two aspects of interviews have been referred to as "conquest or communion," with interviewers exercising power through the use of questioning techniques such as probing to elicit information from interviewees, but also experiencing an emotional interdependence with their interviewees (Ezzy, 2010).

Emotions, as well as cognition or intellect, help people make sense of the social world (Game, 1997; Jagger, 1989). One of us (Holland, 2007) has reviewed discussions of the various ways in which emotions can and do enter the research process, and notes how important acknowledging and reflecting on these emotional dynamics can be for knowledge production.

What to Write in Field Notes

Researchers often find that writing field notes is a challenge. However, the importance of field notes cannot be emphasized enough. Field notes can contain valuable information about the subject of their study. Such information is rarely present in an explicit form. Explicit data is usually recorded in a text, audio, or video format. It is obtained in the participants' replies to the questions asked by the researcher. However, there may be richer and deeper information that may not be explicit. For example, while a respondent might appear to be confident in his (her) answers, the researcher may notice that the body language may indicate otherwise. Such observations must be documented in the researcher's field notes.

To give another example, the respondent might pause for a long time before answering the question. What should the researcher do then? Should they document the reply or record their observations (of the respondent's pause)? Many researchers document only explicit information i.e., the replies given by the respondents but fail to record what they see or sense. Much valuable information and insights can be gained from observations, field experiences and non-verbal communications.

In brief, the researcher must also record the seen and the sensed. They provide deeper information about the phenomenon under investigation. Therefore, researchers should make the effort to document as faithfully as possible signals, cues, and observations. Information captured before and after the interview can be more meaningful and natural than what was obtained in the formal process.

Based on interactions with researchers involved in field research, the author learned that many researchers often find writing field notes a challenge. There are several reasons. It is possible that they do not know what information must be documented; or they may not be aware of the importance of field notes to their study. There are also instances of researchers who maintain field notes but do not know how to use them while analyzing interview data.

According to Okley (1992), researchers should write observation notes, capture, and characterize the participants' responses and reflections. This may include drawings of locations where data was collected, sketches of meetings that depict people seated at the table. Such representations help to comprehend power dynamics, maps of movement, and even stillness.

According to Emerson et al. (2011), writing field notes entails "participating, observing, and taking notes" (p. 21). When taking fieldnotes, Emerson et al. (2011) emphasize the importance of knowing and noting insider terms and concepts.

According to Venkatesh (2008), it is challenging to write field notes during participant observation. It can interrupt the flow of information and cause vital details to be omitted. Writing in such settings can also make the respondents conscious and inhibit the natural flow of information. Both candid and posed photographs are instructive in their own way. But candid photography may produce more naturalistic poses than posed photography. Venkatesh recommended that researchers should note the following points while making field notes.

1. Choose a location where many activities are occurring, if possible.
2. Determine the documentation format, such as audio, video, text, maps, etc.
3. Observe interactions between people, topics of conversation, environment, locations, neighborhood, languages spoken, and describe people and their activities, among other things.
4. What is the justification for your observations and records?
5. What have you chosen to observe and document, and why?
6. What difficulties did you encounter while writing fieldnotes?
7. How will your fieldnotes be analyzed?

Describing Field Experiences

At a workshop on qualitative research on to sexually transmitted diseases, HIV, and tuberculosis among men having sex with men (MSM), the author asked the participants how they collected and processed data. Several participants raised their hands and so, all participants were given an opportunity to speak. Their replies were almost identical:

We go to the field and talk to people who are infected with the disease. We ask respondents a direct question and write down the responses based on their opinions in the research register. Many times, they hesitate to speak, and they avoid looking at us and answering our questions.

The participants were then asked how they analyzed their data. They replied that they are not involved in data analysis. The data collected by them was handed over to another group that is responsible for analysis and interpretation. This was the practice followed in their organization for several years.

In another workshop on a different subject, the reply to the same question was that the respondents themselves collected and analyzed the data. They don't capture their field observations. In fact, they have no clue about capturing observations, documenting, and analyzing them.

These are two examples of common practices. What is missing in them? If the investigators record what respondents have shared with them, what was wrong with that? To understand the situation, we need to examine the question deeper.

The purpose of the examples is to show how people collect and analyze data in different ways, often missing or overlooking crucial information. The purpose was to show how researchers can capture important information in their field notes and make their research more reliable. In the first case, researchers failed to understand participants' behavior and did not make any effort to win their confidence. Rather, they were satisfied with the data they had collected from the interviews, which was handed over to the analysts. For them, documentation of their observations did not seem important. With such an approach to data collection, the chances of crucial insights being missed are high.

In the second case, the researchers only analyzed what their respondents shared with them. Their observations did not seem important enough to be documented. In both examples, the investigators (researchers) failed to record and analyze implicit information, a crucial component of field-based research. Field researchers must document both: words and behaviours.

A key objective of the qualitative researcher is to capture holistic data which is of two types: explicit (explicit) and implicit (latent). When the respondent says something in response to a question, they provide explicit information which the researcher can record in various ways and retrieve as required. But there is also much valuable information that is not expressed but which the researcher can sense. For example, the general manager of an organization may claim that his employees' motivation levels are high, but his body language and facial expression may suggest otherwise. If the researcher captures only the explicit data—the manager's claims—then, obviously, something important has been missed. The researcher must be aware that the body language of the respondents may convey information that may be very different from what was shared in words.

The question that presents itself is whether the (explicit) information shared by the manager is authentic. The answer is, possibly not. This is because the researcher may have overlooked the crucial importance of non-verbal data. Non-verbal data captured by the researcher reinforce the reliability and authenticity of explicit information. Relying only on explicit information is likely to affect the validity of analysis.

Therefore, it is essential that both explicit and implicit data are fully captured. Implicit data is usually documented by the researcher in their field notes. Field notes are also known as research diaries, journals, or field registers. The implicit data must be documented immediately after the interview is over or as soon as possible while the researcher's recall is still strong.

Field notes are an integral part of the data set. If data is analyzed using software, the researcher must write a memo with quotations and codes into the software the same day. Quotations and codes are created from the document (interview data), and the memos are written from the field notes. It is essential for the researcher to understand that field notes contain substantial amounts of tacit information which can be lost if they are not documented and processed quickly. Even if the researcher's field notes and observations are comprehensive, there will still be a significant dependence on recall to make the necessary contextual connections.

Before the Interview

Data collection begins even before the interview. Interviews rarely commence at the scheduled time. Hence, while waiting for the respondent to arrive at the interview site, or get ready for the interview, the researcher should use the time to study the surroundings and make note of what they see, hear or think about the settings. In fact, the interviewer should try to reach the location well before the start of the interview for this purpose. Then, they can make notes on how people behave and interact with each other, or how phones are answered (if the setting is in an office). In fact, anything that can be of importance to the study should be noted. Furthermore, extended waiting times are not unusual. It is important for the researcher to be aware that any show of impatience, annoyance or loss of composure may upset the participants and make data collection more difficult. Figure 3.1 shows the framework for capturing field-notes.

During the Interview

Besides recording the participants' responses, the researcher should also document non-verbal communications in their field notes. Non-verbal communications and cues are present in various forms: display of confidence or discomfort, avoidance, long pauses, or looking away while answering certain questions. Making note of such non-verbal cues serves several purposes. For example, the researcher might notice that opinions stated in the responses may

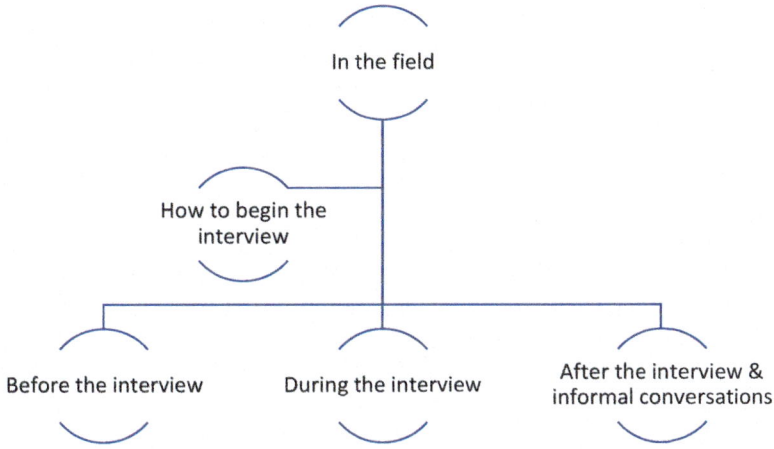

Fig. 3.1 How to begin interview and capture field-notes

be inconsistent with the respondent's expression, offering clues for further probing. The researcher must always be looking for opportunities to probe for richer data.

The researcher's questions should be based on the insights gained from the participants' responses. Questions must be posed in a conversational manner, as naturally as possible, and keeping eye contact with the subject. Looking away, or at the interview schedule, may result in several non-verbal cues being missed. Therefore, the researcher should practice with the questionnaire before meeting the respondents, and only glance at the questionnaire during the interview. One may ask a question that was not in the schedule if it is necessitated by the response to the previous question. Responses to such questions could offer newer dimensions and perspectives on the subject and help the researcher discover the concept in a much more profound manner.

After the Interview

As far as possible, the researcher should not leave the place immediately after the interview. Much useful information can emerge during informal interactions with the respondents. For example, the researcher can use the opportunity to seek answers to certain observations they must have made. Understanding the psychological dimension of these interactions is important. In a formal exchange, such as an interview, the respondent is conscious of his (her) responses being recorded. But after the interview is over and the recording stopped, the respondent is likely to be more relaxed. Warren et al. (2003) call such informal chats as the after-the-interview interactional strips. Natural talks have been used to describe informal conversations (Bernard, 2011). According to some scholars, such as Hammersley and Atkinson (2007), these discussions are still a type of interview, although an informal one.

It is also crucial to remember that qualitative researchers are sometimes social actors who engage in Goffman's (1959) 'impression management,' in which they utilize methods to try to affect others' perceptions of themselves, including, on occasion, deceptive techniques (Bernard, 2011). Sometimes the person (the respondent) is completely ignorant that an interview is taking place, let alone that they are being interviewed. While interviewees may believe they are making comments about seemingly (to them) mundane problems, the researcher is focusing on and digging for components of conversation that answer their study questions (Murphy & Dingwall, 2007). Therefore, to get more and richer information, the researcher should try to prolong informal interactions. One approach to making informal interactions meaningful is by showing sincere appreciation for the respondent's time, knowledge, opinions, and insights.

Often, the researcher is questioned by the audience over the authenticity of information obtained from informal interactions, and its value in data analysis. Another possible query is whether it is ethical to use information shared by the respondent after the formal interview is over. The answer is yes, it is ethical to use such information. In a qualitative study, the researcher is one of the data collection instruments, and hence their thoughts, observations, and experiences are of crucial importance to the objectives of the study.

According to Swain and King (2022), ethical considerations are primarily concerned with consent, which is neither "absolute nor binary, but rather changeable and situational". Obtaining informed consent from respondents to record informal interactions, particularly in written form, in natural settings may not work. Respondents might feel invasive (Akesson et al., 2018; Swain & Spire, 2020).

The researcher must not lose sight of the objective which is to help people and organizations by revealing the truth and then suggest ways of addressing issues.

While Waiting for the Interview

As it is with an important meeting, on the day of the interview, the researcher should reach the location well in advance. This avoids the stress associated with delays and, more importantly, gives the researcher adequate time to study the settings in which the interview will take place. The researcher should carefully observe the environment and people's behaviors and interactions. If there are glass cabins, the researcher should try to note people's movements within them and the expressions on their faces. These observations can help the researcher gain useful insights into the work (or living) environment and people's attitudes.

If the researcher is made to wait longer than one may consider as reasonable, the researcher should also try to understand the reasons. For example, if the location of the interview is an office, is it the office culture to keep people waiting? Or were there other problems? Knowing the reasons will help the researcher to understand the contexts and settings in which the study is conducted. One can decide later as to what extent this information is useful for analyses. While analyzing and interpreting data, researchers should explain the relevance of their observations to their findings.

The researcher can also make use of the wait time to update their journal.

Recommended Readings

Agee, J. (2009). Developing qualitative research questions: A reflective process. *International Journal of Qualitative Studies in Education, 22*(4), 431–447.

Akesson, B., & Hoffman, D. A., "Tony", El Joueidi, S.,& Badawi, D. (2018). So the world will know our story: Ethical reflections on research with families displaced by war. *Forum Qualitative Sozialforschung /Forum: Qualitative Social Research, 19*(3), Art. 5. https://doi.org/10.17169/fqs-19.3.3087. Accessed October 19, 2019.

Anderson, D. G. (1996). Homeless women's perceptions about their families of origin. *Western Journal of Nursing Research, 18*(1), 29–42.

Anderson, D. G., & Hatton, D. C. (2000). Accessing vulnerable populations for research. *Western Journal of Nursing Research, 22*(2), 244–251.

Anderson, J. (2004). Talking whilst walking: A geographical archaeology of knowledge. *Area, 36*(3), 254–261.

Angotti, N., & Sennott, C. (2014). Implementing "Insider Ethnography": lessons from the conversations about HIV/AIDS project in Rural South Africa. *Qualitative Research Online First.* https://doi.org/10.1177/1468794114543402

Baez, B. (2002). Confidentiality in qualitative research: Reflections on secrets, power and agency. *Qualitative Research, 2,* 35–58.

Beauchamp, T. L., & Childress, J. F. (1979). *Principles of biomedical ethics.* Oxford University Press.

Beck, C. (2005). Benefits of participating in Internet interviews: Women helping women. *Qualitative Health Research, 15,* 411–422.

Beneito-Montagut, R. (2011). Ethnography goes online: Towards a usercentred methodology to research interpersonal communication on the internet. *Qualitative Research, 11*(6), 716–735.

Berg, M., & Sigona, N. (2013). Ethnography, diversity and urban space. *Identities, 20*(4), 347–360. https://doi.org/10.1080/1070289x.2013.822382

Bernard, H. R. (2011). *Research methods in anthropology: Qualitative and quantitative approaches.* Thousand Oaks, CA: Sage.

Belousov, K., Horlick-Jones, T., Bloor, M., Gilinskiy, Y., Golbert, V., Kostikovsky, Y., & Pentsov, D. (2007). Any port in a storm: Fieldwork difficulties in dangerous and crisis-ridden settings. *Qualitative research, 7*(2), 155–175.

Bloor, M., Fincham, B., & Sampson, H. (2008). Qualiti (NCRM) commissoned inquiry into the risk to well-being of researchers in Qualitative Research.

Brownlow, C., & O'Dell, L. (2002). Ethical issues for qualitative research in on-line communities. *Disability & Society, 17*(6), 685–694.

Carter, S., Jordens, C., McGrath, C., & Little, M. (2008). You have to make something of all that rubbish, do you? An empirical investigation of the social process of qualitative research. *Qualitative Health Research, 18,* 1264–1276.

Charmaz, K. (2006). *Constructing grounded theory: A practical guide through qualitative analysis.* sage.

Chen, P., & Hinton, S. M. (1999). Realtime interviewing using the world wide web. *Sociological Research Online, 4*(3), 63–81.

Clark, A., & Emmel, N. (2010). 'Using walking interviews', NCRM Realities Toolkit 13. Available at http://eprints.ncrm.ac.uk/1323/1/13-toolkitwalking-interviews.pdf

Cohen, L., & Manion, L. (1994). The interview. L. Cohen & L. Manion (Eds.), *Research methods in education* (4h ed.) Routledge.

Cohn, E. S., & Lyons, K. D. (2003). The perils of power in interpretive research. *American Journal of Occupational Therapy, 57*(1), 40–48.

Corden, A., & Sainsbury, R. (2006). *Using verbatim quotations in reporting qualitative social research: Researchers' views* (pp. 11–14). University of York.

Creswell, J.W. (2003). *Research design: Qualitative, quantitative, and mixed methods approaches* (2nd ed.), CA. Sage: Thousand Oaks.

Crow, G., Wiles, R., Heath, S., & Charles, V. (2006). Research ethics and data quality: The implications of informed consent. *International Journal of Social Research Methodology, 9*, 83–95.

Demi, A. S., & Warren, N. A. (1995). Issues in conducting research with vulnerable families. *Western Journal of Nursing Research, 17*(2), 188–202.

Denzin, N. K. & Lincoln, Y. (Eds.) (2003) *The landscape of qualitative research: Theories and issues* (2nd ed.). Sage.

Dickson-Swift, V., James, E. L., Kippen, S., & Liamputtong, P. (2006). Blurring boundaries in qualitative health research on sensitive topics. *Qualitative Health Research, 16*(6), 853–871.

Dyregrov, K. (2004). Bereaved parents' experience of research participation. *Social Science & Medicine, 58*, 391–400.

Edwards, R., & Holland, J. (2013). *What is qualitative interviewing?* A&C Black.

Ensign, J. (2003). Ethical issues in qualitative health research with homeless youths. *Journal of Advanced Nursing, 43*(1), 43–50.

Emerson, R. M., Fretz, R. I., & Shaw, L. L. (2011). *Writing ethnographic fieldnotes*. University of Chicago press.

Eysenbach, G., & Till, J. E. (2001). Ethical issues in qualitative research on internet communities. *BMJ, 323*(7321), 1103–1105.

Ezzy, D. (2010). Qualitative interviewing as an embodied emotional performance. *Qualitative inquiry, 16*(3), 163–170.

Forbat, L., & Henderson, J. (2003). 'Stuck in the middle with you': The ethics and process of qualitative research with two people in an intimate relationship. *Qual Hlth Res, 13*(10), 1453–1462.

Frick, A., Bächtiger, M. T., & Reips, U.-D. (2001). Financial incentives, personal information and drop-out in online studies. In U.-D. Reips & M. Bosnjak (Eds.), *Dimensions of internet science* (pp. 209–219). Pabst.

Game, A. (1997). Sociology's emotions. *Canadian Review of Sociology/Revue canadienne de sociologie, 34*(4), 385–399.

Goffman, E. (1959). The presentation of self in everyday life. Doubleday.

Goodwin, D., Mays, N., & Pope, C. (2020). Ethical issues in qualitative research. *Qualitative Research in Health Care*, 27–41.

Greenbank, P. (2003). 'The role of values in educational research: the case for reflexivity'. *British Educational Research Journal, 29*(6)

Guillemin, M., & Gillam, L. (2004). Ethics, reflexivity, and "ethically important moments" in research. *Qualitative Inquiry, 10*, 261–280.

Hammersley, M., & Atkinson, P. (2007). *Ethnography: Principles in practice* (3rd ed.). Routledge.
Honebein, P. C. (1996). Seven goals for the design of constructivist learning environments. *Constructivist Learning Environments: Case Studies in Instructional Design, 11*(12), 11.
Hooley, T., Wellens, J., & Marriott, J. (2012). *What is online research?* Bloomsbury Academic.
Holloway, I., & Wheeler, S. (2010). *Qualitative Research in Nursing and Healthcare*. Third edition. Wiley-Blackwell. Oxford.
Holland, J. (2007). Emotions and research. *International Journal of Social Research Methodology, 10*(3), 195–209.
Howe, H., Lake, A., & Shen, T. (2007). Method to assess identifiability in electronic data files. *American Journal of Epidemiology, 165*, 597–601.
Hubbard, J. A. (2001). Emotion expression processes in children's peer interaction: The role of peer rejection, aggression, and gender. *Child Development, 72*(5), 1426–1438.
Humphreys, L. (1970). *Tearoom trade: Impersonal sex in public places*. Aldine.
Hynson, J., Aroni, R., Bauld, C., & Sawyer, S. (2006). Research with bereaved parents: A question of how not why. *Palliative Medicine, 20*, 805–811.
Jagger, A. M. (1989). Love and Knowledge: Emotion in Feminist Epistemolog Feminist Reconstructions of Being and Knowing. *Allison M*, 145–171.
James, T., & Platzer, H. (1999). Ethical considerations in qualitative research with vulnerable groups: Exploring lesbians' and gay men's experiences of health care—A personal perspective. *Nursing Ethics, 6*, 71–81.
Janesick, V. J. (2000). The choreography of qualitative research design. In *Handbook of qualitative research* (pp. 379–399).
Joinson, A. N. (1999). Anonymity, disinhibition and social desirability on the Internet. *Behaviour Research Methods, Instruments and Computers, 31*, 433–438.
Kaiser, K. (2009). Protecting respondent confidentiality in qualitative research. *Qualitative Health Research, 19*(11), 1632–1641.
Kim, J. S. (2005). The effects of a constructivist teaching approach on student academic achievement, self-concept, and learning strategies. *Asia pacific education review, 6*, 7–19.
Keightley, E., Pickering, M., & Allett, N. (2012). Th e self-interview: A new method in social science research. *International Journal of Social Research Methodology, 15*(6), 507–501.
Kozinets, R. V. (2010). *Netnography: Doing ethnographic research online*. Sage.
Kusenbach, M. (2003). 'Street phenomenology: The go-along as ethnographic research tool. *Ethnography, 4*(3), 455–485.
Kvale, S. (1996). The 1,000-page question. *Qualitative inquiry, 2*(3), 275–284.
Kvale, S., & Brinkmann, S. (2009). *InterViews: Learning the craft of qualitative research interviewing*. Sage.
Lewis, J. (2006). Making order out of a contested disorder: The utilisation of online support groups in social science research. *Qualitative Researcher, 3*, 4–7.
Luker, K. (2009). *Salsa dancing into the social sciences*. Harvard University Press.
Maxwell, J. (2005). Qualitative research: An interactive design, (2nd ed.), Thousand Oaks, CA. Sage.
Mann, C., & Stewart, F. (2000). *Internet Communication and qualitative research: a handbook for researching online*. Sage.

McKee, L., Mauthner, N., & Maclean, C. (2000). "Family friendly" policies and practices in the oil and gas industry: Employers' perspectives. *Work Employment and Society, 14,* 557–571.

Mertens, D.M. (2005). *Research methods in education and psychology: Integrating diversity with quantitative and qualitative approaches.* (2nd ed.) Thousand Oaks. Sage.

Murphy, E., & Dingwall, R. (2007). Informed consent, anticipatory regulation and ethnographic practice. *Social science & medicine, 65*(11), 2223–2234.

Okely, J., & Callaway, H. (Eds.). (1992). *Anthropology and autobiography* (No. 29). Psychology Press.

Parkes, C. M. (1995). Guidelines for conducting ethical bereavement research. *Death Studies, 19*(2), 171–181.

Parry, O., & Mauthner, N. (2004). Whose data are they anyway?: Practical, legal and ethical issues in archiving qualitative research data. *Sociology, 38,* 139–152.

Punch, K. F. (1998). *Introduction to social research: Qualitative and quantitative approaches.* Sage Publications.

Punch, M. (1994). Politics and ethics in qualitative research. In N. K. Denzin & Y. S. Lincoln (Eds.), *Handbook of qualitative research* (pp. 83–97). SAGE.

Richards, H. M., & Schwartz, L. J. (2002). Ethics of qualitative research: Are there special issues for health services research? *Family Practice, 19,* 135–139.

Richardson, W.S., & Wilson, M.C. (1997), "On questions, background and foreground", *Evidence Based Healthcare Newsletter,* Vol. 17, (pp. 8–9).

Rosenblatt, P. (1999). Ethics of qualitative interviewing in grieving families. In A. Memon & R. Bull (Eds.), *Handbook of the psychology of interviewing* (pp. 197–209). Wiley.

Rutakumwa, R., Mugisha, J. O., Bernays, S., Kabunga, E., Tumwekwase, G., Mbonye, M., & Seeley, J. (2020). Conducting in-depth interviews with and without voice recorders: A comparative analysis. *Qualitative Research, 20*(5), 565–581.

Ryan, G. W., & Bernard, H. R. (2000). Data management and analysis methods. *Handbook of Qualitative Research, 2*(1), 769–802.

Sieber, J. E. (1998). Planning ethically responsible research. *Handbook of applied social research methods,* 127–156.

Shaw, I. F. (2003). Ethics in qualitative research and evaluation. *Journal of Social Work, 3*(1), 9–29.

Shepherd, N. (2003). 'Interviewing online: qualitative research in the network (ed) society'. In *Paper presented at the AQR Qualitative Research Conference,* 16–19 July, Sydney, Australia. http://espace.library.uq.eduau/eserv/UQ:10232/ns_qrc_03.pdf

Silva, M. C. (1995). *Ethical guidelines in the conduct, dissemination, and implementation of nursing research.* American Nurses Publishing.

Snee, H. (2008). Web 2.0 as a social science research tool, ESRC Government Placement Scheme, The British Library, www.restore.ac.uk/orm/futures/Web_2.0_final_v3.pdf.

Stewart, K., & Williams, M. (2005). Researching online populations: The use of online focus groups for social research. *Qualitative Research, 5*(4), 395–416.

Swain, J., & King, B. (2022). Using informal conversations in qualitative research. *International Journal of Qualitative Methods, 21,* 16094069221085056.

Swain, J., & Spire, Z. (2020). The role of informal conversations in generating data, and the ethical and methodological issues they raise [49 paragraphs]. *Forum Qualitative Sozialforschung/Forum: Qualitative Social Research, 21*(1), Art. 10. https://doi.org/10.17169/fqs-21.1.3344

Tolich, M. (2004). Internal confidentiality: When confidentiality assurances fail relational informants. *Qualitative Sociology, 27,* 101–106.

Tourangeau, R. (2004). Survey research and societal change. *Annual Review of Psychology, 55,* 775–801.

Venkatesh, S. A. (2008). *Gang leader for a day: A rogue sociologist takes to the streets.* Penguin.

Vigo Arrazola, B., & Bozalongo, J. S. (2014). Teaching practices and teachers' perceptions of group creative practices in inclusive rural schools. *Ethnography and Education, 9*(3), 253–269.

Warren, C. A., Barnes-Brus, T., Burgess, H., Wiebold-Lippisch, L., Hackney, J., Harkness, G., & Shuy, R. (2003). After the interview. *Qualitative Sociology, 26,* 93–110.

Watts, J. H. (2008). Emotion, empathy and exit: Reflections on doing ethnographic qualitative research on sensitive topics. *Medical Sociology Online, 3*(2), 3–14.

Weisband, S., & Kiesler, S. (1996). Self-disclosure on computer forms: Meta-analysis and implications. In *Proceedings of CHI96.* Available at http://www.acm.org/sigchi/chi96/proceedings/papers/Weisband/sw_txt.htm

Weiss, R. (1994). *Learning from strangers: The art and method of qualitative interview studies.* Free Press.

Whiteman, N. (2012). Undoing ethics (pp. 135–149). Springer US.

Wiles, R. (2012). *What are qualitative research ethics?* Bloomsbury.

Wiles, R., Charles, V., Crow, G., & Heath, S. (2006). Researching researchers: Lessons for research ethics. *Qualitative Research, 6,* 283–299.

Wiles, R., Crow, G., Heath, S., & Charles, V. (2008). The management of confidentiality and anonymity in social research. *International Journal of Social Research Methodology, 11,* 417–428. Rosenblatt, P. (1999). Ethics of qualitative interviewing in grieving families. In A. Memon, & R. Bull (Eds.), *Handbook of the psychology of interviewing* (pp. 197–209). Wiley.

CHAPTER 4

Codes and Coding

Abstract Qualitative research is built on codes, and researchers must master the processes for creating codes and drawing insights from their analyses. In this chapter, the author discusses the various types of codes and approaches to coding. The quality of the output of qualitative data analysis is dependent on codes and the coding process. Codes may be relevant or irrelevant and the differences between them, and their significance, are explained in this chapter. The chapter discusses code categories and groups, and themes are derived from them. This chapter introduces coding cycle and Computer Assisted/Aided Qualitative Data Analysis (CAQDAS). After a brief overview, the author highlights the advantages and limitations of CAQDAS.

WHAT IS A CODE

A code is typically a word or short phrase that symbolically assigns a summative, salient, essence-capturing, and evocative attribute to a portion of text-based or visual data (Saldaña, 2016). Interview transcripts, participant observation field notes, journals, documents, literature, artifacts, photographs, video, websites, e-mail correspondence, and so on can all be included in the data.

In qualitative research, a code is useful information obtained from data (which is collected from the responses of participants in the study). Codes are assigned through the lenses of the research objectives. They help the researcher study the topic of their interest in greater depth.

All meaningful information in the data, which addresses the research questions, should be coded. It goes without saying that information that is not linked to the research questions should not be coded. The researcher should read the data through the lens of the research question to identify the segments that are significant for the study and code these segments. It is also essential for the researcher to have confidence that the codes contain substantial information from which they can develop themes.

According to Charmaz (2014), coding "creates the bones of your analysis.... Integration will integrate these bones into a functional skeleton" (p. 113). To codify is to organize things systematically, to incorporate anything into a system or classification, and to classify. Codifying is applying and reapplying codes to qualitative data to consolidate meaning and produce explanations by dividing, grouping, and linking data (Grbich, 2012). Analysis, according to Bernard (2013), is "the search for patterns in data and for concepts that explain why those patterns exist in the first place" (p. 338). Coding enables the researcher to organize and classify similarly coded data or "families" because they share a characteristic - the start of a pattern. The researcher utilizes classification reasoning in conjunction with their tacit and intuitive senses to determine which data "look-alike" and "feel like" when they group them (Lincoln & Guba, 1985, p. 347).

Coffey and Atkinson (1996) propose that "coding is usually a mixture of data (summation) and data complication…breaking the data apart in analytically relevant ways to lead toward further questions about the data". The researcher must decide which component of the information gathered by them can be regarded as significant evidence for supporting their findings. A code conveys meaningful information: it may be a word in the text data, a piece of text, an audio or video clip, a photograph or an image, geo-data, graph, or any figure that may reflect information that is directly connected with the research question.

Researchers start the coding process by first reading the data carefully and thoroughly. The data is read and reread as often as necessary to identify meaningful information. When a segment of information is found that is relevant to their inquiry, it is coded (or assigned a preexisting code).

Coding data is a time-consuming exercise. It is an iterative process, requiring multiple readings of raw data, coding, and code review. The researcher must be focused on the task at hand, generating codes, splitting, or linking them, reviewing the codes, deleting and adding codes as required. The process is repeated as many times as is necessary till themes and concepts emerge from the codes with which insights are gained and, depending on the objectives, new theories are proposed.

Coding is also a heuristic process. The word heuristic has its roots in the Greek language (*heureskein*, or "to discover"). Coding explores data to discover information about the phenomenon of study. It is the first stage in the journey of the researcher's understanding of the concepts under investigation. Researchers should not expect to complete data coding in the first coding cycle itself.

Coding involves arranging useful information present in the data in a systematic manner, classifying and categorizing them based on certain characteristics. When the codes are applied (and reapplied) to qualitative data, data can be "segregated, grouped, regrouped and relinked to consolidate meaning and explanation" (Grbich, 2007, p. 21). Where information in another segment is repeated or is similar, a code used earlier can be applied. There is no need to create a new one. However, the researcher has the choice to create the new code and later, during analysis, can merge these 'homogeneous' codes. Codes can be defined in various ways:

They are useful segments of information extracted from responses that helps the researcher to answer research questions.

Codes are significant information from responses that help in deriving concepts to address research objective or research questions.

A code is useful information or idea with reference to research questions and helps in developing variables with reference to research objective.

A code is significant descriptive information identified from responses and written in the form of meaningful words.

Codes are the foundation and the first step towards deriving theme that address the research objectives.

A good code captures the qualitative richness of the phenomenon (Boyatzis, 1998).

In ATLAS.ti, selection of the option 'Apply Code' will display the list of codes already assigned to data segment. If the researcher wishes to assign a code from this list, they can select a code from this list and apply it to the new data segment.

Bernard (2006) states that analysis "is the search for patterns in data and for ideas that help explain why those patterns are there in the first place" (p. 452). After coding is completed, the researcher then tries to discover patterns that are based on various attributes (Hatch, 2002a, 2002b), including, for example,

- similarity (things happening in the same manner)
- difference (they happen in predictably different ways)
- frequency (they happen often or seldom)
- sequence (they happen in a certain order)
- correspondence (they happen concerning other activities or events)
- causation (one appears to cause another).

Code Types

Codes are of several types. Codes based on the first impression are called the initial codes (Saldaña, 2009). Initial, process, in-vivo, focused coding is part of open coding. Coding types have been explained in the next section. Applying multiple codes to a single piece of data is called simultaneous coding. Simultaneous coding is appropriate when the content of the data suggests multiple meanings that necessitate and justify more than one code since complex "social interaction does not occur in neat, isolated units" (Glesne, 2011, p. 192).This happens when two or more codes are applied to, or overlap with a qualitive datum to detail its complexity (Saldaña, 2016). However, two or more significant codes assigned to the same datum may differ in their inferential meanings (Miles et al., 2014). Therefore, researchers should justify the rationale for using simultaneous coding.

Researchers use in vivo codes to capture a key element of information in the respondents' own words. This is also known as verbatim coding. The purpose is to capture deeper meaning of the subject of study derived from the local context. In in vivo code, the segment of the text becomes a code. The main attribute of an in vivo code is that it captures the essence of information present in a segment.

Open Coding

In open coding, the researcher reads the data several times and then begins coding by assigning tentative labels to chunks of data to summarize what was observed or happened. This approach is based on the meaning that emerges from the data and not based on existing theory. The researcher records examples of the words used by the participants and establishes the properties of each code. Then, meaningful information is identified from the responses, which are reflected in certain key words. Thus, it can be said that open coding follows the inductive approach which is explained in the next section. Figure 4.1 shows the approach to inductive, deductive, and abductive coding.

Inductive Coding

Inductive coding is a coding process that does not try to fit data into a pre-existing coding frame, or the researcher's analytic preconceptions. It is also known as bottom-up coding approach (Frith & Gleeson, 2004). The codes are grounded in the data. It is a data-driven coding and is a form of thematic analysis. It uses inductive logic which is often associated with theory-emergent approach in qualitative research. Grounded theory (Glaser & Strauss, 1967) uses the inductive approach of data analysis.

Grounded theory research (Glaser & Strauss, 1967) uses the inductive approach. It relies on observations to establish understandings, procedures, laws, and protocols, and ultimately seeks to develop a substantive and formal theory. Inductive coding is research question-based, data driven and bottom-up process. It is not based on existing theories or concepts.

Glaser and Strauss established an inductive method of qualitative research in which data collection and interpretation are carried out concurrently. Constant comparison and theoretical sampling are employed to aid in the systematic development of theory from evidence. As a result, theories are founded in observations rather than developed in the abstract.

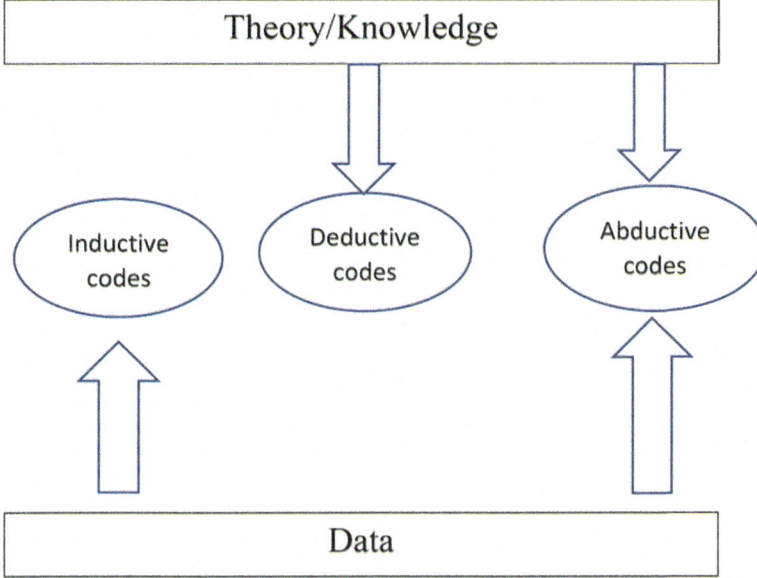

Fig. 4.1 Inductive, deductive, and abductive codes

Inductive analysis is sometimes referred to as the bottom-up method (Frith & Gleeson, 2004). A strong relationship exists between the identified themes and the data itself (Patton, 1990). Inductive analysis entails coding the data without attempting to fit it into a predetermined coding framework or the researcher's analytic ideas.

Deductive Coding

Deductive coding works in the opposite direction. It aims to understand an event by deducing from a general assertion about the conditions. Deductive reasoning is used extensively in quantitative research. A hypothesis is derived from a general law and evaluated against reality by looking for instances that corroborate (or refute) it. Thus, it can be said that deductive techniques are based on a top-down approach (Boyatzis, 1998; Hayes, 1997). Crabtree and Miller (1992) outline a deductive coding approach that uses a priori codes based on the research question and theoretical framework.

Deductive coding is a coding process that tries to fit pre-existing coding frame in the data. This approach is used to explain existing theories, concepts, or knowledge by using the codes available in the data. In other words, researchers do not generate new codes from data. Instead, they examine the data through the lens of pre-existing codes which are assigned to the selected data segment.

Abductive Coding

Abductive logic, first described by Charles Sanders Peirce, is closely associated with pragmatist philosophy. It entails an iterative or dialectical interaction between existing theoretical understanding and empirical data, resulting in fresh interpretation and potentially a modified theoretical framework. To examine theoretical plausibility, mechanisms describing data regularities are compared to well-known regions; they are expanded, and their explanatory value is tested against further data (through their predictive capacity).

In abductive coding, the data is examined using prior knowledge, observations is made and practical wisdom is offered using codes & their description through reflection process. Abductive coding is applied where existing concepts/theories/knowledge from literature is insufficient to understand the phenomena. Or, there data indicate deviations with reference to existing knowledge. The researcher creates codes with the aim identifying new concepts and developing theory that will add to the body of existing knowledge.

Abductive coding helps in theory building, extension and generation of knowledge (Eisenhardt & Graebner, 2007). Abductive methodologies are applied in qualitative and mixed methods,where the aim is to produce theories rather than generalize from a sample to a population, abductive methodologies are widely used.

Hybrid Coding

In some studies, deductive and inductive reasoning are combined in a hybrid approach. The hybrid approach combines the two primary philosophically divergent modes of reasoning. It is the use of both deductive and inductive coding together. The researcher constantly makes choices and selections on how and what to code and how and why data/findings are presented and re-presented (Swain, 2018).

Unlike abductive coding, hybrid coding is simply using inductive and deductive coding to understand the concept emerging from the data.

In Vivo Coding

The phrase *in vivo* is based on a Latin expression that means "with or in the living", which is understood as being within an organism. In science, in vivo refers to experimentation done in, or on, living tissue of a whole, living organism instead of a partial or dead one. In vivo coding is a form of qualitative data analysis that emphasizes the spoken words of the participants. The researcher identifies that section of data in which the respondent has expressed opinions, perspectives and understanding of the topic of the study. The exact words used by the respondent, which offer insightful information about the research, are coded.

The literal meaning of in vivo is "in that which is living" (Strauss, 1987, p. 33, Saldaña, 2016). In vivo coding allows researchers to acquire a comprehensive grasp of the subject under examination using the respondents' own words. In vivo coding is used to allocate a segment of relevant data by utilizing a word, or a brief phrase, that appears in the selected segment of data (Given, 2008). The goal of developing in vivo codes is to capture a key element of information in the respondent's own words. In vivo code is most connected with grounded theory methodology; nevertheless, it is equally pertinent to other approaches that involve discourse and thematic analysis. In vivo coding is coding with the words from the segments of interest. The advantage with in vivo codes is that descriptive codes capture the concepts present in data. It is good practice to create in vivo code categories after the coding process is complete.

Researchers engaged in ethnographic and case study methods use in vivo coding to understand the meaning of slang, jargon, or other expressions that are specific to that language or culture (Manning & Kunkel, 2014a, 2014b). According to Saldaña (2016), in vivo coding helps understand the meanings of words or phrases in a given cultural context. Therefore, it is imperative for the researcher to understand the meanings of context-specific expressions used by participants.

In vivo coding aids academics in generating hypotheses and theories based on the words/short sentences supplied by respondents. In literature on selected approaches, in vivo code is also known as verbatim coding, literal coding, indigenous coding, inductive coding, emic coding, and natural coding (Saldaña, 2016). In vivo codes are used to construct grounded theory (Charmaz, 2014). They help to comprehend the jargon, slang, or other specialized vernacular expressions in a given setting (Manning & Kunkel, 2014a, 2014b). In vivo codes are separated by quotation marks, and researchers code each line of data by aligning it with the terms in the data.

In vivo coding is one type of inductive coding. Codes of this type are not developed by researchers. They are terms used by participants in a certain context, culture, and subculture. In vivo codes reveal the significance of contextual words or phrases that would be missed by other types of coding. In other words, in vivo codes capture the cultural essence of the subject under investigation *in the voices of respondents* in a way that other methods of coding cannot. The meanings inherent in people's experiences are captured by researchers (Stringer, 2014, p. 140). In vivo coding is also useful in action and practitioner research (Coghlan & Brannick, 2014; Fox et al., 2007; Stringer, 2014).

In vivo coding is also used in the study of relational interactions, cultural identities, race, and sexual identities. In fact, this type is used in any study that aims to understand its cultural context. However, since in vivo coding relies on words, it is often criticized for excluding non-verbal communications, information, and symbols. Hence, researchers using in vivo codes should provide a strong rationale for their choice.

Table 4.1 In vivo coding

Quotations	In vivo codes
The human resource manager was transparent in performance appraisal of employees	Transparent in performance appraisal
Employee morale is determined by approach of leadership to equity [in performance appraisal]	Approach of leadership to equity
Management appreciates honest people for their completion of task in time	Appreciates honest people

Researchers should utilize in vivo codes sparingly because the same words/terms/short sentences are commonly coded under in vivo coding and hence may cause problems during the analysis. Examples of in vivo codes are shown in Table 4.1. Meaningful words/concepts in quotations are converted into in vivo codes that appear in the quotations. Writing in vivo codes with an identifying prefix or suffix allows researchers to categorize/group in vivo codes and investigate concepts/phenomena. 'NV' can be prefixed or suffixed in In-vivo codes for easy identification. Researchers must decide how to best use in vivo codes alongside standard codes to investigate the phenomenon.

Axial Coding

Axial coding is used to describe the properties and dimensions of a category and investigates the relationships between categories and subcategories (Saldaña, 2016). According to Boeije (2010), axial coding is used in research to identify which codes in the study are dominant and those that are less significant. In axial coding, synonyms and redundant codes are eliminated, and only the most representative codes are chosen. The objective is to reassemble data that were "split" or "fragmented" during the initial coding process (Strauss & Corbin, 1998, p. 124).

Axial Coding is used for linking codes with each other. It helps to demonstrate causality, or the nature of relationships, between codes. The coding method specifies the properties and dimensions of a category and "seeks to link categories with subcategories and inquiries about their relationships" (Charmaz, 2014, p. 148). Properties and dimensions of a code category refer to contexts, conditions, interactions, and outcomes of a process that enable researchers to determine "if, when, how, and why" something occurs (p. 62) in interview transcripts, field notes, journals, documents, diaries, correspondence, artifacts, video, etc. The example of developing axial codes from emerging code categories is depicted in Table 4.2.

The researcher can connect code categories 'Practicing leadership' and 'Responsible employee' to create an axial code 'Conducive working environment' and explain how they are linked, and what properties they share.

Table 4.2 Axial codes

Axial codes	Code categories	Sub-codes
Conducive working environment	Practicing leadership	Leading by example Demonstrating leadership Encouraging people to grow
Conducive working environment	Responsible employee	Building relations with employees Helping other employees Proving dedication

Figure 4.2 shows a network of Axial codes, code categories, sub-codes, and their associations. They have been linked to express the relationships between code categories, codes and axial codes. Practicing leadership and Responsible employee are code categories. 'Leading by example', 'Demonstrating leadership', and 'Encouraging people to grow' are sub-codes of code category 'Practicing leadership' and their associations with code category is shown. 'Building relations with employees', 'Helping other employees' and 'Proving dedication' are sub-codes of code category 'Responsible employee'. Axial code can be created by linking code categories. The axial code 'Conducive working environment' is shown in Table 4.2 by connecting two code categories.

SELECTIVE CODING

Focused Coding, which is also referred to as "selective coding" or "intermediate coding" in some grounded theory publications, was inspired by the work of Charmaz. In some publications, selective coding is also described as "theoretical coding" or "conceptual coding." After using open and axial coding to identify code categories and subcategories, the researcher uses selective coding to form a core category. According to Ansel Strauss and Juliet Corbin, the purpose of selective coding is to either define a new theory or modify an existing theory based on research. The researcher selects a core category into which other major categories are integrated, and then develops and refines theory.

In grounded theory analysis, a theoretical code functions like an umbrella, encompassing and accounting for all other codes and categories developed to date. The systematic integration of all categories and concepts in theoretical coding around the central/core category suggests a theoretical explanation for the phenomenon (Corbin & Strauss, 2015, p. 13). But, according to Glaser (2005), developing a theoretical code is not always necessary for grounded theory research. In fact, it is preferable not to have one than to have one

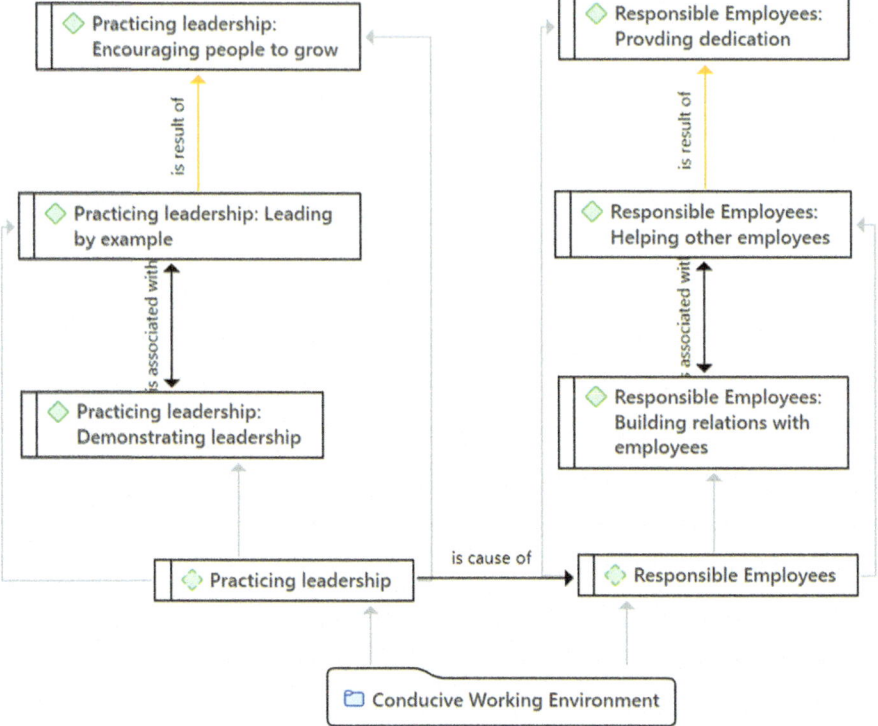

Fig. 4.2 Axial codes network

that is false or improperly applied. A theoretical code specifies the potential relationships between categories and analytic narrative (Charmaz, 2014, p. 150).

Focused coding categorizes coded data using thematic or conceptual similarities. It searches the data corpus for the most significant codes to generate the most prominent categories.

Approaches to Coding in Qualitative Research

There are two major approaches to coding in qualitative research: inductive and deductive. A third, known as abductive, combines the inductive and deductive methods in a back-and-forth process between facts and hypotheses (Pierce, 1978).

Grounded theory research employs three coding strategies: open, axial, and selective (Glaser & Strauss, 1967). According to Saldaña, (2016), grounded theory predominantly considers six coding strategies. All of these can also be utilized in non-grounded theory studies. They include in vivo, process, initial, focused, axial, and theoretical coding. Initial coding is referred to as open coding, while theoretical coding was referred to as selective Coding (Saldaña, 2016).

Open coding is data-driven (Strauss & Corbin, 1998), and the process consists of extracting and categorizing significant information from the data. In axial coding, the researcher seeks to connect code categories to subcategories to determine their relationships. Selective coding is the process of picking a core category related to emerging concepts. The process combines and refines code categories, developing a new theory, or modifying an existing one, based on empirical evidence. Selective coding is used to explain the phenomenon.

Inductive, deductive, and in vivo coding are all types of open coding. In vivo coding allows researchers to understand an idea in the respondent's natural words. It is accomplished by assigning a word/short phrase in the selected data segment (Given, 2008). In some literature, open coding is also known as verbatim, literal, indigenous, inductive emic, and natural coding (Saldaña, 2016). Deductive coding is theory-driven in which researchers generate predetermined codes based on existing theories or concepts. Deductive codes are based on theoretical concepts or themes derived from existing literature. They ensure structure and theoretical relevance from the beginning while allowing for a more inductive exploration of the deductive codes in later coding cycles. Deductive codes are predefined codes used before beginning to code the data (Miles et al., 2014). Closed codes are another name for deductive codes. It is appropriate for counting events based on research questions that follow confirmatory research questions. Closed Coding is widely used in quantitative research.

Major Approaches to Coding in Grounded Theory

In grounded theory research, there are five major coding approaches (Bryant & Charmaz, 2019). Codes are also used in non-grounded theory research. Open, axial, and selective coding are frequently employed in studies relating to grounded and non-grounded theories. Table 4.3 presents an overview of coding approaches and types.

The coding Table 4.4 describes the coding methodologies, grounded theory coding method, and coding cycle proposed by Saldaña (2016), as well as its use in ATLAS.ti. In grounded theory and non-grounded theory research, coding steps and their justifications have been discussed.

Coding approaches in grounded theory i.e., Inductive, deductive, abductive, and hybrid codes have been explained in the next section. Open codings and its types, coding cycle by Saldana and process of theme generation in ATLAS.ti has been explained.

Table 4.3 Major approaches to coding in grounded theory

	Glaser and Strauss	Glaser	Strauss and Corbin	Charmaz and Bryant	Clarke
Research paradigm	Objectivist realist	Objectivist realist Positivist	Objectivist realist Interpretive	Constructivist Interpretivist	Interpretivist Situationist, Constructivist
Approach to coding	Comparisons: incidents; incidents to properties; delimit theory	Substantive Coding: open coding selective coding theoretical Coding	Open coding Axial coding Selective coding	Initial Coding Open Coding Focused Coding Varied coding strategies Axial Coding Theoretical Coding	Open Coding Axial Coding Situational mapping Social worlds/arenas mapping Positional mapping
Analytic tools		Coding families	Conditional matrix	Coding families	
Early coding	Incidents	Line-by-line	Paragraph-by-paragraph; phrase-by-phrase; line-by-line Micro Coding (specific strategic words)	Line-by-line Incident-by-incident Word-by-word	Word-by-word Segment-by-segment

Source Bryant and Charmaz (2019)

Coding Process in Grounded Theory

Figure 4.3 shows the coding process in grounded theory. In open coding, the researcher creates codes based on a first impression that emerges after reading the responses. The researcher starts generating as many codes as possible the information is broken into useful pieces. Open coding is also known as initial Coding.

In the initial coding stage, the researcher's attention will be drawn to many words, phrases, or quotes, that may appear relevant to the research question. They should all be coded.

After the initial coding, the codes are categorized based on their significance to an idea. Code categorization is the process of identifying, selecting, and bracketing codes that share distinct characteristics of the emerging concept.

The next step after code categorization is to describe the properties and dimensions of a category and investigate the relationships between code categories. The researcher identifies the significant codes and their relationships, researchers the relationships between codes/categories are established. This is axial coding, the process of linking codes/categories and describing the nature of their relationship.

Following axial coding, the next step is selective coding, also known as "focused" or "intermediate/theoretical/conceptual coding." Here, the purpose is to define a new theory or modify an existing theory.

Table 4.4 Approaches to coding in qualitative research

Coding approaches	Grounded theory approach to coding method (can be used in non-grounded theory research)	Saldana coding cycle Two coding methods- informant-centric and researcher-centric	Coding and theme Generation in ATLAS.ti
Inductive (data-driven) Deductive (theory-driven) Abductive (dialogical process between data and theory) Hybrid (mix of inductive and deductive)	Open Coding: 1. In-vivo 2. Process 3. Initial 4. Focused	First cycle (Informant-centric) Attribute Initial Concept (analytic) Domain and taxonomic Causation	Coding
	Axial coding	Second Cycle (Researcher-centric) Pattern (inferential) Focused (selective or conceptual) Axial Theoretical (selective or conceptual)	Categorization using axial Coding Code/categories grouping Creating smart codes and smart groups Generating themes and creating reports Creating thematic networks
	Selective coding		Interpreting themes

Selective coding is the process of selecting a core category into which other major categories are integrated and then developing and refining theory.

Selective codes function as an umbrella that encompasses all codes and emerging categories. They systematically integrate the codes/categories/concepts around the code category that explain the phenomenon.

The process of creating selective code, axial codes, sub-codes has been shown in Fig. 4.4. Table and screenshot have also been shown to understand it better.

Table 4.5 helps the researchers to understand codes and their relations in creating code categories, axial codes and selective codes.

In Fig. 4.5, inspirational leadership, practicing leadership, responsible employees and vibrant culture are code categories colored in red. Under each code category, sub-codes are shown. Two axial codes have been created using code categories colored in yellow, i.e., authentic leadership and healthy working environment. Selective codes are created by using axial codes. The

112   A. GUPTA

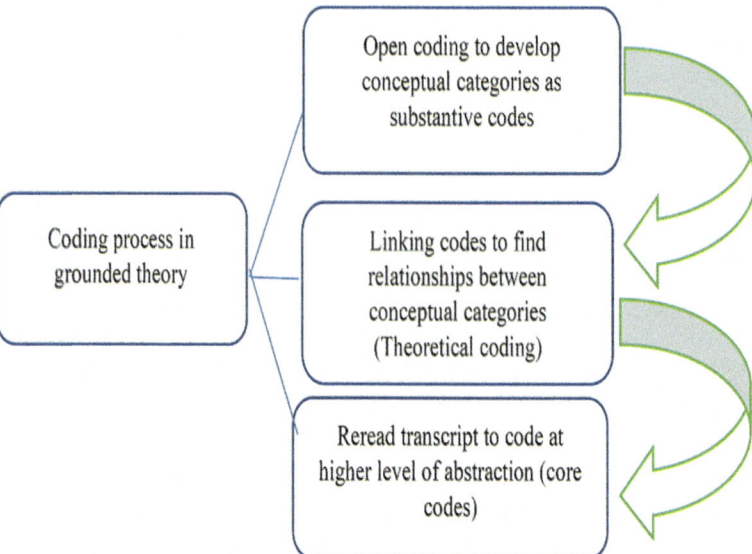

Fig. 4.3 Coding process in grounded theory research

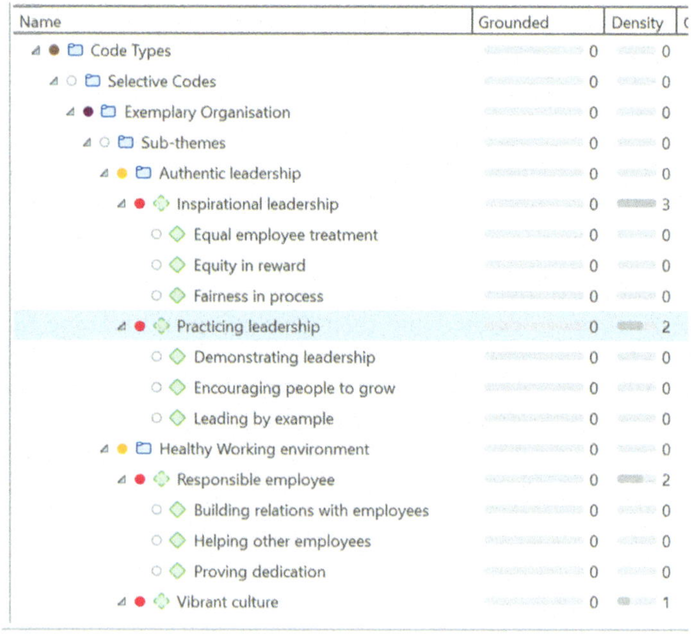

Fig. 4.4 Code types in code manager

Table 4.5 Codes, code categories, axial codes and selective codes

Codes	Code categories	Axial codes	Selective code
Equal employee treatment			
Equity in reward	Inspirational leadership		
Fairness in process			
		Authentic leadership	
Demonstrating leadership			
Encouraging people to grow	Practicing leadership		
Leading by example			
			Exemplary organisation
Building relations with employees			
Helping other employees	Responsible employees		
Proving dedication			
		Healthy working environment	
Conducive working environment			
Harmony among employees	Vibrant culture		
Nurturing employee relation			

selective code- 'Exemplary Organisation' has been created by using axial codes, i.e., healthy working environment and authentic leadership.

The network shows the linkages of sub-codes, code categories, axial codes and selective code.

CODES, CATEGORIES, AND PATTERNS

Codes contain a variety of information. While constructing code categories, researchers should modify and condense the codes. This is done after coding. The data is categorized by the researcher to reflect, each category reflecting a meaning or a way the codes interact. A code represents an individual idea within the data. A category is a collection of codes. Themes representing an overarching concept are derived from code categories. Thus, a code category can be said to represent the collective significance of multiple codes that share information about an idea. Therefore, code categorization can be regarded as the process of recognizing and bracketing codes that share distinct characteristics for the same emerging concept.

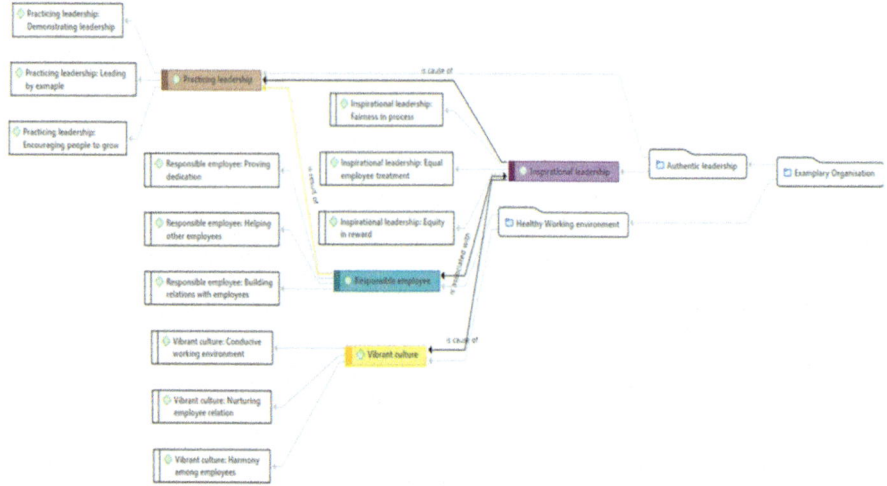

Fig. 4.5 A network of selective code

Richards and Morse (2013) explain, "Categorization is how we move from the diversity of data to *the shapes* of the data, the types of things represented."

The screenshot in Fig. 4.6 shows the codes and code categories. In this example, 'Vibrant culture' and 'Inspirational leadership' are code categories with three codes in each. Table 4.6 shows the codes and categories.

Qualitative researchers must know how to code comprehensively, efficiently, and effectively. The quality of the research is heavily dependent on the quality of the coding. (Strauss, 1987, p. 27). As per Saldaña (2016), coding is merely one way of analyzing qualitative data. Therefore, researchers should determine whether their coding choices are appropriate for their study.

Charmaz and Mitchell (2001) defines coding as the "vital link" between the collection of data and its interpretation. The coding manual of Saldaña (2016) offers several coding methods which researchers should acquaint themselves with, as well as the conditions and contexts in which they must be used.

In the initial coding cycle, researchers can code a single page, a full paragraph, or an entire page of text, depending on their comprehension of the information they have collected.

Researchers must expect their codes and code categories to become more sophisticated and, depending on the methodological approach, more conceptual and abstract as they code and recode. Some of the codes in the initial cycle codes may be subsumed in other codes, renamed, or eliminated in the next iteration. As the researcher proceeds to the second coding cycle, they may reorganize and reclassify coded data into new or different categories. Abbott (2004) compares the procedure to "designing a room; you try it, step back, rearrange a few items, step back again, try a major reorganization, etc." (p. 215). Codes are grouped into categories, and categories to concepts

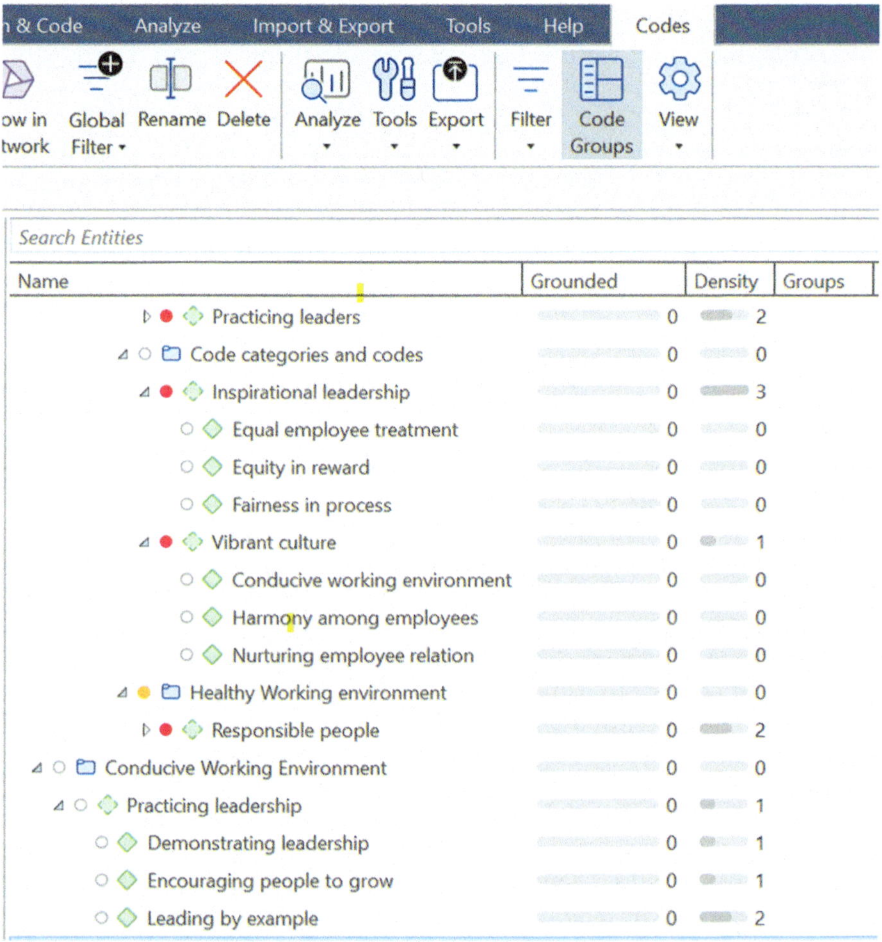

4 CODES AND CODING

Name	Grounded	Density	Groups
▷ ● ◇ Practicing leaders	0		2
▲ ○ 🗀 Code categories and codes	0		0
▲ ● ◇ Inspirational leadership	0		3
○ ◇ Equal employee treatment	0		0
○ ◇ Equity in reward	0		0
○ ◇ Fairness in process	0		0
▲ ● ◇ Vibrant culture	0		1
○ ◇ Conducive working environment	0		0
○ ◇ Harmony among employees	0		0
○ ◇ Nurturing employee relation	0		0
▲ ● 🗀 Healthy Working environment	0		0
▷ ● ◇ Responsible people	0		2
▲ ○ 🗀 Conducive Working Environment	0		0
▲ ○ ◇ Practicing leadership	0		1
○ ◇ Demonstrating leadership	0		1
○ ◇ Encouraging people to grow	0		1
○ ◇ Leading by example	0		2

Fig. 4.6 Codes and categories

Table 4.6 Code and categories

Codes	*Categories*
Conducive working environment	Vibrant culture
Nurturing employee relation	
Harmony among employees	
Equal employee treatment	Inspirational leadership
Fairness in process	
Equity in reward	

(themes), ultimately leading to the development of theory. With categories, "we ascribe meanings; with coding, we compute them", argues Dey (1999). (p. 95).

Patterns are repeated, or regular occurrences of action/data. A "Pattern" is fundamentally concerned with the relationship between unity and multiplicity. A pattern represents a multitude of pieces grouped into a certain configuration" (Stenner, 2014, p. 136). Qualitative researchers seek patterns as they are relatively consistent indications of how humans live and operate to make the world "more understandable, predictable, and manageable" (p. 143). Patterns become more credible evidence for the conclusions of a study since they reveal the routines, importance, and relevance of people's daily lives.

According to Hatch (2002a, 2002b), a pattern is characterized by similarity (things happen the same way), difference (happen in predictably different ways), frequency (often happens or seldom), sequence (happens in a certain order), correspondence (happen concerning other activities or events), and causation (one appears to cause another) (p. 155).

Researchers should remember that the actual process of establishing a theory is considerably more complex than depicted in books and papers. Richards and Morse (2013) explain, "Categorization is how we move from the diversity of data to the forms of the data, the types of things represented." We ascend to more generic, higher-level, abstract constructs through concepts (p. 173). Our capacity to demonstrate how these themes and concepts systematically interrelate contributes to the creation of theory (Corbin & Strauss, 2015). However, Layder (1998) argues that pre-existing sociological theories can inform, if not drive, the initial coding process itself. Although, the formation of an original theory is not necessarily required for qualitative research, it is important to recognize that, whether or not researchers are aware of it, pre-existing theories drive the entire research endeavor (Mason, 2002).

Several texts on qualitative research advocate that researchers begin by "coding for themes." However, Saldana argues that the advice is deceptive because it "muddies the vocabulary". A theme may result from coding, categorization, or analytic reflection, but it is not itself coded (thus, there is no "theme coding" approach in this handbook; nonetheless, there are references to thematic analysis and a section titled "Themeing the Data").

A datum is first and, if necessary, subsequently coded to determine and label its content and meaning following the investigation requirements. "Think of a category as a word or phrase representing an explicit piece of your data, whereas a theme is a phrase or sentence describing more subtle and tacit processes," describe Rossman and Rallis (2003, p. 282, emphasis added). Writing an analytical memo is a crucial step in directing the coding process. The emergence of a group of themes is a positive outcome of the analysis. Still, at the beginning of cycles, coding approaches that study phenomenon such as participant processes, emotions, and values can yield additional interesting discoveries.

Methods in Coding Cycle

According to Saldaña (2016), Coding is done in two cycles: first cycle coding and second cycle coding. Each coding method has a specific purpose, and it is imperative that researchers understand the meaning and methods followed in each coding cycle before they start coding their data.

At the same time, the researcher must also be aware that coding methods are not mutually exclusive. In fact, more than one coding method can be used depending on the research objective and methodological approach. Quinn Patton (2002) rationalizes such an approach saying, "Because each qualitative study is unique, the analytical approach used will be unique" (p. 433). Depending on the nature and goals of the study, the researcher may find that one coding method alone might not suffice and hence two or more are needed to capture the complex processes or phenomena in the data.

Comparing data to data, data to code, code to code, code to category, category to category, category to category, category back to data, etc. suggests that the qualitative analytic process is cyclical rather than linear (Saldaña, 2016). The cyclical collection, coding, and analytic memo writing of data are not linear processes, but "should blur and intertwine continuously from the beginning to the end of an investigation" (Glaser & Strauss, 1967, p. 43). This is one of the major principles developed by the founders of grounded theory, Barney G. Glaser and Anselm L. Strauss, and later elaborated by Corbin (2007).

Lumping and splitting of codes are done in the second cycle of coding. While lumping codes, the researcher collapses (or subsumes) several codes having identical characteristics into a broader code. The researcher may choose to bring many codes together into a code group/smart code or have one code. Or, while analyzing the information, the researcher may find meaningful information in a code that fits better in other codes. In this instance, the code is split. While splitting codes, the researcher moves the associated quotations to the new code.

Saldaña (2016) explained that familiarity with first and second cycle coding methods is necessary for understanding the purpose of codes. The first cycle of coding is known as initial coding, while the second cycle of coding is called thematic coding in which themes-based codes emerge from patterns, theoretical assumptions, and the methodological dimensions.

First Cycle Coding

Attribute Coding Attribute Coding documents pertinent participant data and demographic information for future management and reference. Attribute coding is used in nearly all qualitative studies.

Attribute Codes are assigned to data segments based on what the segment "is about." Thus, data segments are summarized with a label that describes the segment's significance in relation to the research topic. Correct descriptive coding is necessary for a categorized data inventory that provides an overview of the information contained within (Saldana, 2016).

Initial Coding Initial coding is the first major step in grounded theory data analysis. For an initial corpus review, the method is flexible and can incorporate in vivo and process coding. The process of concept coding extracts and labels "big picture" concepts from data. Although several other first- and second-cycle coding methods can generate concept codes, the description demonstrates how this technique can be used independently.

When applied to large amounts of data, in vivo coding is also known as holistic Coding. In holistic coding, a single code captures a sense of the overall data from which the researcher can form code categories.

In vivo codes in data are words or short phrases from the native language of the participant. Folk or indigenous terms from a particular culture, subculture, or microculture can be employed to imply the existence of the group's cultural categories.

Concept Coding Concept coding extracts and labels "big picture" concepts from data. Although several other first- and second-cycle coding methods are capable of generating concept codes, the profile of concept demonstrates how this technique can be used independently. Certain researchers refer to it as analytic coding.

Domain and Taxonomic Coding Domain and Taxonomic coding is an ethnographic method for identifying people's cultural knowledge in order to organize and interpret their behaviors and experiences (Spradley, 1980, pp. 30–31). This knowledge is largely unspoken. Therefore, the ethnographer relies primarily on in-depth interviews comprised of strategic questions to identify these meaning categories.

Causation Coding In this coding method, attribute coding and attribution should not be confused. In attribute coding, attributes are variable descriptions such as age, gender, and ethnicity. Attribution in causation coding refers to reasons or causal explanations.

Causation coding adopts and adapts the premises of the Leeds Attributional coding System (Munton et al., 1999), quantitative applications for the narrative analysis outlined by Franzosi (2010), fundamental principles and theories of causation (Maxwell, 2012; Morrison, 2009; Vogt et al., 2014), and selected explanatory analytic strategies outlined by Miles et al., (2014). The goal is to identify, extract, and infer causal beliefs from qualitative data, such as interview transcripts, field notes from participant observation, and written survey responses.

There is agreement among scholars that at least three factors must be identified when analyzing causality. Munton et al. (1999) outlines the three components of attribution: the cause, the effect, and the linkage between the cause and the effect (p. 9). This is similar to the model presented by Miles et al. (2014), which documents antecedent or starting conditions, mediating variables, and outcomes. Franzosi (2010) argues that "an action has both a cause and an effect" (p. 26), which he refers to as a "triplet."

Second Cycle Coding

Pattern Coding Pattern codes are descriptive or inferential codes that identify an emerging theme, configuration, or explanation. They condense a large amount of data from the initial coding cycle into more meaningful and concise units of analysis. They consist of a "meta code" (Miles et al., 2014, p. 86), the category label identifying similarly coded data. Pattern codes not only organize the corpus, but also attempt to provide meaning for that organization. Pattern coding is utilized to classify these summaries into fewer categories, themes, and concepts. It is comparable to cluster analytic, and factor analytic techniques utilized by quantitative researchers for statistical analysis.

According to Miles et al., (2014, pp. 86–93), pattern coding is appropriate for condensing large amounts of data into a smaller number of analytic units, developing major themes from the data, searching for rules, causes, and explanations in the data, examining social networks and patterns of human relationships, forming theoretical constructs and processes, and examining social networks and patterns of human relationships (e.g., "negotiating," "bargaining").

Focused Coding Various grounded theory publications refer to focus coding as selective, intermediate, or conceptual coding (Saldaña, 2016). Focused coding is applied after In Vivo, Process, and Initial Coding, which are all first-cycle grounded theory coding methods, but it can also be used in conjunction with other data categorization techniques. Focused coding identifies the most frequent or significant codes in a data corpus in order to develop the most salient categories and "requires decisions regarding which initial codes make the most analytical sense" (Charmaz, 2014, p. 138). Focused coding is suitable for nearly all qualitative studies, but particularly for grounded theory studies and the development of major categories or themes from data.

Axial Coding Initial coding and, to a lesser extent, Focused coding are the foundations for Axial coding. The objective is to strategically reassemble data that was "split" or "fractured" during Initial coding (Strauss & Corbin, 1998, p. 124). According to Boeije (2010), the purpose of Axial coding is to "determine which [codes] in the research are dominant and which are less significant… [and] reorganize the data set by crossing out synonyms, removing redundant codes, and choosing the most representative codes" (p. 109).

Axial coding is suitable for grounded theory studies and a wide variety of data types (e.g., interview transcripts, field notes, journals, documents, diaries, correspondence, artifacts, and video). Concept codes can be used as Axial codes for the second cycle.

Theoretical Coding Theoretical coding is referred to in some grounded theory publications as "selective coding" or "conceptual coding." A theoretical code functions like an umbrella, encompassing and accounting for all other codes and categories developed in grounded theory analysis to date. Integration of codes begins with identifying the primary theme of the research, also known as the central or core category in grounded theory, which "consists of all the products of analysis condensed into a few words that seem to explain what 'this research is all about'" (Strauss & Corbin, 1998, p. 146) via "what the researcher identifies as the major theme of the study" (Strauss & Corbin, 1998, p. 146).

According to Stern and Porr (2011), the central/core category identifies the major conflict, obstacle, problem, issue, or concern for participants. The systematic integration of all categories and concepts in theoretical coding around the central/core category suggests a theoretical explanation for the phenomenon (Corbin & Strauss, 2015, p. 13). The theoretical code, such as survival, balancing, adapting, and personal preservation while dying, is an abstraction that represents the theory's integration (Glaser, 2005, p. 17). It is a key phrase or keyword that initiates a discussion about a theory.

Theoretical coding is a necessary final step in the development of grounded theory. However, theory development is not always necessary in qualitative research. According to Hennink et al. (2011), research that applies pre-existing theories in different contexts or social situations or that elaborates or modifies pre-existing theories can be just as substantive. However, the most crucial aspect of this cycle of theory development is to answer the "how" and "why" questions to explain phenomena in terms of how they operate, develop, and compare, and why they occur under specific conditions (pp. 258–61, 277).

Computer Assisted/Aided Qualitative Data Analysis (CAQDAS)

The application of computers in data analysis has been in practice since the 1980s. Muhr (1991), the founder of ATLAS.ti, was among the earliest advocates of using computer-assisted software for analyzing qualitative data. Researchers can organize and analyze their data (Kelle & Bird, 1995)—interviews, field notes, audio, video, images graphs and geodata—with the help of computer-assisted qualitative data assessment software (CAQDAS). Students and researchers may initially find the CAQDAS challenging to learn. However, with practice and experience, CAQDAS becomes easy to use.

ATLAS.ti uses artificial intelligence to quickly retrieve and display words, concepts, and phrases that the researcher searches in the data. ATLAS.ti 23 has AI (artificial intelligence) coding features that researchers can use to code their data with much precision and organize codes based on significance. Further, the auto-coding function can filter, link groups, and enable researchers to perform various tasks such as comparing, relating, interpreting, and building theory from the data (Lewins & Silver, 2007). Saldaña (2021) highlighted the advantages that CAQDAS has over the manual paper and pencil data coding and analysis method. Researchers can use CAQDAS across research methods and disciplines.

Advantages and Limitations of Computer-Assisted Analysis of Qualitative Data

Computer assisted qualitative data analysis software facilitates several functions such as storing, classifying, indexing, filtering, retrieving, and reporting data to help people across disciplines and firms.

ADVANTAGES

- CAQDAS-based qualitative data analysis saves time on data analysis, research refinement, and report generation from standard to customized.
- Enables for more rigorous data analysis, establishing credibility and validity by allowing for data analysis with visible audit trails.
- Reduces bias by allowing self and peer review of the analysis process.
- Replaces file cabinets, saving space and allowing for more systematic data management.
- Allows for the graphical representation of themes, models, and concepts.
- Examines complex relationships in data and engages in theory development.
- Enables collaborative research on the same project.
- Analyzes a variety of data, including text, images, audio, video, geodata, and images, using various social media platforms.
- Provides a systematic approach to coding, data analysis, and data review.
- Refines findings in relation to research questions, respondent attributes, and demographics.

LIMITATIONS

- Qualitative data analysis using CAQDAS is perceived as difficult in comparison to quantitative research, resulting in avoidance. However, its effective use depends on the researcher's ability to learn and apply CAQDAS techniques.
- Adaptability to CAQDAS is a challenge.
- Challenges with learnability, as qualitative data analysis tools require a significant amount of effort and time to master. However, this is true for most other software (i.e., Python, STATA, R, etc.)
- Poses difficulties in terms of data analysis, presentation, and reporting

- CAQDAS is meant primarily for qualitative data analysis. The researcher must be familiar with methodologies, research design, and the philosophical paradigms of their study to be able to use CAQDAS to their benefit.

What the Researcher Should Know About CAQDAS

Computer Assisted Qualitative Data Analysis Software is a tool for analysing qualitative data. Its utility and effectiveness depend on the researcher's understanding of how the software works. Researchers intending to use CAQDAS should know what it can do and cannot.

What the Software Does

- Software helps to structure the analysis. The researcher is provided instant access to all project components like data files (i.e., transcripts), analytical tools, etc.
- The researcher can 'explore' the data, such as searching a large volume of text for one word or phrase.
- Qualitative analysis software generates codes and retrieves sections of the coded text.
- It manages projects and organises data.
- The researcher can carry out searches and interrogations of the database, looking for correlations between codes,
- Several writing tools are available: the researcher can write notes, comments, and annotations.
- The researcher can generate various types of reports for printing or export to another package.

What the Software Does not Do

- It is not capable of analytical thinking. However, the researcher can use the various tools and functionalities to help in analysis.
- In general, the researcher must decide what can and cannot be coded. While some software allows for automatic coding of text search results, the researcher must verify what has been automatically coded.
- It does not reduce bias, increase reliability, or simply improve the quality of the analysis on its own.
- Qualitative analysis software will not tell the researcher how to conduct an analysis of your data.
- It does not perform statistical calculations, although some programmes will generate simple counts and percentages.

Recommended Readings

Abbott, A. D. (2004). Methods of discovery Heuristics for the Social Sciences.
Bernard, H. R. (2006). *Social research methods: Qualitative and quantitative approaches*. Sage.
Bernard, H. R. (2013). *Social research methods: Qualitative and quantitative approaches*. Sage.
Boeije, H. (2010). *Analysis in qualitative research*. Sage Publications Ltd.
Boyatzis, R. E. (1998). *Transforming qualitative information: Thematic analysis and code development*. sage.
Bryant, A., & Charmaz, K. (Eds.). (2019). *The SAGE handbook of current developments in grounded theory*. Sage.
Charmaz, K. (2014). *Constructing grounded theory* (2nd ed.). Sage.
Charmaz, K., & Mitchell, R. G. (2001). Grounded theory in ethnography. In *Handbook of ethnography* (pp. 160–174).
Coffey, A., & Atkinson, P. (1996). *Making sense of qualitative data: Complementary research strategies*. Sage Publications, Inc.
Coghlan, D., & Shani, A. B. (2014). Creating action research quality in organization development: Rigorous, reflective and relevant. *Systemic Practice and Action Research, 27*, 523–536.
Coghlan, D., & Brannick, T. (2014). *Doing Action Research in your own organization* (4th ed.). London. Sage.
Corbin, J. (2007). Strategies for qualitative data analysis. *Journal of Qualitative Research*, 67–85.
Corbin, J., & Strauss, A. (2015). *Basics of qualitative research: techniques and procedures for developing grounded theory* (4th ed.). Sage.
Crabtree, B. F., & Miller, W. F. (1992). A template approach to text analysis: Developing and using codebooks.
Creswell, J. W. (2015). Revisiting mixed methods and advancing scientific practices. In *The Oxford handbook of multimethod and mixed methods research inquiry*.
Dey, I. (1999). Grounding grounded theory: Guidelines for qualitative inquiry. *No title*.
Eisenhardt, K. M., & Graebner, M. E. (2007). Theory building from cases: Opportunities and challenges. *Academy of Management Journal, 50*(1), 25–32.
Fox, M., Martin, P., & Green, G. (2007). *Doing practitioner research*. Sage.
Franzosi, R. (Ed.). (2010). *Quantitative narrative analysis* (No. 162). Sage.
Friese, S. (2019). Qualitative data analysis with ATLAS. ti. Sage.
Frith, H., & Gleeson, K. (2004). Clothing and embodiment: men managing body image and appearance. *Psychology of Men & Masculinity, 5*(1), 40–48.
Gioia, D. A., Corley, K. G., & Hamilton, A. L. (2013). Seeking qualitative rigor in inductive research: Notes on the Gioia methodology. *Organizational Research Methods, 16*(1), 15–31.
Glaser, B. (2005). *The grounded theory perspective III: Theoretical coding*. Sociology Press
Glaser, B., & Strauss, A. (1967). Grounded theory: The discovery of grounded theory. *Sociology the Journal of the British Sociological Association, 12*(1), 27–49.
Glesne, C. (2011). *Becoming qualitative researchers: An introduction* (4th ed.). Pearson Education Inc.

Grbich, C. (2007). *An introduction: Qualitative data analysis*. London, UK: Sage.
Grootenhuis, M. A., & Last, B. F. (1997). Predictors of parental emotional adjustment to childhood cancer. *Psycho-Oncology, 6*(2), 115–128.
Grbich, C. (2012). Qualitative data analysis: An introduction. *Qualitative Data Analysis*, 1–344.
Gupta, A. K. (2014). A comparative study of middle managers morale in two public sector banks in India.
Hatch, J. A. (2002a). Doing qualitative research in education settings. Suny Press.
Hatch, J. A. (2002b). *Doing qualitative research in educational settings*. State University of New York Press.
Hayes, N. (1997). Theory-led thematic analysis: social identification in small companies. In N. Hayes (Ed.), *Doing qualitative analysis in psychology*. Psychology Press.
Hennink, M., Hutter, I., & Bailey, A. (2011). *Qualitative research methods*. Sage Publications.
Kelle, U., & Bird, K. (Eds.). (1995). *Computer-aided qualitative data analysis: Theory, methods and practice*. Sage.
Layder, D. (1998). Sociological practice: Linking theory and social research. *Sociological Practice*, 1–208.
Lewins, A., & Silver, C. (2007). Qualitative coding in software: principles and processes. Using software in qualitative research. *Using Software in Qualitative Research*. 10.9780857025012.
Lincoln, Y. S., & Guba, E. G. (1985). *Naturalistic Inquiry*. Sage.
Manning, J., & Kunkel, A. (2014a). Making meaning of meaning-making research: Using qualitative research for studies of social and personal relationships. *Journal of Social and Personal Relationships, 31*(4), 433–441.
Manning, J., & Kunkel, A. (2014b). *Researching interpersonal relationships: Qualitative methods, studies, and analysis*. Sage.
Mason, J. (2002). *Qualitative researching* (2nd ed.). Sage.
Maxwell, J. A. (2012). The importance of qualitative research for causal explanation in education. *Qualitative Inquiry, 18*(8), 655–661.
Merton, R. K. (1987). The focussed interview and focus groups: Continuities and discontinuities. *The Public Opinion Quarterly, 51*(4), 550–566.
Miles, M. B., Huberman, A. M., & Saldaña, J. (2014). *Qualitative data analysis: A methods sourcebook* (3rd ed.).
Morrison, K. (2009). *Causation in educational research*. Taylor & Francis eBooks DRM Free Collection.
Morrison, K. (2012). *Causation in educational research*. Routledge.
Muhr, T. (1991). ATLAS/ti—A prototype for the support of text interpretation. *Qualitative Sociology, 14*(4), 349–371.
Munton, A., Silvester, J., Stratton, P., & Hanks, H. (1999). *Attributions in action*. Wiley.
Patton, M. Q. (2002). Two decades of developments in qualitative inquiry: A personal, experiential perspective. *Qualitative Social Work, 1*(3), 261–283.
Pierce, C. (1978). Pragmatism and abduction. In C. Hartshorne & P. Weiss (Eds.), *Collected papers* (Vol. 5, pp. 180–212). Harvard University Press.
Quine, S., Bernard, D., & Kendig, H. (2006). Understanding baby boomers' expectations and plans for their retirement: Findings from a qualitative study. *Australasian Journal on Ageing, 25*(3), 145–150.

Richards, L., & Morse, J. M. (2007). Coding. In *Readme first for a user's guide to qualitative methods* (pp. 133–151).
Richards, L., & Morse, J. M. (2012). *Readme first for a user's guide to qualitative methods*. Sage publications.
Richards, L., & Morse, J. M. (2013). *Readme first for a user's guide to qualitative methods* (3rd ed.). London, England. Sage.
Rossman, G. B., & Rallis, S. F. (2003). *Learning in the field: An introduction to qualitative research* (2nd ed.). Sage Publications.
Saldana, J. (2016). Saldana-coding manual for qualitative research-Introduction to codes & coding. *The coding manual for qualitative researchers*, 1–39.
Saldaña, J. (2009). *The coding manual for qualitative researchers*. Sage.
Saldaña, J. (2021). *The coding manual for qualitative researchers*. sage.
Spradley, J. P. (1980). *Making an ethnographic record*. Participant observation.
Spradley, J. P. (2016). *Participant observation*. Waveland Press.
Stebbins, R. A. (2001). What is exploration. *Exploratory Research in the Social Sciences*, 48, 2–17.
Stern, P. N., & Porr, C. J. (2011). *Essentials of grounded theory*.
Stenner, P. (2014). Pattern. In Lury, C., & Wakeford, N. (Eds.), Inventive methods: The happening of the social (pp. 136–146). New York: Routledge.
Strauss, A. L. (1987). *Qualitative analysis for social scientists*. Cambridge University Press.
Strauss, A., & Corbin, J. (1998). Basics of qualitative research techniques.
Stringer, E. T. (2014). *Action research* (4th ed.). Sage Publishing.
Swain, J. (2018). *A hybrid approach to thematic analysis in qualitative research: Using a practical example*. SAGE Publications Ltd.
Timmermans, S., & Tavory, I. (2012). Theory construction in qualitative research: From grounded theory to abductive analysis. *Sociological Theory*, 30(3), 167–186.
Vogt, W. P., Gardner, D. C., Haeffele, L. M., & Vogt, E. R. (2014). *Selecting the right analyses for your data: Quantitative, qualitative, and mixed methods*. Guilford Publications.
Wolcott, H. F. (1994). *Transforming qualitative data: Description, analysis, and interpretation*. Sage.

CHAPTER 5

Data Analysis Methods

Abstract This chapter introduces analysis of qualitative data. After a brief overview, coding in ATLAS.ti, including auto-coding, redundant coding, code merging, code splitting, and when and how to use them, are explained with the help of coding frameworks and supported by screenshots. This is followed by an introduction to ATLAS.ti and its functioning. Then, with the help of examples and screenshots, the author explains quotations, coding data and field notes, comments, and memo writing. Codes are the crucial first step in the analysis of qualitative data, which is followed by their grouping and categorization and eventually, identification of themes. The process of identifying themes is explained with the code-theme hierarchy and the frameworks for data analysis.

An Overview of ATLAS.ti

ATLAS.ti is a computer assisted qualitative data analysis software for analyzing text, graphical, audio, video, geo, social media, and other types of data from primary and secondary sources. It offers tools to manage, extract, compare, explore meaningful segments of information to help the researcher gain insights from data. It visualizes, integrates, discovers, and explores data to build new knowledge and theory. ATLAS.ti can be used by a single user or several people working in groups to work on a project. ATLAS.ti facilitates extraction categorizing, filtering, and interlinking data segments, and discovering patterns and themes from a large variety and volume of source documents.

ATLAS.ti is used in social sciences, engineering, business administration, tourism, media, architecture, graphology, psychology, linguistics, law

© The Author(s), under exclusive license to Springer Nature Switzerland AG 2024
A. Gupta, *Qualitative Methods and Data Analysis Using ATLAS.ti*, Springer Texts in Social Sciences,
https://doi.org/10.1007/978-3-031-49650-9_5

and criminology, theology, history, medicine, literature, and geography etc. ATLAS.ti analyses field-based (primary), secondary, observation, transcripts, screenshots, diagrams, pdf, audiovisual, and locative data.

Before using ATLAS.ti, the researcher must become familiar with the meaning of certain important concepts in qualitative research. These concepts are not new; they are rooted in the qualitative research paradigm and hence their understanding is essential, whether the researcher uses computer-assisted software for analysis or manually analyses data. The researcher should know how to identify these concepts and their meanings from interviews, documents, articles, or other material. The researcher must also be conversant with the terminologies associated with the various types and methods of qualitative research.

This chapter discussed the four basic entities (documents, quotations, codes, and memos) in qualitative research, and how they are used in ATLAS.ti software. We will also discuss how to identify these concepts and locate them in the documents that will be analyzed. There are other concepts also, but they will be explained and discussed as and when the need arises.

About the Examples Used in the Discussions in This Book

The book uses several examples from studies spanning several methods and disciplines to explain qualitative data analysis in ATLAS.ti. These examples will help researchers to understand the challenges they must face in data collection, analysis, and reporting. Frameworks for generating codes, quotations, writing memos, forming networks, themes, code categories and code groups have been explained with the help of examples.

The author has made a special effort to give theoretical and conceptual clarity for working with ATLAS.ti. Data analysis processes have been explained step wise using appropriate screenshots of operations in ATLAS.ti. Researchers can learn more by practicing with their data while using the book. Researchers should keep themselves updated with theoretical underpinnings and concepts associated with their study for better understanding of the subject of their study.

The frameworks for data analysis are meant to help researchers to understand the data analysis process and visualize the linkages between concepts. With a better understanding, they should be able to plan the analysis of their research data in ATLAS.ti.

A Comprehensive Framework for Data Analysis Using ATLAS.ti

A comprehensive framework for qualitative data analysis in ATLAS.ti is shown in Table 5.1. The framework is based on qualitative research methodologies and can be used across disciplines. The framework can also be used even if the researcher is not using ATLAS.ti or other software. Some variations in the framework are possible depending on the research methods and the terminology used, but its basic structure will remain unchanged. It is imperative for researchers to be thorough with the theoretical concepts and methodological approaches in qualitative research.

Methodology is the backbone of qualitative research. ATLAS.ti is a powerful tool for analyzing qualitative data and its value is dependent entirely on the quality of input data. Therefore, it is the methodology used and the rigor with which the researcher conducts fieldwork that determines the outcomes of the study, and not their expertise with the software.

In one study, the author collected data from 20 respondents (Table 5.1). There were also 20 field notes in addition to the data, one for each respondent. The interviews were categorized in 5 groups.

Data analysis began with a thorough reading of the responses to identify the informative segments. Coding and memo writing followed. The researcher created 400 codes and grouped them. There were 30 groups based on the homogeneity of the codes, and their attributes. Memo groups were also formed for the memos. The author then created smart codes and smart groups from the code groups and codes. The smart groups and smart codes became the themes which were interpreted in the study's context. Quotations and memos provided the justification for the researcher's interpretations.

Table 5.1 A framework for data analysis framework using ATLAS.ti

Document and groups level	Thematic Interpretation			Within and across documents		
	Themes (5–8)					
Evidences				*Evidences*		
Respondents' statements	Smart codes	Code categories	Smart groups	Researcher's field experiences		
Clusters of quotations (50–100)	Code categories (less than 10) (Grouping code groups)			Memo groups (40–50)	Memo comments	Coding & Code grouping
Naming quotations	Code groups (30) (homogeneity)			Linking to codes	Converting into documents	
Source Target Verbatim Heuristic Quotations (200)	Merging Splitting Codes (400) Inductive Abductive Grounded theory/ Phenomenology/Case study/ Ethnography/Narratives/ Thematic		Linking Deductive	Methodological Memos (50) Observations	Theoretical Experiences	Descriptive Insights
Document groups (05)	Documents (20)			Field notes (20)		

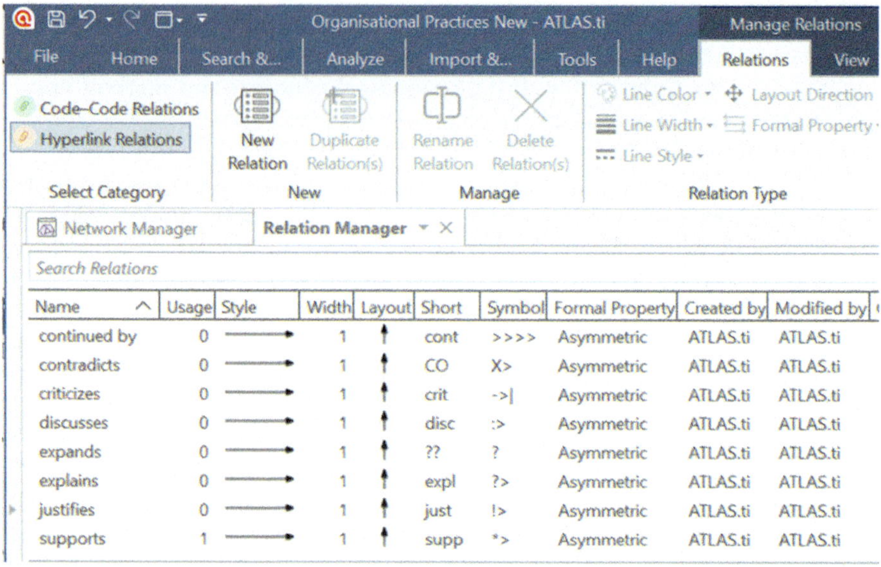

Fig. 5.1 Hyperlinks for quotations

Thematic interpretations should be supported by researcher's field experiences and quotations from respondents' answers. Themes can be found throughout the data. Therefore, researchers should search themes within a category of respondents or across categories of respondents to understand how themes have been created and their significance for that category of respondents. Researchers can generate themes across categories of respondents, based on several parameters like demographic, socio-economic, professional, contexts, social and economic etc. The themes identified by the researcher should be supported by relevant quotations and memos.

Quotations are evidence of themes. They are significant segments of information found in the statements of the respondents, which are linked to the research questions. Memos are notes of the researcher's field experiences. Quotations and memos provide evidence to support the insights into the phenomenon under investigation.

The connections between quotations are shown by hyperlinks. In a hyperlink of a causal relationship, one quotation become the source and another one becomes the target. Figure 5.1 shows the hyperlinks list used to link quotations.

Analysis of Qualitative Data

Data analysis in qualitative research is the methodical search and organization of qualitative data for understanding the phenomenon of study. Qualitative data is mostly subjective. The information is contained in the transcripts of interviews and focus group discussions, the researcher's observations, field notes, pictures taken during the study, and various other sources.

The data analysis process involves coding or categorizing the data for which the researcher must read the transcripts, often multiple times, and look for similarities and differences. Coding is followed by categorizing them after which themes are identified from significant patterns in the data. Data analysis concludes with interpretation of the themes and the researcher's conclusions.

Analysis of qualitative data requires thorough understanding of the importance of codes and coding processes, comments, quotations, memos, and themes. Lack of familiarity and their improper use is a major cause of insufficient rigor in data analysis and weak interpretation of results. This chapter discusses codes, comments, quotations, memos, and themes with the aim of providing the necessary foundations for qualitative data analysis in ATLAS.ti.

The author has endeavored to maintain the logical sequence followed in computer-aided qualitative data analysis in the discussions. However, at certain places the reader may find it necessary to retrace steps for a better understanding of the qualitative data analysis process, as well as the interconnections between the steps.

Figure 5.2 shows the data analysis process that researchers should follow in data analysis across methods. Table 5.2 shows the explanation of the data analysis process. The process is useful for qualitative data analysis with or without CAQDAS.

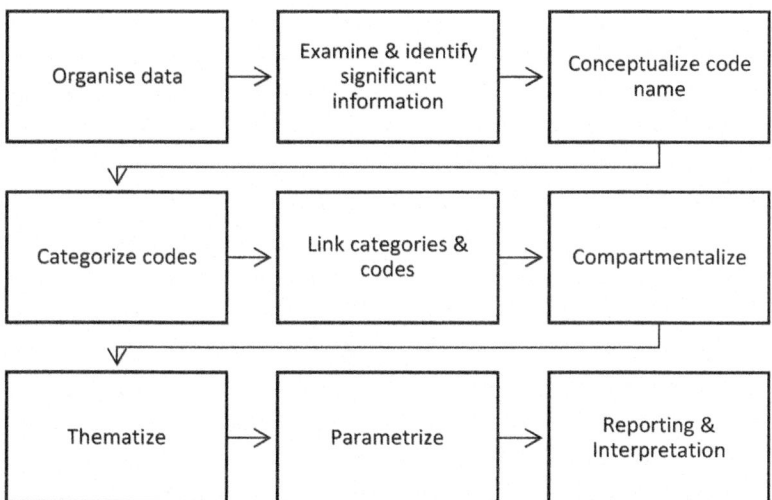

Fig. 5.2 Data analysis steps

Table 5.2 Data analysis process

Process	Description
Organise data	Data should be organized according to need. For example, according to respondents' personal, professional, social, economic, and demographic profiles
Examine and Identify significant information	Probe data using the research question, apply the appropriate coding approach and identify the significant information from responses. The step refers to reading, identifying and selecting relevant segment of information for coding
Conceptualise code name	Create a suitable code and attach it to the selected data segment. Repeat for all data segments
Categorise codes	Categorise codes based on shared traits/ attributes/sharing relations. Name the code categories
Link categories and codes	Establish relations between code categories using the axial coding process
Compartmentalize	Group code categories/codes based on specific attributes/ relations/preferences. The step refers to grouping code categories and codes
Thematize	Identify themes using code categories/code groups and organize them in significant order
Parametrize	Refine themes based on respondent categories: personal, professional, contextual, demographic, socio-economic, etc
Reporting and interpretation	Report the themes, explain their significance and interrelationships, and provide interpretations with reference to the research questions. Support the interpretations with evidence for your report. Conclude with suggestions for suitable actions and scope for future study

CODES IN ATLAS.TI

Researchers using ATLAS.ti, or those who intend to, must be familiar with meaning of codes, the types of codes, and coding methods. Only then will they be able to apply the coding method appropriate to their study.

ATLAS.ti offers several coding options from which the researcher can select the one that is best suited for their research objectives, the nature of their data, and coding preferences. Figure 5.3 shows the coding options in ATLAS.ti, while Fig. 5.4 shows the auto-coding window under 'Search & Code' option.

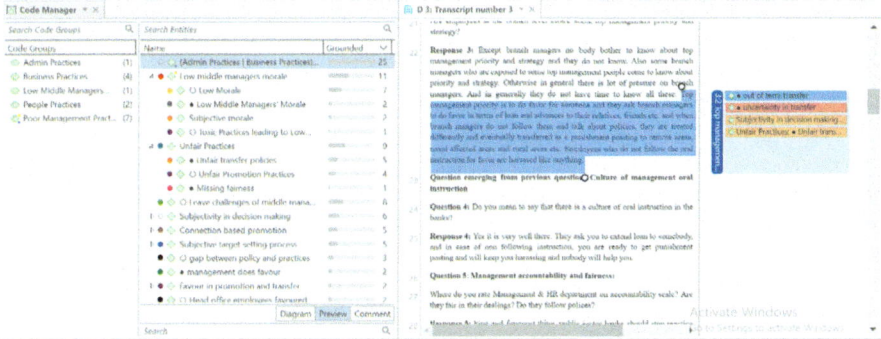

Fig. 5.3 Coding windows In ATLAS.ti 22/23

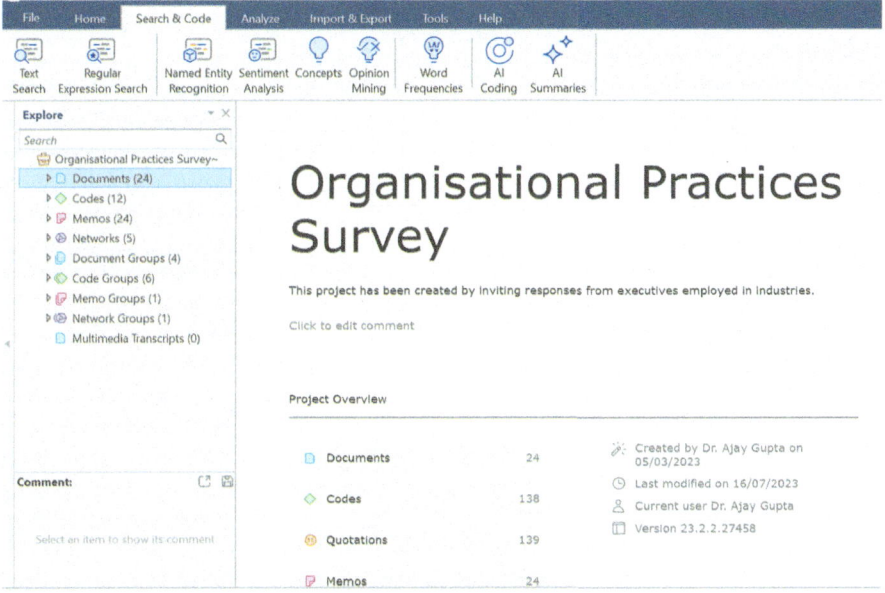

Fig. 5.4 Auto-coding windows In ATLAS.ti 23

The Auto-coding options available in ATLAS.ti are Text Search, Regular Expression Search, Named Entity Recognition, Sentiment Analysis, Concepts, Opinion Mining, Word Frequencies, and AI coding. There are slight differences in terminology in various versions of ATLAS.ti. For example, 'Open Coding' in ATLAS.ti 8 is the same as 'Apply Codes' in ATLAS.ti 9 and 22. The Auto-coding option in ATLAS.ti 8 does not have a 'Sentiment Analysis' option and opens in a different window than ATLAS.ti 9 and 22.

Using the 'Home tab', the researcher can create 'Free Codes' under the 'New Entities' option in all three versions of ATLAS.ti. Free codes are used to create a list of codes that can be used during analyses. The researcher creates 'Free codes' based on their experience during their literature review of the topic of their study and data collection.

The 'Open Coding' or 'Apply Codes' option is used to code data with either an inductive or deductive coding approach. Using the 'Quick Coding' option, ATLAS.ti applies the preceding code to code the next data segment. Here, the researcher does not write the code. A previously used code becomes the default code for the next data segment. 'Code in Vivo' is verbatim coding where the selected data segment becomes the code.

The Auto-coding option in ATLAS.ti 9 and 22 has two additional features: Named Entity Recognition and Sentiment Analysis (Fig. 5.4). Researchers use these codes for creating code categories, code groups, smart codes & smart groups to develop themes.

Free Codes

As the name suggests, free codes are free and floating. They are not attached to any data segment. Free codes may be regarded as "dead" unless attached with relevant information. Free codes are the ideas, information, insights, and concepts that emerge during the data collection process (Fig. 5.5).

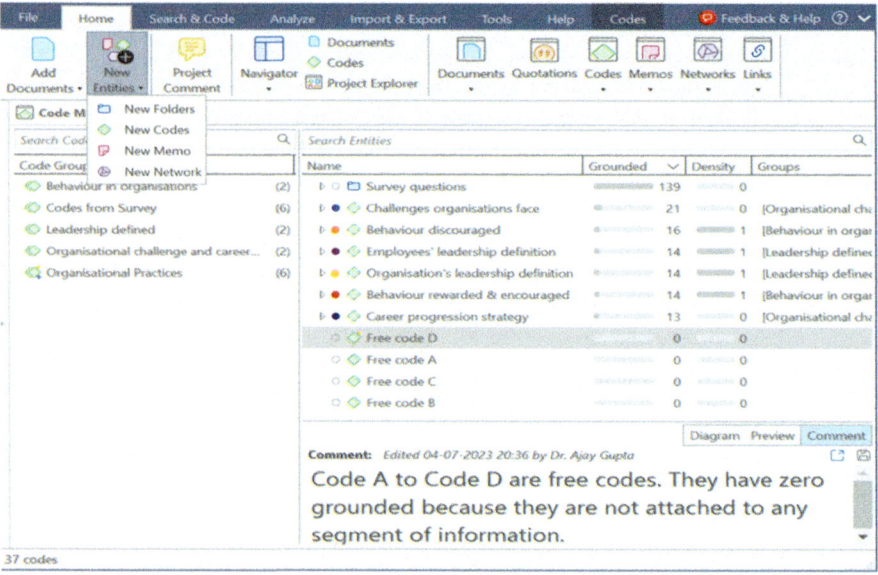

Fig. 5.5 Free coding option windows in ATLAS.ti 22

Open Codes

As described earlier in this chapter, there are two approaches to coding: inductive and deductive. Inductive codes emerge from the data, while deductive codes are developed from a theoretical perspective. Thus, when the researcher wants to understand the information present in the data, an inductive approach is used for coding. Inductive coding is not influenced by the researcher's preconceived ideas, concepts, and understanding.

On the contrary, deductive coding is explanatory and attempts to justify/reject already available theory/concepts. While the inductive approach to coding is data-driven, a deductive approach is theory-driven.

> **Important** Quantitative data can also be explored and verified. The researcher should describe the nature and scope of particular research as qualitative–exploratory, confirmatory, quantitative–exploratory, confirmatory.

In an exploratory inquiry, researchers may think deductively, largely within their emerging theoretic framework rather than within established theory and the hypotheses deduced from it. Moreover, they engage in a sort of verification; that is, they (tentatively) confirm their emergent generalizations rather than an ensemble of a priori predictions (Given, 2008).

After coding and memo writing, the researcher "steps back" to view the data in their entirety. The themes are based on the patterns that emerge from the data and then a theory is developed. Themes identified are strongly linked to the data (Patton, 1990). New knowledge is created that becomes the basis for proposing a hypothesis to generalize the findings. In this case, one does not use a codebook; the researcher builds from scratch using the data. Since the purpose of qualitative research is to understand the phenomenon, and the underlying factors influencing the phenomenon, the researcher develops themes that are emerging variables. If the researcher wants to generalize the phenomenon, they must create a hypothesis based on emergent themes, use a large sample size and survey methods to collect and analyse quantitative data which is then analyzed, and a generalization is proposed with the results.

In deductive Coding the researcher develops a codebook which will guide the coding process. The codebook must be ready before commencement of data collection. Some methodologists advise that the choice of coding method(s), or even a provisional list of codes, should be determined beforehand (deductive) to harmonize with the study's conceptual framework, paradigm, or research goals. But emergent, data-driven (inductive) coding choices are also legitimate (Saldaña, 2016a, 2016b). This is refined and reorganized as data collection progresses. In its final form, the codebook should reflect the structure of the data. A codebook is a document that contains the code concept scheme. It describes the codes, categories, and themes used in qualitative data analysis.

In the deductive approach to coding, the codes are based on specific theoretical or epistemological positions (DeCuir-Gunby et al., 2011). Codes summarizing the surface meaning of the data can be identified as semantic codes, and those that dig deeper into the data and prioritize the analytical framework can be termed as latent codes (Clarke & Braun, 2014).

Crabtree and Miller (1999) adopted a theory-driven, deductive approach to coding in which someone else's theoretical framework(s) is applied to develop the codebook(s) and then codes are attached to the texts. An integration of inductive and deductive Coding reflects a balanced, comprehensive view of the data, instead of purely relying on the frequency of codes decontextualized from their context (Xu & Zammit, 2020).

Quick Coding

Quick Coding, as the name suggests, does not take much time. Quick Coding takes the previous code and assigns it to the next data segment. For example, in ATLAS.ti, if a data segment is coded as Code A, the next data segment that needs to be coded is highlighted and then, with a right-click and selecting the quick coding option, Code A is assigned to it.

Alternatively, the user can highlight the segment for coding and click on Quick Coding at the top menu bar. The selected data segment is assigned the previous code.

Figure 5.6 shows coded responses in open, and in vivo codes. In addition, the figure shows axial codes using networks, and free codes in the code manager window. The open codes are 'No HRM practices', 'Promotion and performance disconnect', and 'Long tenure rewarded'. The corresponding in vivo code is 'HRM creates division between morale and mindsets', a verbatim copy of the data segment. As the reader can see, the in vivo code has taken a selected response segment and made it a code.

The quotation 1:4 in the screenshot has been coded as 'Long tenure rewarded (an open code)' which is a quick code that used the preceding code. When the researcher uses the quick code option, the preceding code becomes the code for the next selected segment.

Axial coding helps to establish the link between codes using relationships shown in the network. For example, the quotation 1:1 has been coded with two codes: 'No HRM practice', which is the cause of the second code, 'Promotion-performance disconnect'.

Free or floating codes are not attached to quotations. They have zero groundedness. In other words, they are not associated with, or supported by, a quotation because they are 'floating' codes generated by the researcher for later use. In code managers, there are four codes with zero groundedness. They are 'Free code, free code A, free code B, and Auto-coding'. These are not linked to any segment of information.

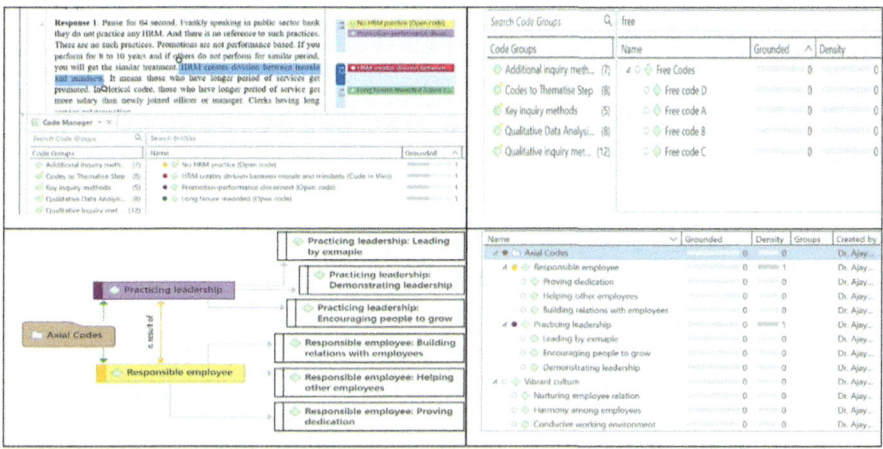

Fig. 5.6 Open, free, in vivo, and axial codes

Auto-coding

Auto-coding is used when the volume of data is large, and it is difficult to read every line or word of the transcript. As the name indicates, codes are automatically assigned to the selected segments in the entire data set. The following example explains how this is done.

In one study the researcher has to review a 50-page document (i.e., a research paper). Auto-coding allows the researcher to search the document for segments containing a specific type of information using the key concepts of the paper. In the case of qualitative content analysis, the researcher is interested in searching contents based on their significance from the document. The researcher can use the word cloud, or a word list function to search for contents. After selecting the document of interest, the researcher can use the options in ATLAS.ti to determine the frequencies and order of the words present in the data. The words can be arranged in descending or ascending order of their frequencies, which will help the user identify the key concepts.

Order of concept means words that are most frequently used in the document. By using the auto-coding option, the researcher can arrange their order either in ascending or descending order. ATLAS.ti provides researchers the option to arrange the codes according to their density by clicking on Grounded and Density in Code Manager (Fig. 5.7). This functionality in ATLAS.ti helps the researcher to understand the usage of concepts linked to the phenomenon of study in the document. The auto-coding function is then used to code the selected documents.

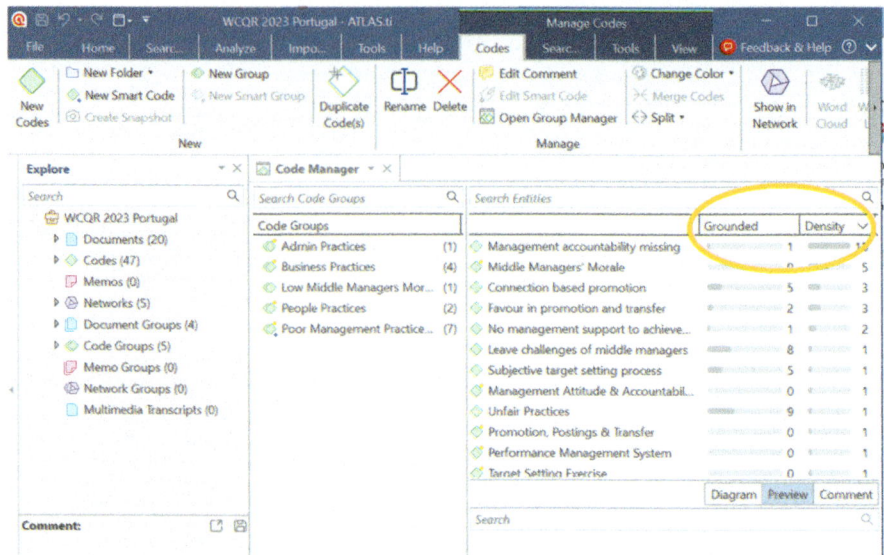

Fig. 5.7 Code manager, grounded and density

Splitting Codes

Split coding is needed when a coded text can contain more than one type of information for a single code is not sufficient. Alternatively, the researcher may assign one code to many text segments, which is then split into subcodes to reflect the precise meaning of each text. In code splitting, a code is broken into subcodes. Code splitting begins when the researcher groups several quotations under a broad code. This process of grouping (or clubbing) of many quotations under a broad code is called lumping. Splitting codes into subcodes help researchers to better understand the concept they are investigating.

Figures 5.8 and 5.9 explain how codes are split in ATLAS.ti.

The process can be explained with the help of an example. In one study, the researcher has coded all their data. Several quotations are attached to a code. While reviewing the quotations, the researcher realizes that some of the quotations need a separate code because they have a slightly different meaning from the initial code. The researcher then removes these quotations from the existing code and links them to another code. In effect, the researcher has split the quotation from an existing code and assigned them to another, more suitable code.

In Figs. 5.8 and 5.9, the code 'Leave issues of middle managers' was split in the following steps:

5 DATA ANALYSIS METHODS   139

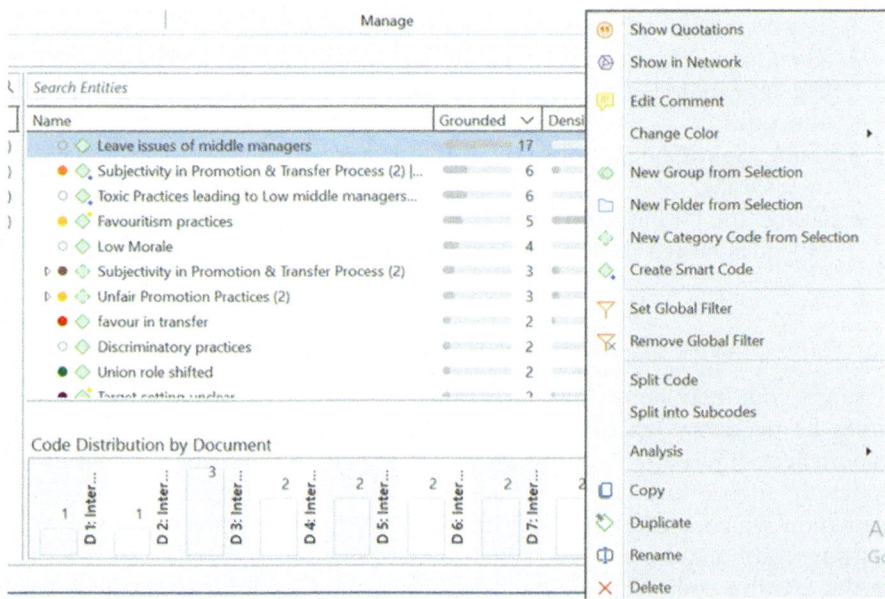

Fig. 5.8 Split Code options in ATLAS.ti 22

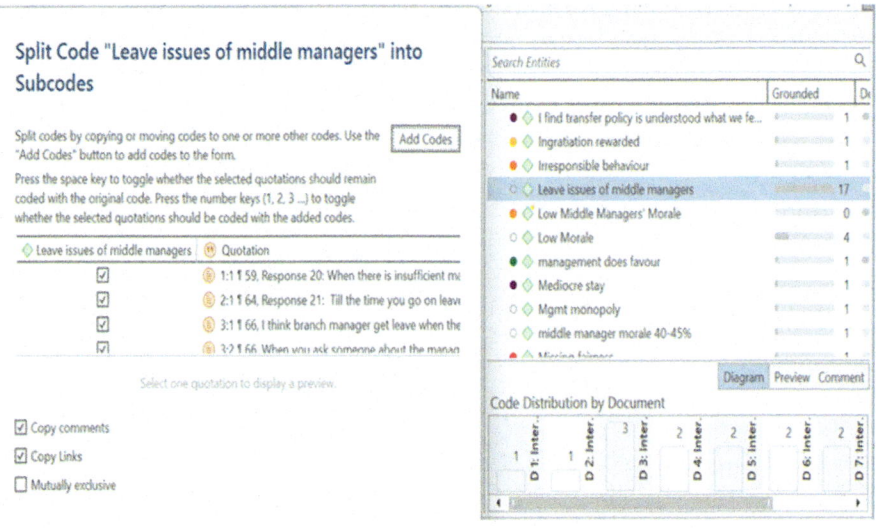

Fig. 5.9 Split Code window in ATLAS.ti 22

1. The code list in code manager option was opened.
2. Then, the selected code was highlighted followed by a right-click. A flag appeared (Fig. 5.8) showing two options: Split Code and Split into Subcodes.
3. Using 'Split Code' option, the researcher created separate codes. whereas by selecting
4. Using the "Split into Subcodes" option, the researcher can create subcodes nested in the selected code.

When and How to Split Codes

A single code may be divided into two or more codes. Splitting may sometimes be necessary when the researcher feels that quotations linked with a code reflect different characteristics and hence should be attached to separate codes. In such a case, the researcher may decide to delink the code from a quotation which is then linked with a new code. Or a sub-code nested in an existing code may be created, which is then linked to the quotation as well as the existing code (Fig. 5.10). To do this, the user must check both code boxes. If this is not done, the existing code will get unlinked automatically.

Codes that are merged can also be split this way. For example, the researcher can, if required, split merged codes into two or more codes. For example, if the code for motivation has six connected quotations for which the researcher may feel the need for two more codes from the existing code for motivation, the split code option can be used to create two new codes. Once the new codes are created, the quotations linked to motivation will shift to new codes.

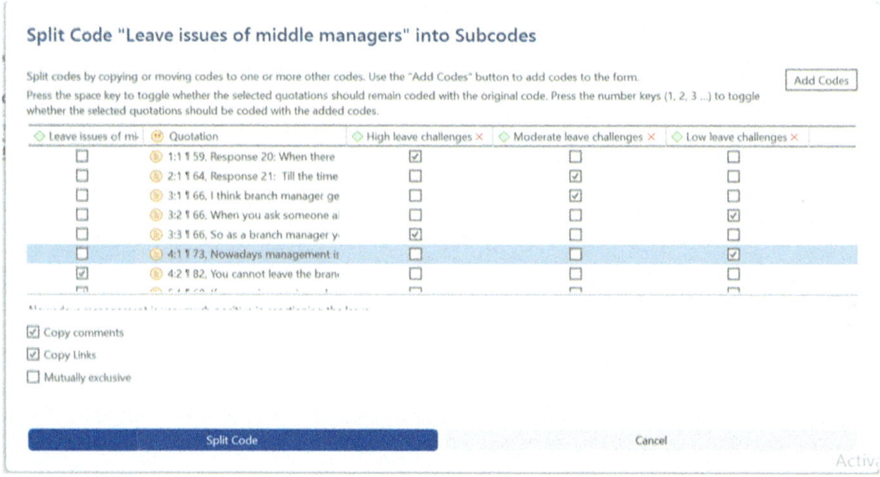

Fig. 5.10 Splitting codes

Figure 5.10 shows an example of a split code. The code "Leave issues of middle managers" has been split into three sub-codes "High leave challenges", "Moderate leave challenges", and "Low leave challenges". Quotations for the checked codes in the figure were unlinked from the original code and then linked with the checked codes. If the researcher wants to check more than one code for the same quotation since both codes fit, the quotation gets linked with all checked codes. If the researcher wants to keep quotations with the original code, they can check both codes. This means that the quotations linked with the original code are also linked to the new code. Quotations can be linked or unlinked based on the researcher's understanding of the relevance of the quotations linked with a code.

Steps for Splitting Codes in ATLAS.ti

1. Open the codes in Code Manager (Fig. 5.8).
2. Select the code for splitting and right-click. A flag will be displayed. Select the 'Split' option at the bottom of the flag. Click.
3. A flag will be displayed.
4. Create new code/codes as required using the New Codes option (Fig. 5.9).
5. Check the box of the original or new code as required. By checking the box of the new code, the quotation will be linked to the new code (Figs. 5.9 and 5.10).
6. On clicking Create, the original code will split.
7. For creating mutually exclusive codes, only one box must be checked.
8. To remove a quotation from the code, uncheck the box before the original code.

Redundant Coding

Redundant coding is superfluous or surplus coding. It does not create value for the information that is being coded. A redundancy might occur while a paragraph is being divided into various sub-segments, each having the same code. Redundant codes only cause confusion. Therefore, when a paragraph indicates one concept, the entire segment must be coded.

Redundant codes can be identified by "overlapping or embedded quotations" having the same code, meaning that two different bits of information have the same code. A quotation may be embedded in another quotation, or two quotations may have overlapping features (Fig. 5.11). Such redundancies can result from the normal coding process; but they can also go unnoticed during a merging of codes, or when researchers work in a team on the same project. Redundant codes should be removed before data analysis. ATLAS.ti helps researchers to address the issue of code redundancy.

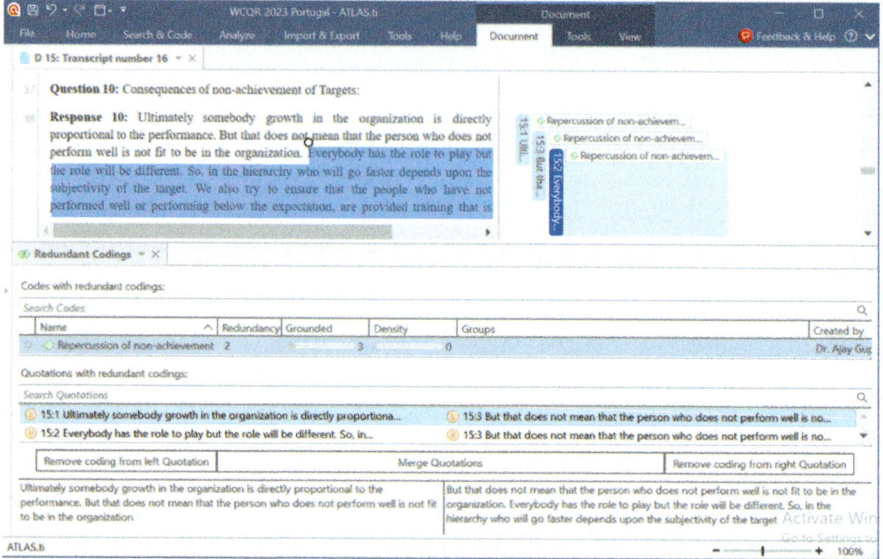

Fig. 5.11 Redundant coding

There are two ways of removing redundant codes: by using the Tools and Support options or opening Code Manager and selecting the Tool tab option.

Figure 5.11 shows the names of redundant codes, their groundedness, and density. The redundancy column displays the number of pairs of redundant quotations associated with the codes. The user selects any code from the list to find redundant quotations listed in pairs in the middle pane. A preview of the quotations for the code is displayed in the bottom pane. If the reader wants to see the quotations' original positions in the document, a double click will display the text.

There are various ways of correcting code redundancy, such as by unlinking left quotations, merging quotations, and unlinking right. By clicking on "unlink" from the left quotation, the selected code from the quotation is shown on the left. Unlinking from the right quotation removes the selected code from the quotation shown on the right. By clicking on merge quotation, the two quotations shown on the right and left are merged with the code. All references to the merged quotations are "inherited" by the code. If the two quotations overlap, the resulting quotation includes all data from both quotations.

On "hoovering" over a quotation, information about the "Connectivity" of the quotation can be seen. This includes the number of connections to other codes (= Coding), the number of connections to other quotations (=hyperlinks), and the number of connections to memos (= memo links). Connectivity information provides additional clues to the researcher about code redundancy. In the author's opinion, unlinking the quotation connected with the code is the best option because it removes uncertainty and ambiguity over quotations liked with redundant codes.

The three buttons for removing redundant codes—unlink from left quotation (unlink left), Merge quotations, and unlink from right quotation (unlink right), either at the bottom or top pane —become active only after the user selects and clicks on the quotations.

Figure 5.11 explains removal of redundant codes with an example. Quotations numbers 15:1, 15:2 and 15:3 have the same code: Repercussion of non-achievement. These quotations overlap. The passage can be coded using the code once. In such cases, it is good practice to merge the quotations. However, in other cases, where removing the quotation is the only option, the researcher should remove the code. This will leave the quotation an orphan, without any link to a code. It is good practice to remove orphaned quotations.

When and How to Merge Codes

Codes reflecting similar characteristics can be merged into one code. For example, 'head office employees favoured', 'transfer based on trends', and 'performers leave' reflect similar traits and thus, if required, these codes can be merged with an existing code. The merged code can be identified separately if the new code reflects a broader meaning. With the merger of codes, all linkages of the codes are also merged with the new code. Thus, for example, if three codes with a total of 20 linked quotations are merged, the new code will display 20 quotations.

Figure 5.12 shows the option for merging codes in ATLAS.ti 22. The researcher can select the codes for merging and use either option: right click and select Merge or select the Merge option as shown by the arrow.

Auto-Coding in ATLAS.ti

The auto-coding function helps to quickly identify the words the researcher is interested in and then codes them. ATLAS.ti offers a wide range of options to the researcher for auto-coding data. The auto-coding function searches a single document or all documents, depending on the researcher's selection. If one document is selected for auto-coding, the function searches for a specific word, or words, in the document. The researcher can then select the data segments—a word, phrase, sentence, or a whole paragraph—that need to be coded.

Auto-coding performs three tasks: text search, automatic segmentation, and coding. It searches the text for words of interest, and they are found, the words are coded.

However, the auto-coding option must be used with caution. Researchers should avoid total dependence on auto-coding for data analysis. They should review the codes at the end of an auto-coding cycle. Scholars and experienced

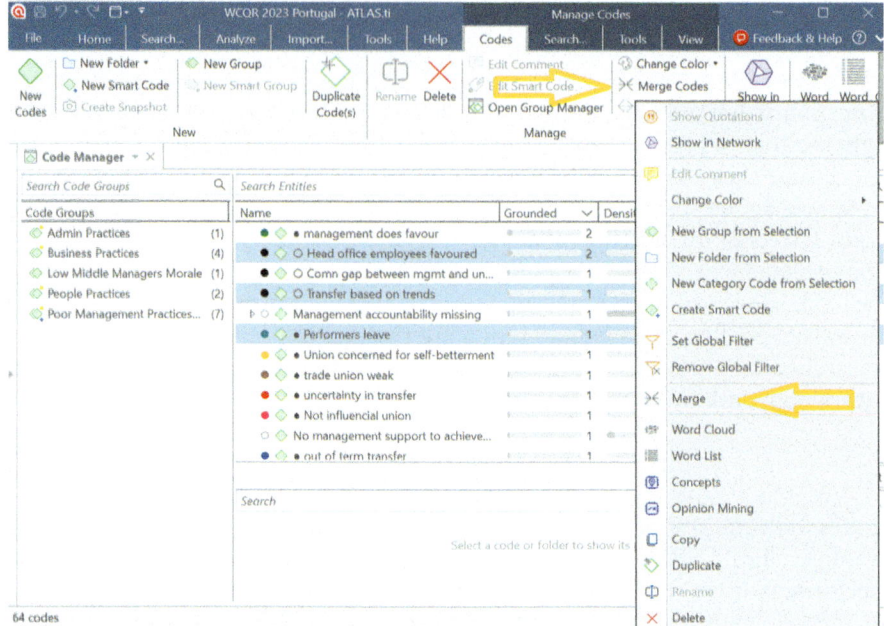

Fig. 5.12 Merging codes in ATLAS.ti 22

users advise that the auto-coding option must be used only if it is necessary and unavoidable, and not just because it is available.

The auto-coding function is useful when the researcher knows the ideas or concepts that are likely to emerge from their data, as it sometimes becomes evident during field work. The auto-coding function can also be used to search for a specific content, concept, or variable in the data, or when the researcher wants to test theories. For testing a theory, or explaining one, the researcher uses the auto-coding function for aligning the variables in the data with theory. In such a case, the variables are used as keywords that are searched for in the document. With the hits achieved by the search, the researcher identifies the segments that need to be quoted, as well as the codes that should be linked to the segmented data. This depends on the methodology used in the study and data analysis methods.

Researchers can also use auto-coding along with other utilities like word clouds or word list. A word cloud lists all the words in the documents along with the frequencies of their occurrence in the text. A word list arranges all the words in the text in the ascending or descending order of word frequency and their prevalence (percentage of total) in the document. From this information, the researcher can get an indication about the focus of the documents and accordingly use the auto-coding function to identify the most frequently used words. The auto-coding function saves time by offering a faster way to analyze large volumes of data in qualitative research. If the researcher wants to discover the concepts emerging from a collection of books on a specific topic, the word cloud is used to get information about the word frequencies after which the sentences or paragraphs containing the keywords are selected and coded.

Researchers can create a word cloud or word list by a right-click on Document, Document Group, Code, Code Group, Smart Code, and Smart Group (Fig. 5.13). Then, a double-click on a word appearing in the word list will take one to the context menu. The context menu will appear by selecting the word appearing in the word cloud, and right-clicking on Search in Context.

> **Important** Although auto-coding is useful for locating key concepts, a thorough reading by the researcher is necessary to have deeper and better understanding of concepts in their contexts. This is because words with the highest frequencies may not necessarily indicate a paper's key idea and concepts. Relying only on word frequency may result in the researcher's focus on only a few words, which increases the risk of key concepts being overlooked. Thus, it should be always remembered that, at best, word frequency can only provide clues about the concepts in a research paper; the researcher must probe further.

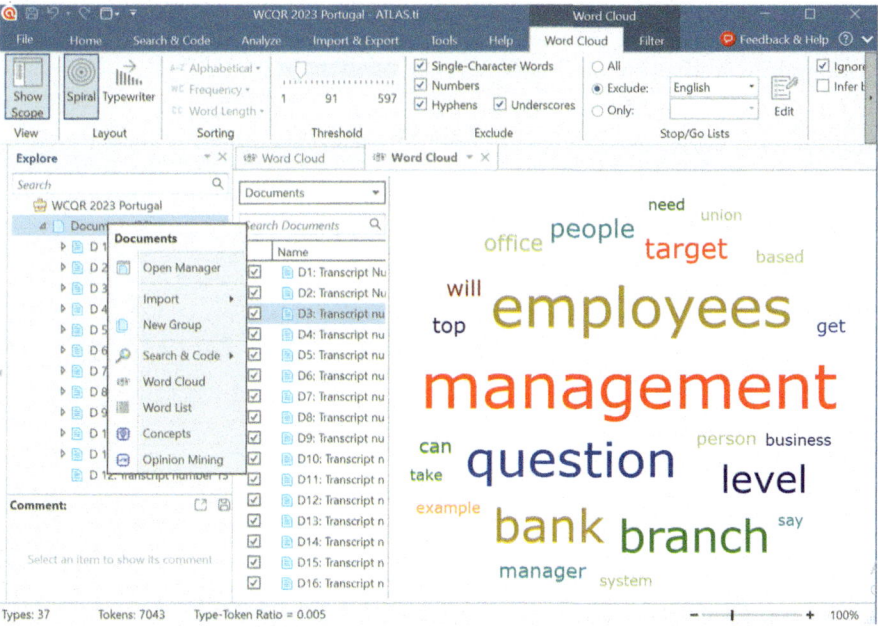

Fig. 5.13 Word cloud and word list

The details about auto-coding have been explained in Chap. 15.

The Auto-Coding Window

The Auto-coding function in ATLAS.ti 22 and 23 has several advanced features (Figs. 5.14, 5.15 and 5.16). It has an integrated search and code option which one can use and select the document/documents that are to be searched and coded. Alternatively, the user can also first select the document/documents and then right-click to locate the search and code option or click on the search and code option in the main menu.

Every auto-coding option serves a specific purpose. Text search searches words and synonyms in 'Paragraphs', 'Sentences', 'Words' and 'Exact matches' options. Several combinations of search are possible using "OR" and "And" option. Regular expression search option searches words in same option as in text search. Researchers can use auto-coding option based on objective and appropriateness.

The search and code function in ATLAS.ti uses machine-learning concepts to search for words, and finding exact matches in sentences and paragraphs, as well as synonyms. An expert search option is also available with which the user can search and code specific words or concepts. It uses natural language processing techniques to help researchers fine-tune their search.

Named Entity Recognition finds sentences or paragraphs to find specific persons, locations, organizations, while using the search and code option. This is helpful while searching for a specific context associated with the various parameters. The researcher can search for information based on various demographic and contextual parameters.

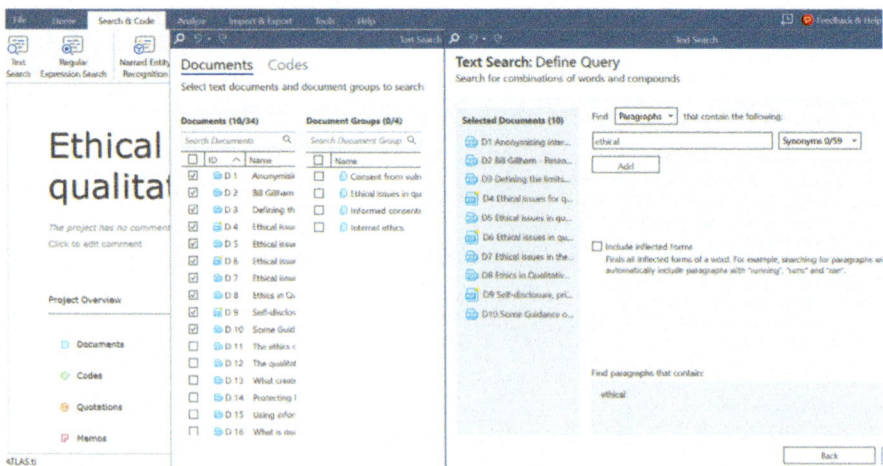

Fig. 5.14 Auto-coding text search option in ATLAS.ti 23

5 DATA ANALYSIS METHODS   147

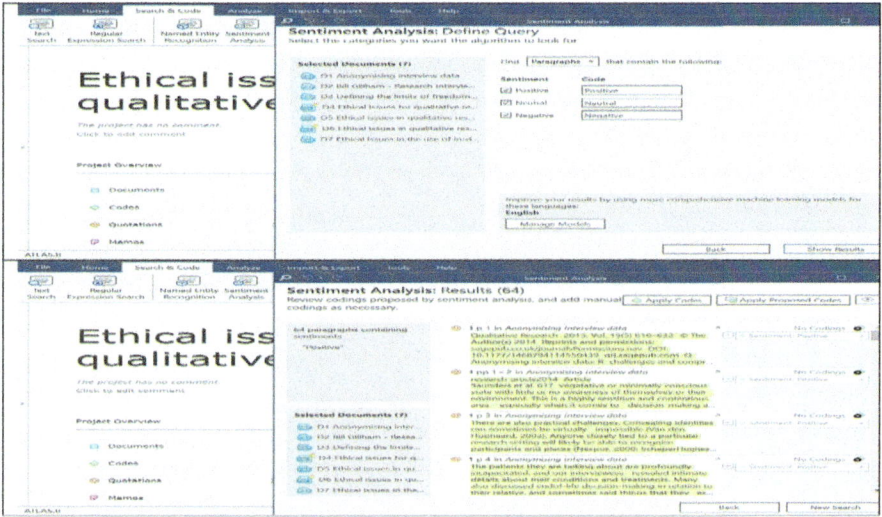

Fig. 5.15 Auto-coding sentiment analysis option in ATLAS.ti 23

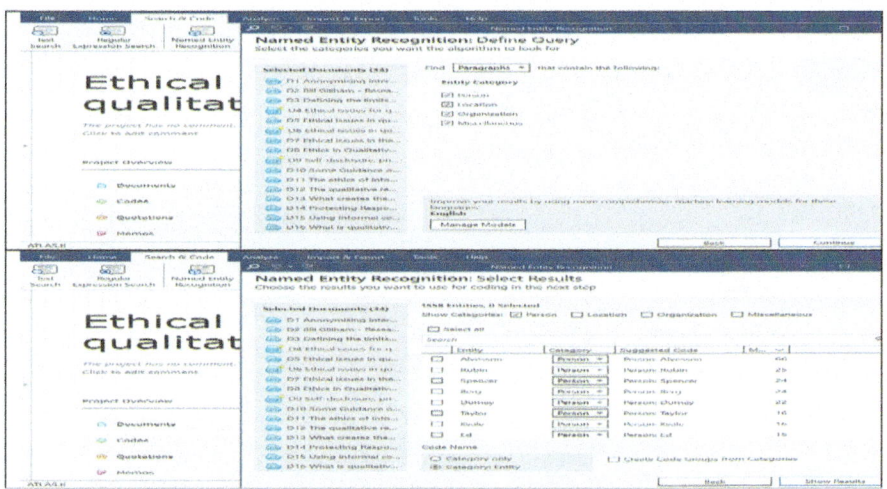

Fig. 5.16 Auto-coding named entity option in ATLAS.ti 23

Sentiment analysis option is an information extraction technique using a machine-learning technique. It is also known as entity identification, entity chunking, and entity extraction. Sentiment analysis classifies text segments by sentiment and suggests suitable codes. The user can search and code positive, negative, and neutral sentiment in paragraphs or sentences in the selected document/documents. ATLAS.ti extracts and groups the selected information based on the researcher's preference.

Creating a Research Project in ATLAS.ti

1. Create a new project by clicking on + sign and give it a name (Fig. 5.17)
2. Open Add Documents and locate Add Files (Figs. 5.17 and 5.18).

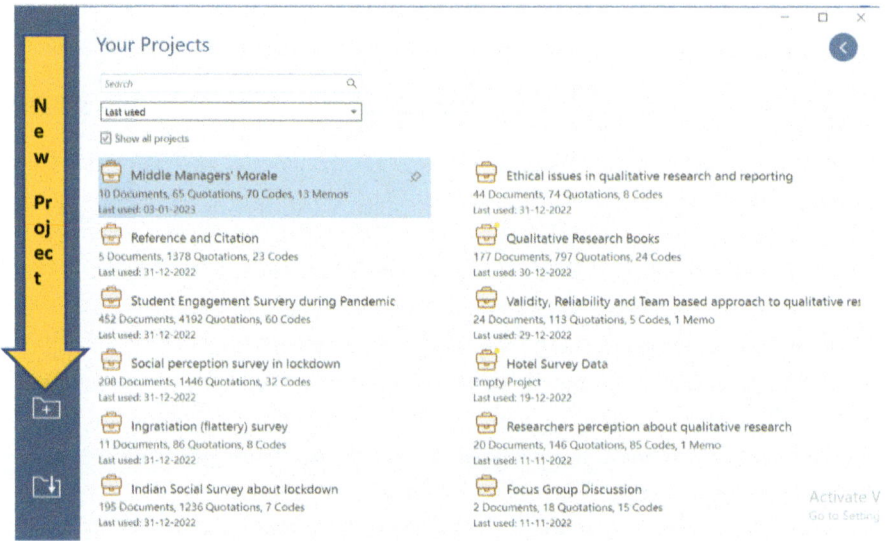

Fig. 5.17 Creating a project

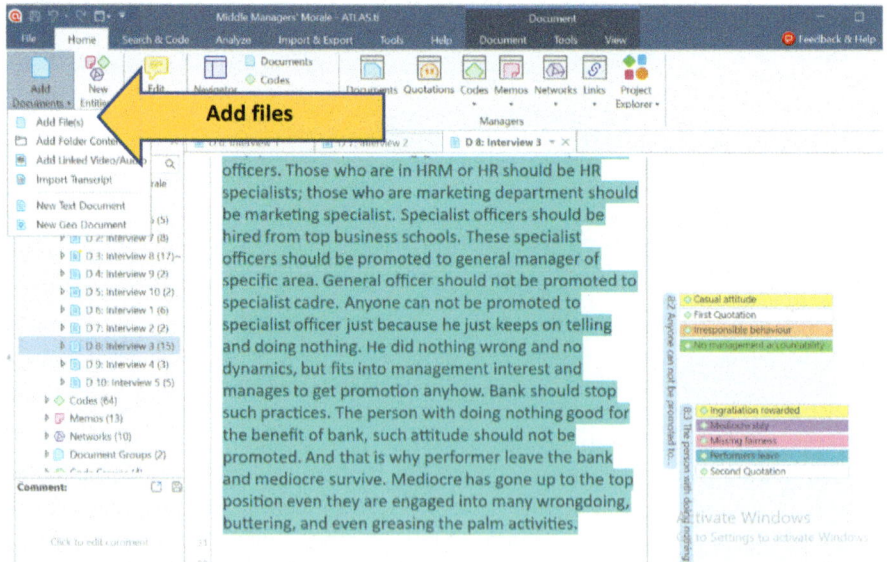

Fig. 5.18 Adding files in ATLAS.ti

3. Locate the document for analysis. Double-click on the Document or click Open. The document will open.
4. Read the document thoroughly, select segments containing useful information by using the research question lens.
5. The next step is coding. Select a segment of interest and right click. The coding window will open. Write the code name and click on the + sign. The code will be shown before the selected segment (the quotation).
6. The quotations for the code will be displayed in the left margin of the codes in the image.

Understanding Documents in Qualitative Research

Documents are a collection of several data forms that have meaningful information for the topic of research. They may contain the answer(s) to the research questions. The documents being analyzed contain the information collected from primary or secondary sources or both. Any material relevant to the research, and which the researcher intends to work with, is a document. Documents are important to researchers for analysis. The results of analysis are used to help the researcher to present the findings of the research and make logical conclusions.

Audio or audio-visual data must be transcribed verbatim. This means that the researcher converts the spoken word into a written form with the exact words of the speaker without any omission, not even a pause. The researcher is not permitted to summarize or assume what the speaker *intended* to say while transcribing. Transcripts are crucial records and subject to scrutiny and review.

Meaning of Quotations

While reading a book, the reader's attention is drawn to a particularly forceful point made by the author, or to the manner vital information or key points are presented. The reader marks, underlines or highlights the point because, in their opinion, this is a critical part of the text: the narration is built around that part, or from which the reader can draw important inferences or make conclusions.

What do we do when we read a storybook, newspaper article, magazine, or journal and come across some information that catches our attention? In some instances, we might find an entire paragraph to be important. The usual reaction would be to highlight the part with a pen, pencil, or highlighter. The purpose is to show that this part is crucial to the understanding of what we read, or it contains the gist of the topic. Other information in the next may be of secondary interest or may only support the highlighted portion.

Similarly, in a qualitative research setting, the researcher looks for information of relevance and value that can be interpreted and from which inferences can be drawn. That portion of text data of interest and relevance to the study's research questions is coded. The coded portion is then known as a quotation.

In qualitative studies, the researcher looks for informative segments or expressions that will help them to uncover meaningful information from documents. These are called quotations. A quotation is helpful information gathered from interviews, opinions, or information that the researcher uses. A quotation can be a single sentence, a few words, or a paragraph. Local words reveal he contextual meaning that should be coded using in-vivo coding. Responses may include native words that reflect the meaning applicable in the specific context. Coding them with other words may result in the loss of the meaning of the selected segment of information. As a result, the researcher selects the segment of information as a code, which is known as in vivo code.

Quotations are important because they support the researcher's findings in their study. Quotations are evidencing the researcher must present to justify the findings. They are important for thematic presentation. They are the meaningful lines in data that address the research questions, which the researcher finds informative and valuable. Quotations show the specific forms of a general phenomenon (Weiss, 1994) and are shreds of evidence about the researcher's claim about participants' responses. In other words, quotations are necessary to convince readers that the researcher's claim is valid.

According to Blauner (1987) and DeVault (1990a, 1990b), quotations are crucial to qualitative analysis and reporting. Researchers must have a clear idea of what information they are looking for in the data. A cluster of quotations strengthens the researcher's claim about their findings. Therefore, the research report should support their findings with as many quotes as possible (Wolcott, 1994).

Often, researchers tend to create an abundance of quotations to make their descriptions appear thick. According to Wolcott (1994), thick description is not about "heaped data" but, rather, they represent information on a cultural context. *Therefore, researchers should avoid offering multiple quotations representing the same idea. Quotations should reflect the variety present in the information.* This makes the themes rich, informative, and insightful.

Identifying Quotations in ATLAS.ti

After the data is exported to ATLAS.ti, the project is saved with a name after which each document in the project is opened and read thoroughly (Figs. 5.19 and 5.20). While reading the document from the perspective of the research questions, the researcher becomes a seeker of information that speaks about research questions, directly or indirectly. Any information found to have relevance to the research question or objectives becomes a quotation.

Quotations in a data set will be different for different research questions. This is because different research questions concern different aspects of the subject of study and hence the responses to different questions will not be similar. An example will help explain this point:

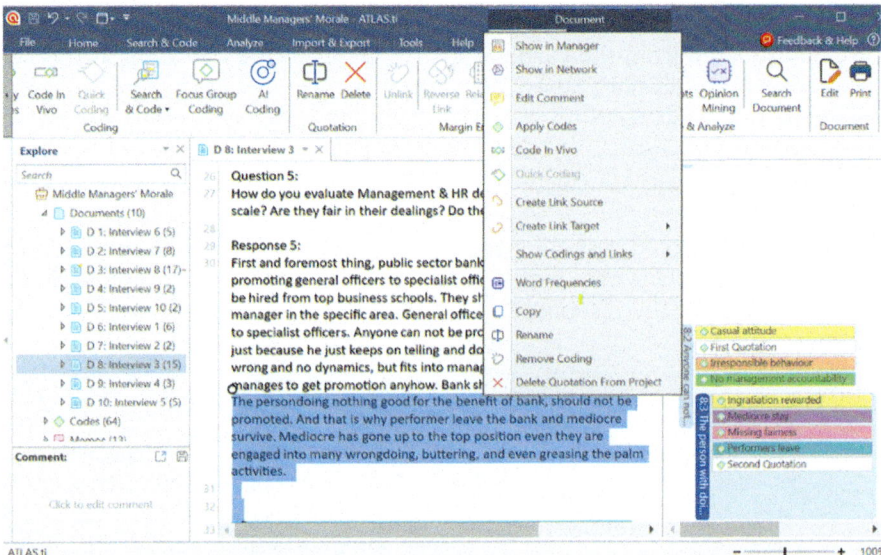

Fig. 5.19 Identifying quotations and coding them

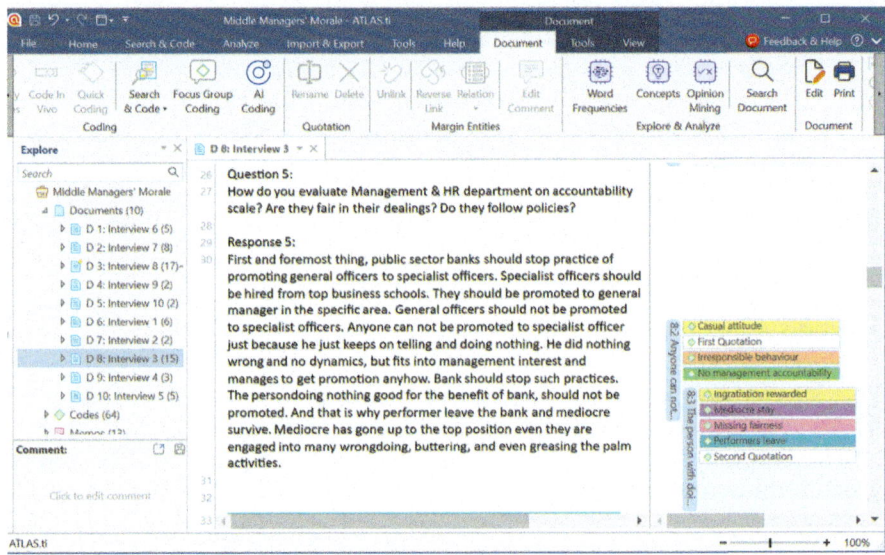

Fig. 5.20 Quotations and codes

In one study, responses to a set of questions were documented. Quotations were be extracted from the participants' responses to the questions as follows.

Question 1

How do you evaluate Management and HR department on the accountability scale?

(It should be noted that the respondents in this study were free to express their thoughts in their own words. The researcher did not try to regulate the content or tone of the responses).

> Response
> First and foremost, public sector banks should stop promoting general duty officers to specialist roles. Specialists should be hired from top business schools; they should be promoted to general managers in a specific area. The general officer should not be promoted to the specialist cadre. *Anyone cannot be promoted to a specialist rile because he just keeps on telling and does nothing. However, he fits [well with] the management's interests and manages to get a promotion. The bank should stop such practices. For the person doing nothing good for the benefit of the bank, such attitudes should not be encouraged. This is why performers leave the bank and [the] mediocre survive. Mediocrity has risen [even] to the top positions. They (the seniors) are engaged in many wrongdoings, buttering, and even greasing the palm activities.*

Question 2

What is your opinion about management fairness in the system? Could you express your opinion about whether they follow policies?

(The responses to Questions 1 and 2 are slightly edited (for clarity) verbatim transcripts from recorded data)

> Response
> Management in many cases is not fair. They favor employees who are close to them. They follow policies according to their convenience. They also interpret policies the way they want.

Identifying Quotations from Responses

The first question relates to management and HR's accountability towards middle managers. On a scrutiny of the response, the researcher found that the remark "anyone can not be promoted to….. should not be promoted" is an insightful one in the context of the research question and, therefore, is marked as a quotation. The second question relates to fairness in management's dealings with employees. The researcher finds the respondent's view that "the person with doing nothing to….palm activities" to be relevant. This segment of the participant's response was also identified as a quotation. The screenshots in Figs. 5.19 and 5.20 show these quotations.

Figure 5.21 shows a quotation that has been coded. After coding, the quotation gets linked with its code.

There are five quotations linked to the code 'Favouritism practice' (Fig. 5.21). When needed, the researcher can generate a report about this association in text or network form. Such reports help the researcher to support their interpretation of the results.

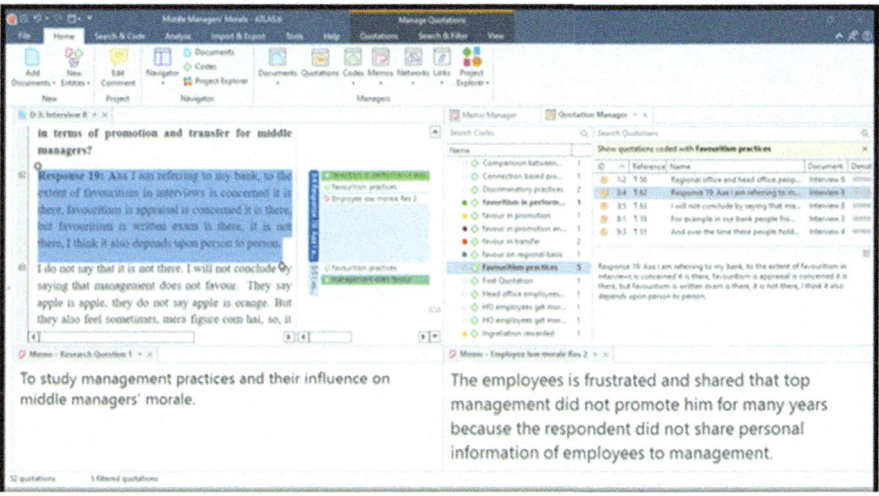

Fig. 5.21 Quotations, codes and memos

Coding Window in ATLAS.ti

Figure 5.22 is a screenshot in ATLAS.ti version 22. The coding window with coding options is shown in Fig. 5.23. After selecting a data segment, the researcher can either code by clicking on "Apply Codes" on the top bar or by selecting "Apply Codes" with a right click (shown with a yellow arrow in Fig. 5.23). There are three options in "Apply Codes": All Codes, Applied Codes, and Search Options (shown by a green arrow in Fig. 5.23). The system default is to show all codes in the window, applied and free.

Free codes are not applied to any data segment. When applied to a data segment, they become applied codes. Free codes are a 'bank of free codes' from the researcher selects a suitable one for applying to a data segment of interest. Researchers may generate and apply as many codes as possible based on their data and field experiences. By typing the code name in the search field, the user can search and retrieve any code, and by selecting the +sign, the code can be assigned to the selected data segment. The researcher can also use an applied code for another data segment. In that case a code is selected and with the +sign, it is assigned to the new data segment.

By clicking on Applied Codes, only applied codes are displayed. If needed, the researcher can write a comment for the code by clicking on the code and writing in the edit comment section.

It is also possible to give a coding preference. Two signs are provided with which the user can add or remove codes. The plus (+) sign is selected for adding the code, and the minus (−) sign for removing it. For auto-coding, one can use the "search and code" option.

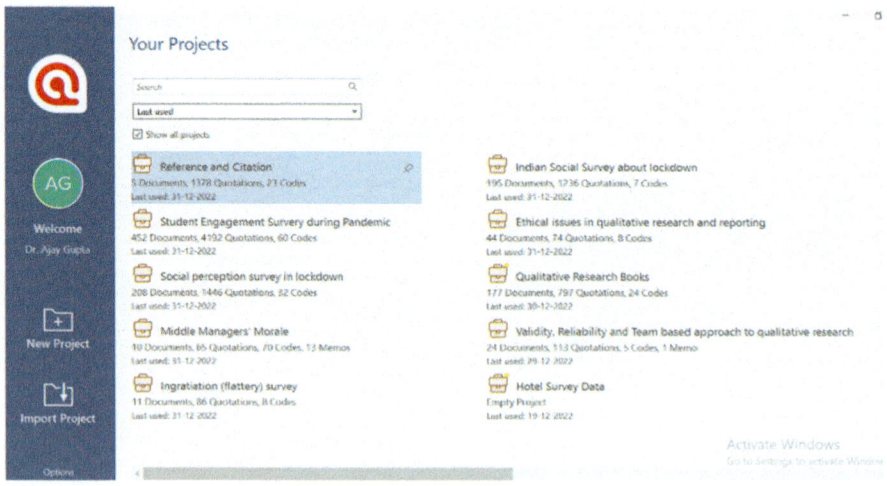

Fig. 5.22 Window for projects in ATLAS.ti 22/23

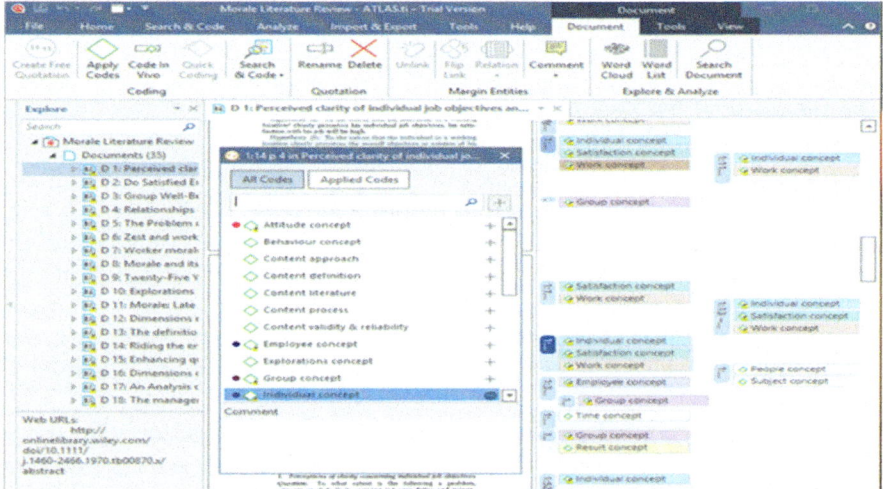

Fig. 5.23 Coding window in ATLAS.ti

Generating Codes from Quotations

Coding is like highlighting (or underlining) important or interesting words, lines, and passages in a text. After highlighting (or underlining) the parts (or segments) of the text that the reader has found interesting or important, the next step is to is to write something related to the highlighted segment which can be a word or several words, sentence, or paragraph. The approach to coding is similar to highlighting important parts of a text.

Codes are meaningful bits of information which are important for addressing research questions or objectives. Coding is the purposeful extracting of information from a section of data (or quotations) from the responses. Codes are not sentences; rather, they are a word or a set of words that are of value to the researcher in their search for insights into the topic of study. The researcher must also understand that although coding is a crucial step in the analysis of data, it is not synonymous with analysis (Basit, 2003). Codes help to create variables for the research. The help to make code categories from which code groups are formed. Code categories and code groups are variables that emerge from the data analysis. Themes emerge from code groups.

Coding is a cyclical and iterative process. Researchers generate codes based on their experiences and understanding of the research context. It is also an idiosyncratic process as different researchers code the same data set differently (Glesne, 2006). According to Coffey and Atkinson (1996), "coding is usually a mixture of data (summation) and data complication … breaking the data apart in analytically relevant ways to lead toward further questions about the data". This "breaking" of the data should be driven by the purpose of the research. Therefore, as a practice, the researcher should always ensure that while coding, the research concern, theoretical framework, central research question, study goals, and other significant aspects of the project are always visible. If necessary, these should be written on a page that the researcher can keep referring to. This practice helps keep the researcher focused and take the right coding decisions (Auerbach & Silverstein, 2003).

Coding is regarded as being more of an art than a science. Coding and codifying are not precise sciences that require the researcher to follow specific algorithms or procedures. Codes are not pre-decided before commencing the study. Coding evolves as the analysis progresses. Although there are no thumb rules for coding, it demands diverse skills of the researcher, such as those of induction, deduction, abduction, synthesis, evaluation, critical thinking, and several others (Inductive, deductive, and abductive coding are discussed in Chap. 4).

Approaches to coding depend on the researcher's experience and understanding. But it is essential to be aware that codes help researchers in several ways. They help in building new concepts or theories or extend the validity of existing ones. Coding techniques cannot be learned from textbooks. The process is driven by curiosity and the lens from which a phenomenon is studied. Coding is not a search for the occurrence of specific words; rather, codes help to identify concepts or meanings in the contexts of the study. Researchers categorize concepts at different levels of abstraction and then compare them across cases. It is essential for the researcher to keep code names concise and meaningful.

Figure 5.10 shows the coding of a text done through the lens of the research objectives. The researcher identifies the words which describe the meaning of the quotation. For example, the first quotation points to missing management accountability, responsibility missing from the management side, and a casual attitude. We can modify, rename, or reframe these concepts using keywords to make them codes, such as, in the present example, *No management accountability, irresponsible behavior, casual attitude,* etc.

It needs to be mentioned here that there are no hard and fast rules for the number of codes that must be generated. The number of codes depends upon the amount of information present in the selected segment of data. Researchers should avoid judging the quality of data analysis based on the number of codes. The number of codes generated depends on the amount of information that can be extracted from the data. The codes for responses to the second questions in the example under discussion may include **"pleasing practices rewarded"**, **"mediocre stay"**, **"missing fairness"**, **"performers leave"**, and others. Figure 5.24 shows the codes generated for the first and second quotations in the example. They are circled in the right margin; they also appear in the code list for each quotation.

Steps for Generating a Code

1. Open the document that needs to be coded (Fig. 5.24).

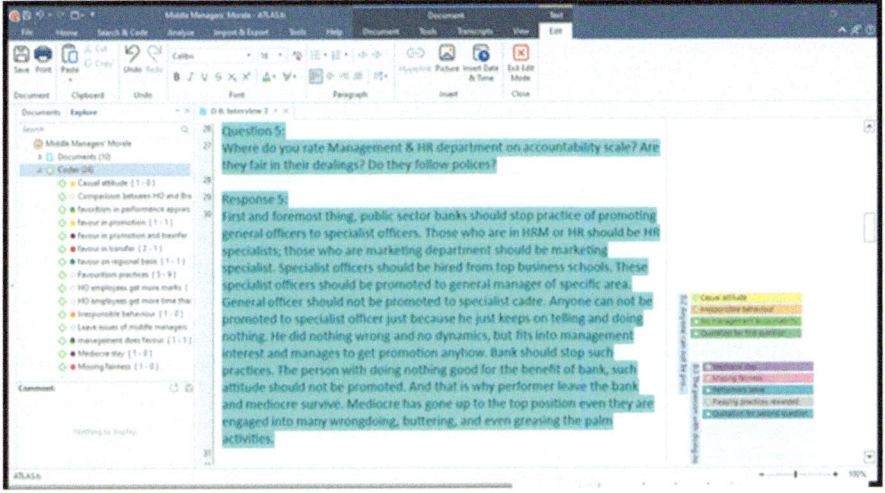

Fig. 5.24 Generating codes

2. Read the responses through the lens of the research question.
3. Identify the segment of interest in the document, highlight and right click. The coding window will open.
4. Select "Apply code" option, write code name and press +sign
5. The code will attach to the selected segment (the quotation)

Memos

Memos help to explore the data. They describe the researcher's understanding and observations about the entity or concept. Memos perform several functions in the research process. They are analytic or conceptual notes. One can write memos about the methodology of a study, its conceptual framework, or about a memo group, etc. Memos help the researcher to add insightful information. The information could be anything: a quotation, code, code groups, memos, etc. A memo written for codes is attached to a specific code or code groups. A memo written for a particular response can be attached to its associated quotation.

Memos may also contain a critical analysis of the methodology of the study. There could be several reasons for writing a memo. The purpose should be to enrich the information. While presenting the research findings in a network form, the research memo helps the researcher show the logical connections among quotations, codes, code groups, and their relationship with the findings. For making the connections more obvious, it is good practice to write the research question in a memo. Here, the memo serves to remind the researcher to "discover information" through the lens of the research objectives and follow the appropriate coding processes.

According to Glaser (1978), memos are "the theorizing write-up of ideas about codes and their relationships as they strike the analyst while coding" (p. 83). Glaser urges the researcher to prioritize memo writing to ensure the retention of ideas that may otherwise be lost. Therefore, researchers should consider all tasks in the conduct of the study as subordinate to the writing of memos to record ideas, musings, and reflections. The most important contribution made by memoing is that it initiates and maintains productivity in the researcher (Charmaz, 2006). Regardless of how inconsequential these thoughts, feelings, and impressions may initially seem, creating a record in the form of memos ensures the preservation of such ideas that may later prove significant (Polit & Beck, 2006).

Memos are based on field notes maintained by the researcher. However, this is just one possible use of a memo. Memos can be categorized as theoretical, methodological, and descriptive. Theoretical memos are a specific subset of memos that focus on the theorizing aspect of research. A theoretical memo helps researchers to develop reflexivity, make decisions with respect to data and research design; and make connections between evidence and abstract ideas to support assertions and interpretations (Given, 2008). Commentary memo contains the general observations of researcher for the research.

Memos can also be defined as "the narrated records of a theorist's analytical conversations with him/herself about the research data" (Lempert, 2007, p. 247). The researcher documents their insights, observations, and experiences of field work. Field notes also include information shared during interactions with the respondent before and after the interview. They include the researcher's notings on, for example, the waiting time for an interview, and the overall data collection experience at both individual and collective levels. The importance of memos has been stressed in various studies.

In short, the researcher writes memos from his field experiences. Memos form the basis for the final written product: a report, article, or thesis. The researcher who can realize the significance of memoing from the outset of the study will find the final stages of their research much less onerous than the one who fails to make this critical investment in what will ultimately prove to be 'intellectual capital in the bank' (Clarke, 2005, p. 85).

The value of memos in qualitative studies can be appreciated by the mnemonic in which the **M** stands for Mapping (the activities in the study); the **E** for Extracting meaning from the data; the second **M** for Maintaining the momentum of research; and **O** for Open communication. Researchers use memos while doing critical analysis of theories or research methodology, the.

Regardless of the research method employed, field notes and memos are indispensable to qualitative research. Ideas captured in memos are often the "pieces of a puzzle" that are later put together in the report writing phase.

In ATLAS.ti, there is a provision to classify memos based on their characteristics. Like codes, memos can also be classified in groups. If memos contain valuable information, they can be converted into documents and analysis. In ATLAS.ti entity means codes, quotations, code groups, smart codes, and smart groups. The researcher should write about field experiences while investigating a phenomenon or event, and then link them to the segments of responses known as quotations. While coding a quotation, the researcher recalls experiences from their field notes to write a memo about the answer, question, or incidence.

A memo can also be free and not attached to any entity. Sometimes, the researcher may want to document their observations and insights and create memos for analysis at a later stage. If the research is a team project, memos can be used to exchange information among team members. While interpreting the results of analysis, memos help to link the research questions to the findings, as well as with quotations, comments, codes, code groups, smart codes, etc.

Writing a Memo

In qualitative studies, researchers write memos to document their field experiences. Memos capture the experiences, moods, sentiments, and observations of researchers and respondents during data collection in the field. These experiences are generally difficult to capture in the interactions between researchers and their subjects. The title of memo depends on the type of memo and observations and analysis written in it (Fig. 5.25).

In ATLAS.ti, the research objectives are written in a memo. A memo in ATLAS.ti serves the following purposes:

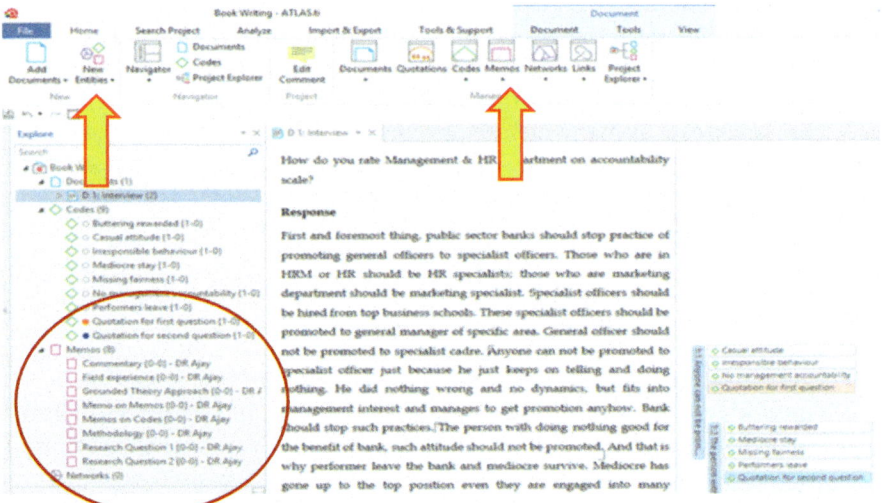

Fig. 5.25 Writing a memo

1. Recording the researcher's moods and sentiments
2. Recording the respondents' moods and sentiments
3. Recording the researcher's experiences and observations
4. Other information that the researcher may want to include, such as significant observations and descriptions, specific incidences, etc. that may be considered worthy of mention.

The information gathered in the field is of two types: latent and manifest. Latent information is that which cannot be directly recorded. It includes respondents' expressions, such as eye movement, facial expression, body language indicating various feelings and emotions, facial expressions indicating frustration, happiness, or excitement, etc. The researcher must capture as much latent information as possible. Manifest information, on the other hand, consists of quotations, jargon, analogies, the manner of use of local language, or other words that have richer meaning concerning the research questions. Manifest information can be recorded, while latent information is captured by the researcher in field notes.

Steps for Writing a Memo in ATLAS.ti

1. Open the Home tab.
2. Open New Entities or Memos on the taskbar at the top.
3. Select New Memo on New Entities or create a free memo in Memos on the taskbar at the top.
4. Select the option and write the name of the memo and save. The name given to the memo name should be informative and relevant to the research topic.
5. The user can place the memo in a category defined them (Fig. 5.25).
6. The user then describes their field experiences, methods, instruction, description, etc. and saves the memo.
7. One can create as many memos as required. The researcher can also write memos during data analysis when the need arises. But, to save time, it is good practice to create as many memo titles as possible. The descriptions can be written later.

Steps in Writing Memo Types

1. Create a memo and check if it is set in the correct type (Fig. 5.25).
2. Open Memos on the top taskbar between codes and networks. The left arrow points to "New Entities" which contains "New Memo". The right arrow points to "Memos" option that can also be used to write the memo.
3. Click once on the memo title.

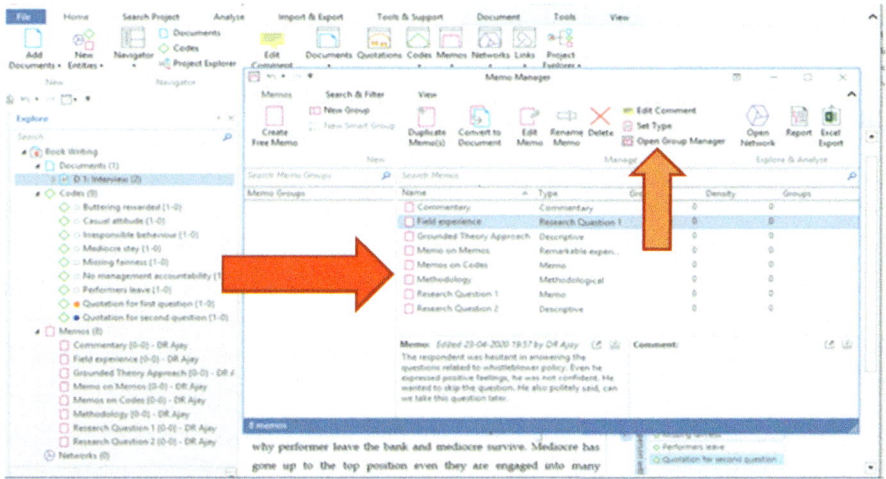

Fig. 5.26 How to set a memo type

4. Open set type on the top taskbar and 'Set Memo Type' will display. The user can select from the existing ones or create new ones.
5. As many memo types can be written as required.
6. The user can change, modify, add, delete as many memos as are required.
7. To change each memo title, click once on the selected title, set the type, and change type (Fig. 5.26).

Coding Field Notes

Despite awareness of their importance, researchers are often unsure about working with field notes. Field notes are sources of rich data; they cannot be excluded from analysis because of the researcher's lack of understanding or confidence in using them. Researchers must, first and foremost, understand that they are outsiders to their study. Excessive immersion in the study's context and settings may affect perceptions and objectivity, which they may be required to explain from an insider's perspective. Therefore, researchers should be careful about how they perceive their role. They should observe, document, and report. Showing indulgence to the participants might change their perceptions, eventually developing a bias instead of offering insights. Hence, researchers should be clear about how they will be using their field notes in their analyses and interpretations.

Field notes serve many purposes. They help to develop theory, methodologies, write memos, and coding data. Most researchers describe their experiences in their field notes, documenting their observations and insights about the phenomenon they are studying. Despite their value, researchers avoid using them for coding and analyses.

Table 5.3 Guidelines for making field notes

- What are people doing? What are they trying to accomplish?
- How, exactly, do they do this? What specific means and strategies do they use?
- How do members talk about, characterize, and understand what is going on?
- What assumptions are they making?
- What do I see going on here? What did I learn from these notes?
- Why did I include observed information?

According to Emerson, Fretz, and Shaw (1995), coding field notes serve to capture contextual and situational information, regardless of research purposes. Field notes capture much holistic information, including context, culture, and the environmental dimensions. Such information may be of great value from the perspective of the research objectives. Therefore, it is necessary to code field notes for capturing holistic information.

Coding field notes are essential to qualitative data analysis because they often contain deeper and more nuanced information. Researchers must code their field notes, write memos, and analyze them with primary data. Not documenting latent information or omitting it from analysis may lead to incomplete understanding of the subject of study, even when advanced data analysis tools are used. Therefore, researchers should treat the need to maintain field notes with utmost seriousness. Emerson et al. (1995) have suggested following the order while coding the field notes (Table 5.3).

Researchers can create codes for behaviors, goals, strategies, insights, interests, and other information on their field experiences and observations.

After coding their field notes, they should categorize them as memo codes. As memos are attached to codes, code groups, smart codes, and smart groups, researchers can import the codes linked to memos while creating networks.

A report on the connection of codes, quotations, code groups, memo, memo groups, document, document groups with the themes can be generated. Themes appear on the right side and their sources on the left side (Fig. 5.27). Four themes and their associations show how memos categories and categories of respondents contribute to creating themes. Creswell et al. (2007) advises researchers to prepare a list of critical research queries that should be addressed during coding and data analysis. This is necessary to discover the knowledge emerging from data, including what is unusual, interesting, or conceptually informative.

Coding Field Observations

Table 5.4 shows the stages of field observations and coding guidelines. The framework helps researchers to capture field observations and code them using several contextual parameters. Researchers should note that information that brings insights and exploration in their research is vital information.

Fig. 5.27 Visualization of themes from field notes and transcripts

Comments

Comments are annotations. They are used for identifying and describing an entity and its attributes. There are six entity types in ATLAS.ti: documents, quotations, codes, memos, networks, and links. With each entity, the researcher can perform several functions. Thus, a comment for code describes the code. Similarly, a comment for a quotation or document are their descriptions.

The researcher can also write comments about the categories of respondents and their backgrounds (i.e., social-economic, cultural, or other characteristics) to make data analysis more precise by using them as document groups and creating code-document tables. Similarly, comments can be added to memos, memo groups, codes, code groups for identification and explanations. Comments are usually written quickly without much attention to their order and detail, somewhat like jotting one's thoughts on paper. For example, while writing a comment, the project may be described as, "This project is about studying employees' morale in public sector banks where the researcher has collected data from management and employees".

Writing Comments

Figure 5.14 shows written comments for the document, quotation, code, and memo. For each section, the comments are shown by a differently colored arrow. The comment for the document says that it is a transcription of an audio-recorded interview. If required, a researcher can also add relevant information in the comment pane or change, modify, shorten, or delete the comment as needed.

Table 5.4 Coding field observations

How to capture, code, and report field observations from formal organisations

Stage	Observations	Contents to code	How to note and report
The Wait Period (for interviews, discussion, etc.)	Observe the settings, people's behaviour, appearances, work, interactions, what they talk about, how they regard the researcher	People appearance/ interaction/ surrounding/ movement/ environment	Write showing the contexts to help the reader to understand and visualize the study settings
Before the interview	What the respondent talks about, their appearance, suggestions, office settings, papers, walls, tables, telephones, confidence, happiness, unhappiness etc. How does the respondent regard the presence of the researcher	Significant and relevant information/office setting/respondent confidence/ acceptance/avoidance Mention details in code description	Describe the work settings, observations about the respondent, and other significant information
During the interview	Body movement, smiles, appearance of uneasiness, eye movement, anxiety, excitement, happiness, tones, volume, avoidance, interest, focus, etc	Excitement/anxiety/ pauses/ avoidance moments etc Assign memos written from field notes to the responses that showed contrasting signals Explain in code description	Make notes of the moments when the respondent seemed uneasy, or showed emotions that were not consistent with the response. Also describe the body language of the respondent. Use symbols/short words
After the interview	Engage in informal conversation. The researcher should politely raise questions or points that were either not answered, avoided, or those that the respondent skipped or appeared uneasy while answering,	Significant, relevant, new information/ suggestions/cautions/ insider information/ comfort and concerns Describe the information which emerged, including suggestions, cautions, insider information, etc. in the memos for supporting themes	Observe the moments during the interviews and conversations. Report insights that emerged during informal interactions

Figure 5.28 also shows a comment about a quotation. The comment says that the respondent was unhappy about the promotion processes in the organization. The comment elaborates and clarifies researcher's thoughts. It also explains the respondent's response. Although the researcher can write as many comments as possible, there should be a justification for writing them. In the same figure, the researcher has written a comment on a code named "Buttering (a colloquialism for sycophancy) rewarded" and explained what it meant—the need to please their superiors to improve their chances of promotion—in the comment. The comment indicated by the blue arrow is associated with a memo named "field experience." It mentions that the respondent was visibly angry while answering the researcher's questions. It appeared (to the researcher) he was not motivated by the organization's culture.

Steps for Writing Comments

The screenshots in Figs. 5.29 and 5.30 show the comment section in ATLAS.ti. The user must follow these steps while writing comments.

Researchers can use codes, memos, networks, and links to write comments (Fig. 5.29). If the researcher wants to include a comment about a code or code groups, memo or memo groups, network or network groups, or links, they should select the entity, click once & write. Alternatively, they can right click on the entity, open edit comment, write their comments and save.

Writing Comments for Documents

1. Open Home or Project Explorer. Figure 5.30 shows the resulting screen.
2. Click on Open Documents. A list of all project documents will be displayed.
3. Open Document Manager.
4. Select the document to which a comment is to be added.
5. Write a comment in the bottom right and save.
6. Repeat for other documents or groups.

Writing Comments for Quotations

1. Open Home or Project Explorer (Fig. 5.31).
2. Open Quotations. The quotations file will be displayed.
3. Open Quotation Manager, select a quotation (i.e., 1:1, 1:2, 2:1, 2:2, etc.)
4. Write a comment in the field at bottom right and save.
5. Repeat for other quotations or groups.

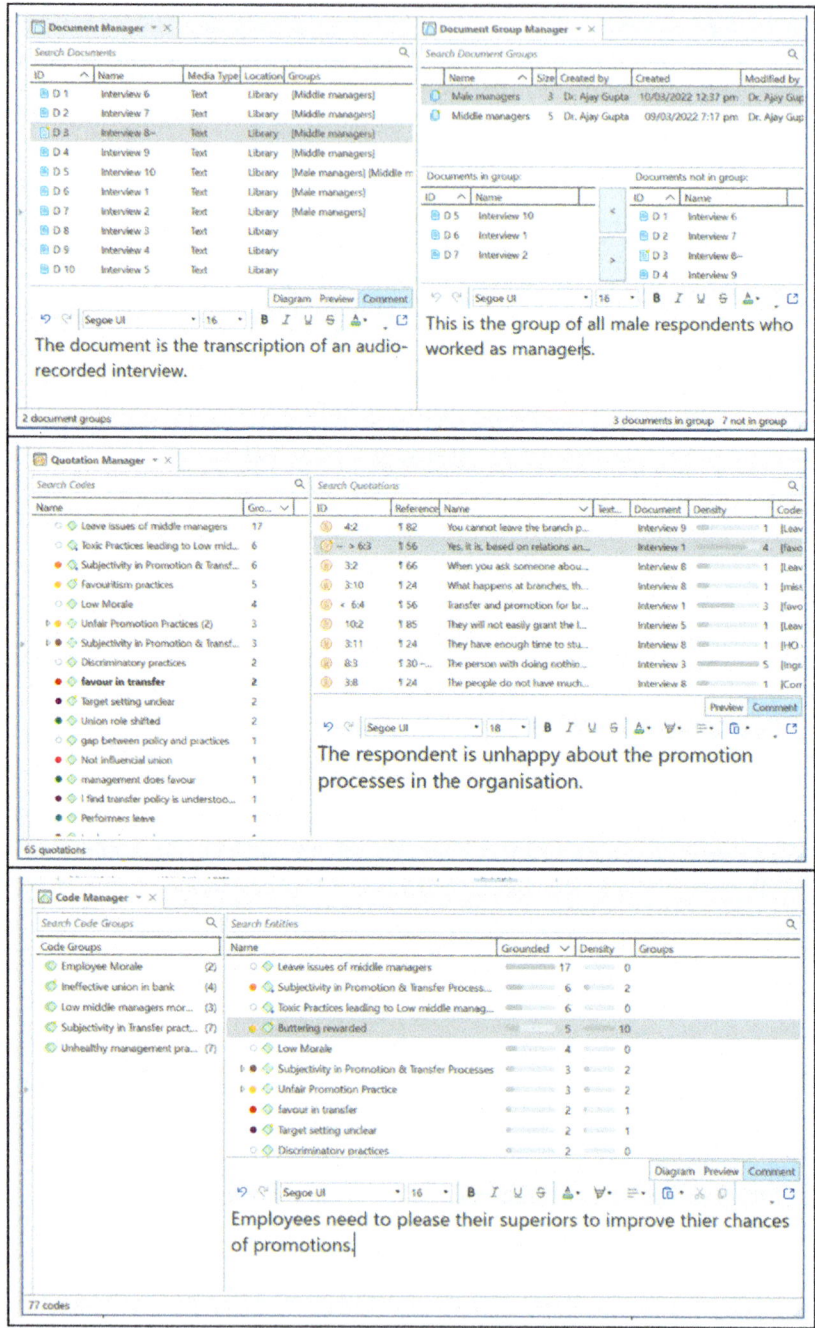

Fig. 5.28 Comments for entities

5 DATA ANALYSIS METHODS   167

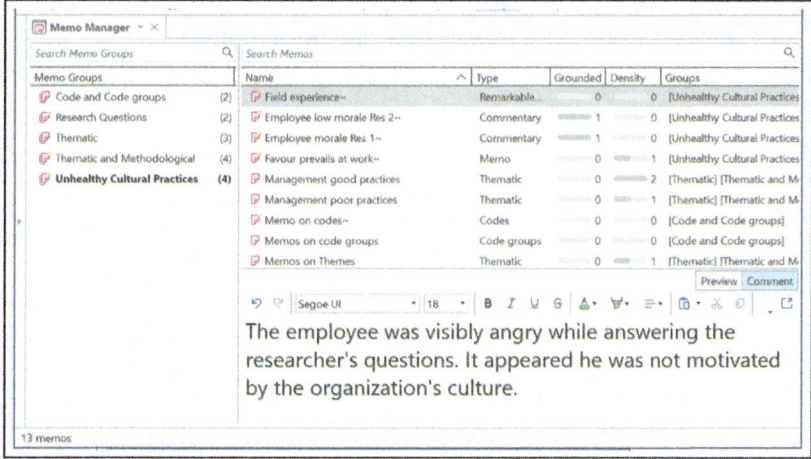

Fig. 5.28 (continued)

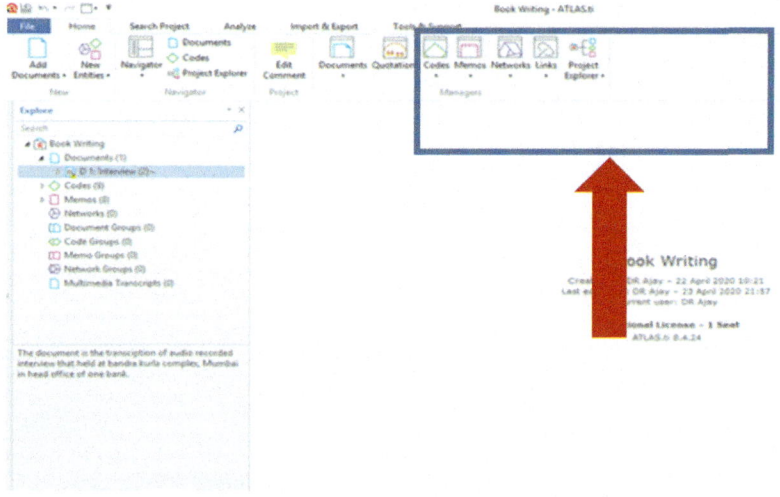

Fig. 5.29 Writing comments

Writing Comments for Codes

1. Open Home or Project Explorer (Fig. 5.32).
2. Open Codes. All codes will be displayed.
3. Open Code Manager, select code and click.
4. Write a comment in the bottom right and save.
5. Repeat process for other codes or groups.

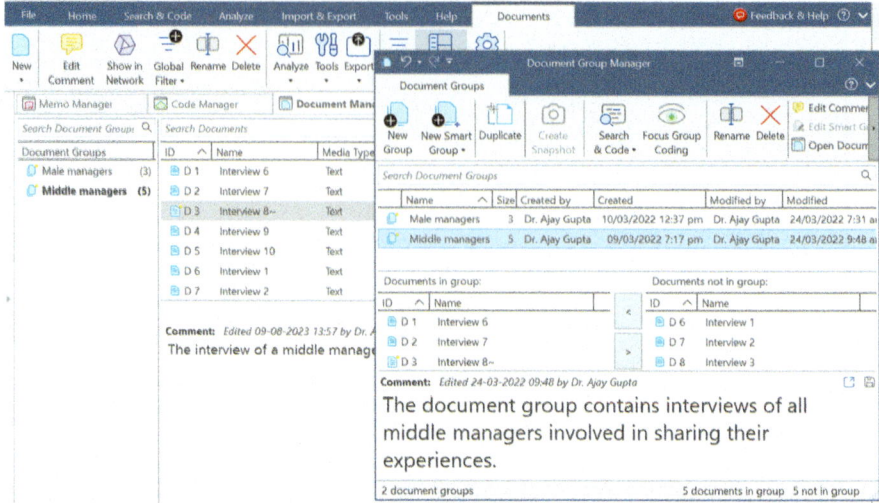

Fig. 5.30 Writing a comment for document

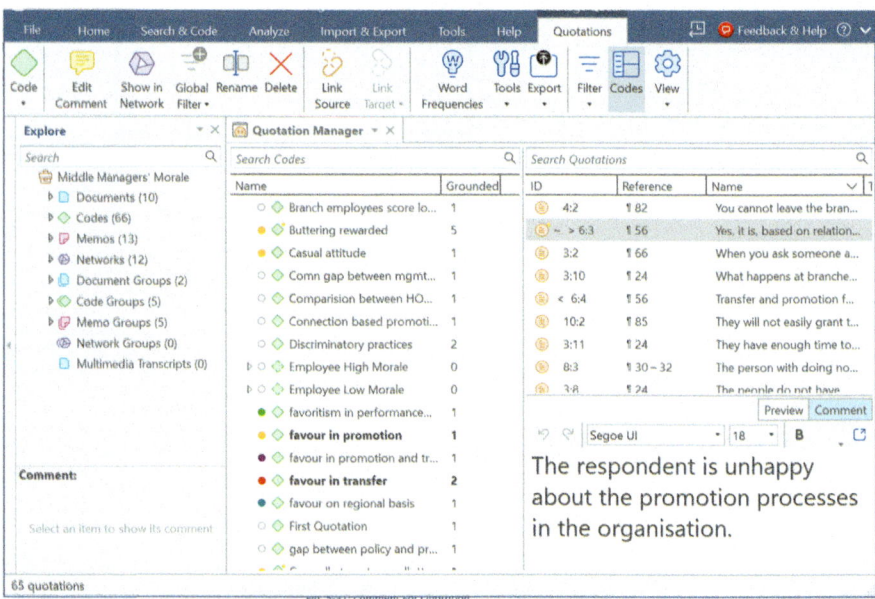

Fig. 5.31 Comment for quotation

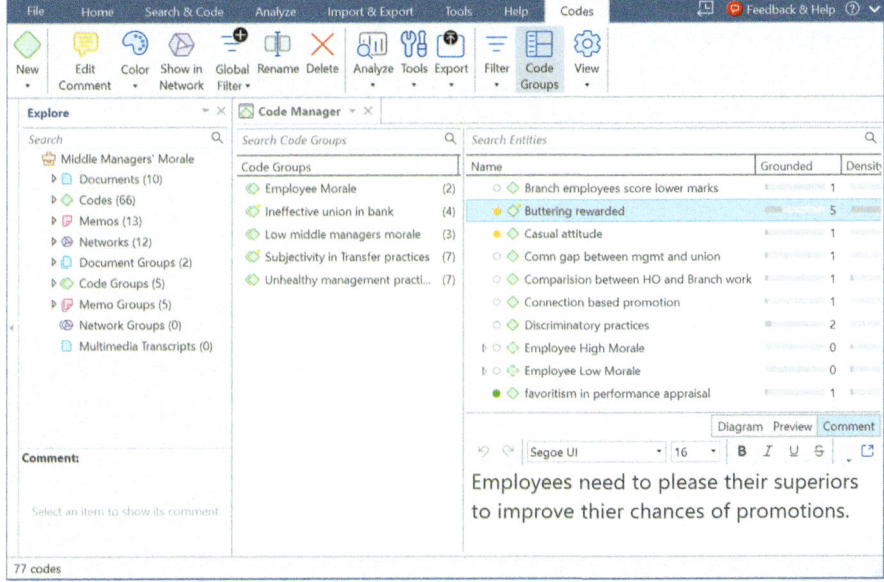

Fig. 5.32 Comment for code

Writing Comments for Memos

1. Open Home or Project Explorer (Fig. 5.33).

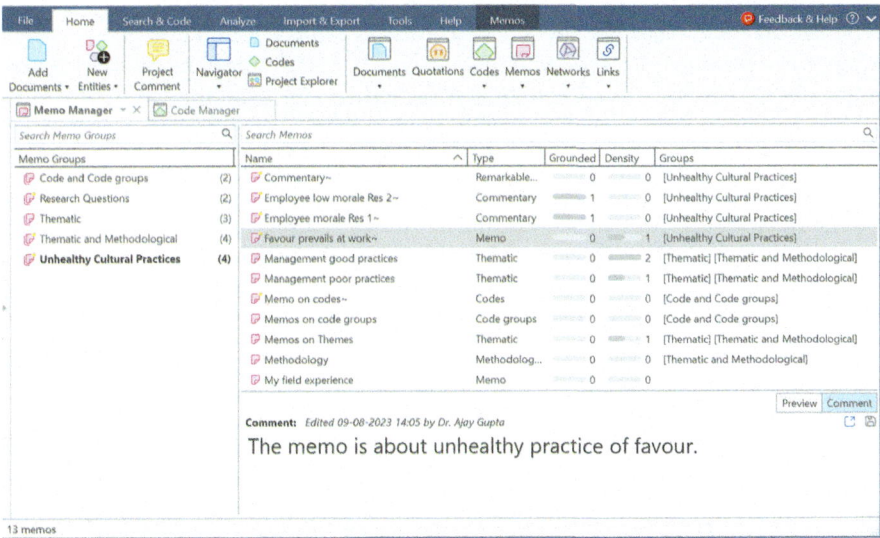

Fig. 5.33 Comment for memo

2. Select Open Memo. All memos will be displayed.
3. Open Memo Manager
4. Select memo and write comments in the field at the bottom right of the screen.
5. Save
6. Repeat for other memos or groups.

Comments for networks and links can also be written in a similar manner.

Understanding Comments and Memos

A comment is a description of the object. It describes the concepts, segments, field experiences and the researcher's reflections. A comment can be connected to any entity: projects, documents, quotations, codes, code groups, memos, networks, links, relations, etc. It describes the entity's attributes, characteristics, and other information. For example, a comment about a document group may be about the characteristics of the respondents, about quotations indicating a positive perception, codes with a high density, and so on.

Comments are usually written quickly and need not have a structure and much detail. Their main importance is to capture the researcher's impressions about the entity for which the comment is written. Thus, a project on employees' morale can have the comment, "This project is about studying employees' morale in the public sector banks, where the researcher has collected data from management and employees".

While writing a comment about a quotation, the researcher may describe their interpretation. Whether all memos must be commented on or not depends on their contents and the researcher's impression. Thus, for example, when the researcher describes a respondent's body language that indicated discomfort, such observations should be written in a memo. *But the memo can be described in a comment*, such as, for example, "This was the experience of the researcher while observing the respondent's body language which indicated discomfort" in the comment field.

The network, "Unhealthy Cultural Practices" in Fig. 5.34 shows the memo and its content. At the bottom, a comment describes the memo, "The memo is about the experience of the respondent. He is very enthusiastic about sharing his good experience about the organisation while the comment on another memo 'Favour prevails at work' is also written. Thus, every movement, detail, and observations about the interview can be described in a memo.

Comments and memos are different. A comment only describes the entity (i.e., codes, quotations, memos, etc.). They are helpful because the researcher can describe, define, or explain their understanding of a code, code groups, smart code, etc. ATLAS.ti allows the user to edit, add, modify, or shorten the comment with the help of the text editor.

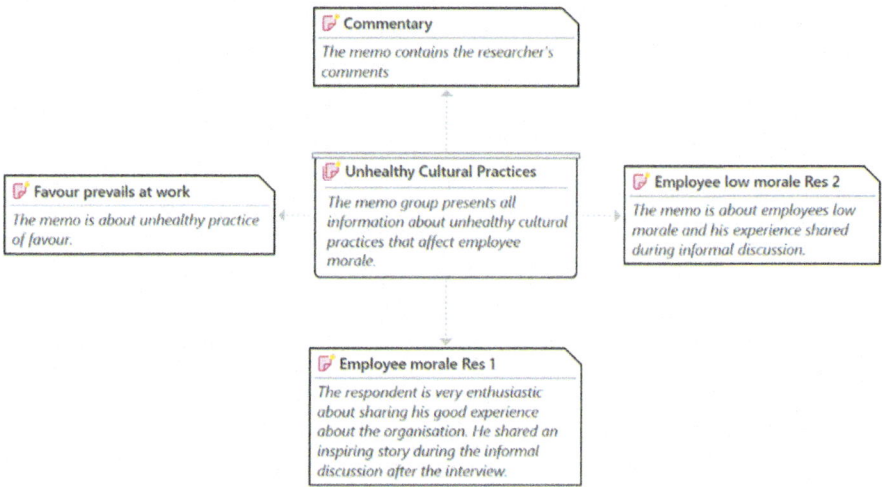

Fig. 5.34 Memo and comment network

Memos are described in comments. If a memo is about the methodological aspects of the research, the researcher can explain the methodology and provide related information in the memo. There can be memos about networks of documents, codes, other memos, quotations, etc.

The basic difference between a memo and its comment is that a memo is a document that describes the respondent and researcher's observations. Memos can be converted into documents for further analysis, whereas comments only define and clarify quotations, codes, memos, networks, etc. Memos support findings, while comments elaborate on the concepts and provide rich meaning to the entities.

Figure 5.34 shows quotations, codes, and memos in the network. The top half of the memo describes the field experiences of the researcher. In the bottom half, the researcher has described the memo with a comment. The title of the memo is "Employee morale Res 1", which is Respondent 1's reply to Question 1 about unfair practices. It is always advisable to give a memo a title for easy identification (of the quotation, code, or memo it seeks to explain).

THEMES

According to Rubin and Rubin (2011), a theme is a process of discovery in which the concepts that support themes are embedded in the data (from interviews, questionnaires, focus group discussions, and so on). Themes are necessary for capturing the important features of qualitative data. They are directly related to the research question and represent some level of patterned response or meaning within the data set. In the context of coding, it is necessary to understand what can be considered as a pattern or theme.

The researcher must understand that mere occurrence of an incident does not decide a theme. A theme might be given considerable space in some data items, little or none in others, or it might appear as a relatively insignificant part of the data set. A theme is decided not by data but more by the researcher's judgment about the data. Therefore, rigid rules do not work; instead, the researcher must adopt a flexible approach to identifying themes and analysing them.

In qualitative research, themes emerge from data through the researcher's understanding of the research question. After collecting data, the researcher studies the responses through the lens of the research question (s) to make meaning of the data. While doing so, the researcher may also refer to her/his field observations and experiences.

As explained earlier in Chap. 3 (Writing Field Notes), the observations are recorded in field notes from which memos are created. Themes emerge from codes and memos. Memos play an important role in identifying themes. Since the researcher is usually immersed in the study, the researcher is, in a sense, also a respondent. Hence, their observations and opinions are important. While memos document the researcher's observations and experiences, memos will be absent while working with secondary data.

Themes emerge from meaningful information and the researcher's observations and field experiences. However, they can be questioned, or even misinterpreted, by the audience who might question the themes proposed by the researcher, as well as the language in which they are presented. Ely (1997) argues that if themes reside anywhere, they reside in the researcher's head from thinking about data and creating links. Therefore, the researcher should make all efforts to ensure that themes are presented and interpreted in such a way to appear natural and logical, emerging from the data and researcher's field experiences. The researcher should provide a detailed explanation for the data collection and field-observation processes and, most importantly, support the emerging themes with key quotations and memos, and their relationships with research objectives.

Using codes and memos, and code-coefficients, code densities, groundedness values while identifying themes will help in instilling confidence in the researcher's audience and give greater credibility to the research's outcomes. The researcher should constantly engage with their audience, explaining the rationale for qualitative approach, the methods used for data collection and analysis while developing themes.

Code-Theme Relationships

Data coding is foundational to qualitative research. It is central to qualitative data analysis and hence it is imperative for researchers to have superior coding skills. In fact, as several scholars have pointed out, process rigor and quality of coding directly determine the quality of the research.

Coding is not just about selecting keywords and levelling data; it also involves creating meaningful and insightful information out of large volumes of segmented data. As stressed several times in this book, qualitative data is obtained in various forms—text, audio, video, images, pictures, geo maps, and so on. The approach to coding each data type is almost the same; however, text data is richer and denser. But the study of data in text form is time-consuming. Coding helps to "index" or "map" critical information from disparate data, which allows the researcher to find insights which address the research questions.

Opinions about the number of codes that a researcher should create from their data vary and are often the cause of disagreements and ambiguity. There is also considerable divergence in views on the approach to coding. However, these should not confuse the researcher. According to Lichtman (2006), the researcher should generate 80–100 codes that can be organized in 15–20 categories. The categories can be further grouped under five to seven major concepts. Creswell et al. (2007) suggests that the researcher begins with a small list of five to six provisional codes, following the principles of "lean coding", and then expanding the list to 25–30 categories which can be grouped under five to six major themes. According to Wolcott (1994), three of anything major (themes) is good for reporting qualitative work. The number of themes/concepts should be kept to a minimum to avoid complexity in analysis and for ensuring coherence in reporting.

Despite the diversity of opinions, scholars are in general agreement that the number of themes reported should be kept to a manageable number. Besides the need to avoid unnecessary complexity, researchers must also be mindful of the need to fit their arguments in limited journal space. However, a study can have a larger codebook for multiple research articles/transcripts i.e., a qualitative PhD study. Therefore, the number of codes, categories and themes should also be treated as an exercise in optimization aimed at the requirements of the study and its reporting.

While creating code groups, the researcher should know how to identify patterns and homogeneity of traits and attributes. Many codes—as much as 10 to 20 percent of the total number—may not fit in a pattern or reflect homogeneity in certain characteristics. These codes should be, as a rule, excluded from analysis unless the study has adopted the grounded theory approach. Codes not fitting in a pattern may indicate the direction for further study, which the researcher could elaborate on while presenting the findings.

Table 5.5 shows the relationship between codes, code categories, smart codes, code groups and smart groups. This is an example of a project. Codes are categorised based on shared relationships. Using code categories researchers can create Smart codes using set or proximity operators. Smart codes can be considered as theme. And further, smart codes can be used to create code groups and smart groups. In such cases, code group or smart group will become theme.

Table 5.5 From codes to smart groups

Smart groups	Code groups	Smart codes	Code categories	Codes (sub-codes)
Selection and association code groups using set & proximity operators to develop themes	Grouping code categories under an emerging concept	Selecting and associating code categories using operators to develop themes	Linking codes with shared attributes using code-code relations	Discovering meaningful information from response through research objective lens
		Low Employee Morale (operators) unhealthy employee transfer practice (Unhealthy transfer practices and employee low morale have been used to create the smart code)	High Employee Morale	Satisfied Motivated Committed
Subjectivity in promotion and transfer (operators) unclear leave practices	Employee morale		Low Employee Morale	Dissatisfied Frustrated
		Unhealthy transfer practice (operators) positive experience	Unhealthy transfer practice	Unfair transfer Relation-based transfer
Employee morale (operators) unclear leave practices	Subjectivity in promotion and transfer	High Employee Morale (operators) positive experience	Dysfunctional promotion practice	Connections-based promotion Favoritism in promotion
			Negative experience	Ask to postpone leave Ask to cancel leave
	Unclear leave practices	Low Employee Morale (operators) dysfunctional promotion practice	Positive experience	Easily get leave No issue with leave

Table 5.6 shows the stages of coding. The description of each stage has been explained. Axial and selective coding are generally used in grounded theory; however, coding can also be used in non-grounded theory research. Researchers can use the coding framework for data analysis across disciplines.

Figure 5.35 shows the code-theme relationships. Codes sharing identical information are categorized into several codes. Further, code categories are used to create sub-themes (axial codes in grounded theory). Sub-themes are created from code categories and codes. Further, sub-themes can be used to create core theme (Selective code in grounded theory). Since every code is linked with quotations, every theme and sub-themes contain selected quotations which are used to report and interpret the findings. In field-based research, memos are used to support themes along with quotations.

Figure 5.36 shows the relationship between theme, sub themes, code categories and codes in ATLAS.ti 23. Sub-themes are shown by blue arrow. Code-categories are shown by red arrow. Exemplary organization is the theme (Fig. 5.36). Authentic leadership and a Healthy working environment are sub-themes. Practicing leadership, Inspirational leadership, Responsible employees, and Vibrant culture are dimensions of sub-themes. Codes are components of sub-themes.

Table 5.6 Stages of coding and description

Stage	Description
Coding	The research objective determines the coding approach. Researchers justify their coding strategy. It could be inductive, deductive, abductive, in-vivo, hybrid, or a combination of several coding methods. Codes provide fragmentary and meaningful information about the phenomenon under investigation
Categorizing	Codes are classified according to their homogeneity. Codes that share similar information should be combined to form a code category. Positive and negative code categories can also be formed based on sentiments. Codes with similar attributes should be categorized. Many codes may be excluded from the code category due to this process. Information is classified based on attributes, sentiments, and identical relationships between them
Identifying sub-themes (Axial coding)	Axial codes are sub-themes that are created using code categories. The axial codes share the meanings of code categories within it. Several code categories are typically used to generate lesser axial codes
Creating themes (Selective coding)	Selective codes can be created if axial codes can be made more meaningful by developing relationships between them. The core theme is represented as selective code in such cases

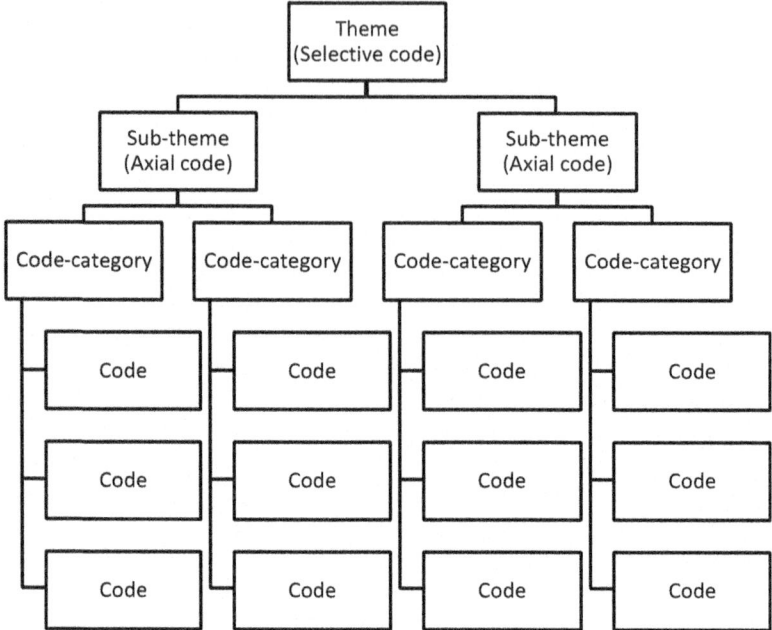

Fig. 5.35 Codes-theme relationships framework

Report for theme, sub-themes and code categories will show connected quotations and codes.

Language of Qualitative Research

In qualitative research, the researcher is the medium through which the meanings made by the subjects of the study are transmitted to the audience. The researcher reports the findings which are a social construction in which the author's choice of writing style and literary devices provide a specific view on the subjects' lived experiences (Kvale, 1996, p. 253). In this process, researcher influence is significant (Alvesson & Skoldberg, 2017). The researcher decides how specific information or respondent should be given voice based on research objective. The researcher also decides what aspect of the context and other relevant information should be presented to the reader.

Presenting Results and Findings

The results of a qualitative study are presented as themes. Themes capture the vital information present in the data in relation to the research questions. The researcher should present the themes and show the relationships among them and their relevance to the research question. The theme is a discovery process (Rubin and Rubin, 2011) and is necessary for capturing important

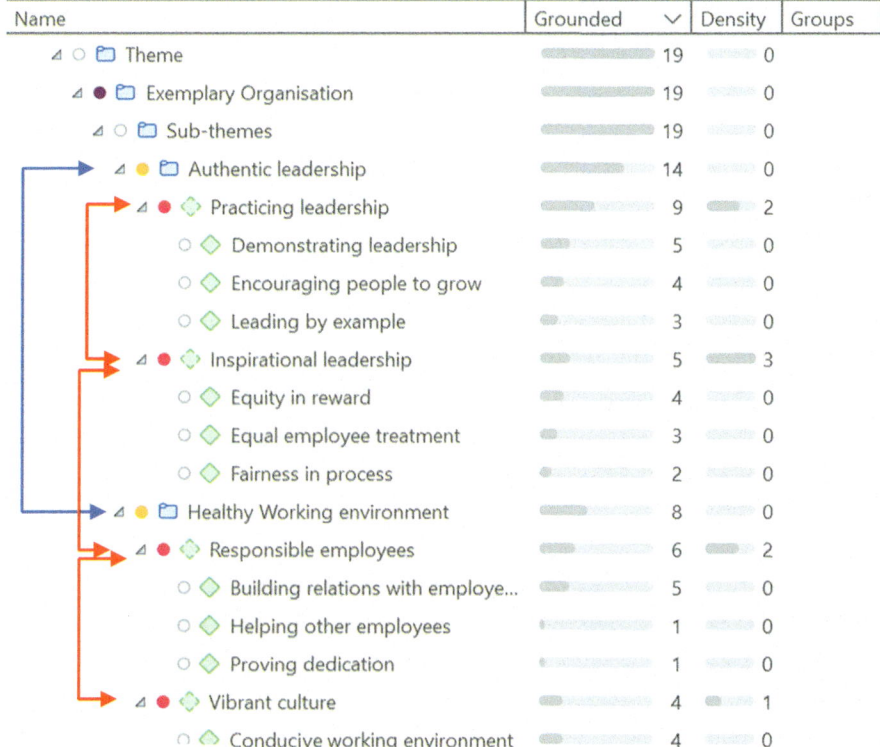

Fig. 5.36 Codes-theme relationship in ATLAS.ti

qualitative data features. A theme directly relates to the research question and represents some patterned response within the data set. The themes identified are strongly linked to the data (Patton, 1990).

Each theme represents a concept, theory, reflecting the new knowledge that emerges from data analysis. The themes are supported by significant statements (Quotations) made by the respondents and in the memos written by the researcher (Figs. 5.26 and Tables 5.1 and 5.2). Figure 5.36 shows the theme, sub-themes and their components. The numbers under 'grounded' show the significance of the theme and associated components. Researchers should organize themes based on grounded, generate reports and interpret with reference to research objectives. Co-occurrence analysis, Code-document analysis table, code manager should be preferably shown in presenting results and findings. Results should be supported by appropriate tables, screenshots, networks and text reports.

Recommended Readings

Agar, M., & MacDonald, J. (1995). Focus groups and ethnography. *Human Organization, 54*(1), 78–86.
Alvesson, M., & Sköldberg, K. (2017). *Reflexive methodology—New vistas for qualitative research*. Sage.
Anzul, M., Downing, M., Ely, M., & Vinz, R. (2003). *On writing qualitative research: Living by words*. Routledge.
Auerbach, C., & Silverstein, L. B. (2003). *Qualitative data: An introduction to coding and analysis* (Vol. 21). NYU Press.
Barnett, J. M. (2002, November). Bloor, M., Frankland, J., Thomas, M., & Robson, K. (2001). Focus groups in social research. In *Forum qualitative Sozialforschung/forum: Qualitative social research* (Vol. 3, No. 4).
Basit, T. (2003). Manual or electronic? The role of coding in qualitative data analysis. *Educational Research, 45*(2), 143–154.
Becker, P. H. (1993). Common pitfalls in published grounded theory research. *Qualitative Health Research, 3*(2), 254–260.
Berente, N., Seidel, S., & Safadi, H. (2019). Research commentary—data-driven computationally intensive theory development. *Information Systems Research, 30*(1), 50–64.
Bernard, H. R. (2006). *Research methods in anthropology* (4th ed.). Altamira Press.
Blauner, B. (1987). Problems of editing 'first-person' sociology. *Qualitative Sociology, 10*, 46–64.
Boeije, H. (2010). *Analysis in qualitative research*. Sage Publications Ltd.
Boyatzis, R. E. (1998). *Transforming qualitative information: Thematic analysis and code development*. Sage.
Braun, V., & Clarke, V. (2014). What can "thematic analysis" offer health and wellbeing researchers? *International Journal of Qualitative Studies on Health and Well-Being, 9*(1), 26152.
Butina, M. (2015). A narrative approach to qualitative inquiry. *Clinical Laboratory Science, 28*(3), 190–196.
Charmaz, K. (2005). Grounded theory in the 21st century: A qualitative method for advancing social justice research. *Handbook of Qualitative Research, 3*(7), 507–535.
Charmaz, K. (2006). *Constructing grounded theory: A practical guide through qualitative analysis*. sage.
Clarke, A. (2005). *Situational Analysis: Grounded Theory after the Postmodern Turn*. Thousand Oaks, Calif. Sage Publications.
Clarke, M., Seng, D., & Whiting, R. H. (2011). Intellectual capital and firm performance in Australia. *Journal of Intellectual Capital*.
Coffey, A., & Atkinson, P. (1996). *Making sense of qualitative data: Complementary research strategies*. Sage Publications, Inc.
Cohen, S. M., Gravelle, M. D., Wilson, K. S., & Bisantz, A. M. (1996, October). Analysis of interview and focus group data for characterizing environments. In *Proceedings of the Human Factors and Ergonomics Society Annual Meeting* (Vol. 40, No. 19, pp. 957–961). SAGE Publications.
Corbin, J. M., & Strauss, A. (1990). Grounded theory research: Procedures, canons, and evaluative criteria. *Qualitative Sociology, 13*(1), 3–21.
Crabtree, B. F., & Miller, W. F. (1992). A template approach to text analysis: developing and using codebooks.

Crabtree, B. F., & Miller, W. F. (1999). A template approach to text analysis: Developing and using codebooks. In BF Crabtree & WL Miller (eds). *Doing qualitative research*. Newbury Park, CA: Sage.

Creswell, J. W., & Poth, C. N. (2016). *Qualitative inquiry and research design: Choosing among five approaches*. Sage Publications.

Creswell, J. W., Hanson, W. E., Clark Plano, V. L., & Morales, A. (2007). Qualitative research designs: Selection and implementation. *The Counseling Psychologist, 35*(2), 236–264.

DeCuir-Gunby, J. T., Marshall, P. L., & McCulloch, A. W. (2011). Developing and using a codebook for the analysis of interview data: An example from a professional development research project. *Field Methods, 23*(2), 136–155.

DeVault, M. L. (1990a). Novel readings: The social organization of interpretation. *American Journal of Sociology, 95*(4), 887–921.

DeVault, M. L. (1990b). Talking and listening from women's standpoint: Feminist strategies for interviewing and analysis. *Social Problems, 37*(1), 96–116.

Ely, J. W. (Ed.). (1997). *Main themes in the debate over property rights* (Vol. 6). Taylor & Francis.

Emerson, R. M., Fretz, R. I., & Shaw, L. L. (1995). Writing up field notes I: From field to desk._____. In *Writing ethnographic field notes*. The University of Chicago Press, Chicago e Londres.

Evers, J. C. (2016, January). Elaborating on thick analysis: About thoroughness and creativity in qualitative analysis. In *Forum: Qualitative social research* (Vol. 17, No. 1, pp. 152–172). Freie Universität Berlin.

Frith, H., & Gleeson, K. (2004). Clothing and embodiment: Men managing body image and appearance. *Psychology of Men & Masculinity, 5*(1), 40.

Given, L. M. (Ed.). (2008). *The Sage encyclopaedia of qualitative research methods*. Sage Publications.

Glesne, C. (2006). Becoming Qualitative Researchers, 3rd Edn. Pearson Education, New York, NY, USA.

Glaser, B. G. (1978). *Theoretical sensitivity*. University of California.

Glaser, B. G., & Strauss, A. L. (2017). *The discovery of grounded theory: Strategies for qualitative research*. Routledge.

Glesne, C. (2016). *Becoming qualitative researchers: An introduction*. Pearson.

Hsieh, H. F., & Shannon, S. E. (2005). Three approaches to qualitative content analysis. *Qualitative Health Research, 15*(9), 1277–1288.

Knodel, J. (1993). The design and analysis of focus group studies: A practical approach. *Successful Focus Groups: Advancing the State of the Art, 1*, 35–50.

Kvale, S. (1996). *InterViews—An introduction to qualitative research interviewing*. Sage.

Lakoff, G., & Johnson, M. (1980). Conceptual metaphor in everyday language. *The Journal of Philosophy, 77*(8), 453–486.

Lempert, L. B. (2007). Asking questions of the data: Memo writing in the grounded. *The Sage handbook of grounded theory* (pp. 245–264).

Lichtman, M. (2006). *Qualitative research in education: a user's guide*. Sage Publications.

Patton, M. Q. (1990). *Qualitative evaluation and research methods* (2nd ed.). Sage.

Polit, D. F., & Beck, C. T. (2006). The content validity index: Are you sure you know what's being reported? Critique and recommendations. *Research in Nursing & Health, 29*(5), 489–497.

Polkinghorne, D. E. (2005). Language and meaning: Data collection in qualitative research. *Journal of Counseling Psychology, 52*(2), 137.

Polkinghorne, D. E. (2007). Validity issues in narrative research. *Qualitative Inquiry, 13*(4), 471–486.

Richards, L., & Morse, J. M. (2013). *Readme first for a user's guide to qualitative methods* (3rd ed.). Sage.

Rubin, H. J., & Rubin, I. S. (2011). *Qualitative interviewing: The art of hearing data*. Sage.

Saldaña, J. (2016a). *The coding manual for qualitative researchers* (3rd ed.). Sage.

Saldaña, J. (2016). *Ethnotheatre: Research from page to stage*. Routledge.

Scott, K. W., & Howell, D. (2008). Clarifying analysis and interpretation in grounded theory: Using a conditional relationship guide and reflective coding matrix. *International Journal of Qualitative Methods, 7*(2), 1–15.

Smithson, J. (2000). Using and analysing focus groups: Limitations and possibilities. *International Journal of Social Research Methodology, 3*(2), 103–119.

Stewart, D. W., Shamdasani, P. N., & Rook, D. W. (2007). Analyzing focus group data. *Focus groups: Theory and practice* (Vol. 20).

Strauss, A. L. (1987). *Qualitative analysis for social scientists*. Cambridge University Press.

Van Nes, F., Abma, T., Jonsson, H., & Deeg, D. (2010). Language differences in qualitative research: Is meaning lost in translation? *European Journal of Ageing, 7*(4), 313–316.

Weiss, R. S. (1994). *Learning from strangers: The art and method of qualitative interview studies*. Free Press.

Wolcott, H. F. (1994). *Transforming qualitative data: Description, analysis, and interpretation*. Sage.

Xu, W., & Zammit, K. (2020). Applying thematic analysis to education: A hybrid approach to interpreting data in practitioner research. *International Journal of Qualitative Methods, 19*, 1609406920918810.

CHAPTER 6

Using Networks and Hyperlinks

Abstract Nodes and links help to make networks more informative, allowing the researcher to directly link the research questions with the findings. A network is a relationship of codes, memos, quotations, documents with research questions. It is shown by intersecting horizontal and vertical lines that represent a collection, or system, of connected concepts. Networks assist in locating the origin of concepts and their links with other concepts, showing causal relationships, and the research outcomes. They are a more sophisticated way of organizing data and information as compared to groups. Networks enable researchers to visualize the structure of concepts by connecting groups of related elements. They help to express the relationships between nodes. Hyperlinks show the connection between quotations and citations, helping to strengthen the researcher's claim. The chapter shows, with the example of a research project involving multimedia data, how to create hyperlinks, as well as demonstrate their utility.

Overview

The findings of analysis, insights, and ideas can be presented using networks or as text reports. A network is a visualization tool that contains concepts, objectives, ideas, situations, or sets of information such as a chart, images, audio or videos, etc. Visualization helps to discover connections between objectives, concepts, findings, and interpretations, and to communicate the researcher's research outcomes effectively, as well as for easy comprehension by the researcher's audience.

Networks are used to locate the origin of information, linkages between concepts, causal relationships, and output of data analysis. A theme can

© The Author(s), under exclusive license to Springer Nature Switzerland AG 2024
A. Gupta, *Qualitative Methods and Data Analysis Using ATLAS.ti*, Springer Texts in Social Sciences,
https://doi.org/10.1007/978-3-031-49650-9_6

be presented as a network showing its links to codes, categories, memos, comments, and quotations. ATLAS.ti uses quotations, codes, memos, documents, and groups to present networks. They are a more sophisticated way of organizing data and information as compared to a group. Networks enable researchers to visualize the structure of themes and concepts by connecting groups of related elements.

Nodes are the entities in a network (Fig. 6.1). They are used to add information to the network to show the origins of information (in documents) and themes (outcomes of analysis), and how they are connected. Themes are generated from codes and memos (Chap. 5). As explained in Chaps. 4, 5 and 7, codes, memos, documents, networks are grouped based on traits and their connections with the objectives of the study. Every entity that makes the network informative is called a node. Documents, document groups, quotations, codes, code groups, memos, memo groups, networks, and network groups are nodes. The interconnections between nodes in a network are shown by using relations (code to code relations and hyperlinks).

A very useful property of graphs is their intuitive presentation, usually in a two-dimensional layout of labeled *nodes* and *links*. In contrast with linear, sequential representations (i.e., text), representing knowledge in a network form closely resembles how human memory and thought are structured. ATLAS.ti uses networks to help represent and explore conceptual structures. But first, it is necessary to understand the meanings of the key terms used while using networks.

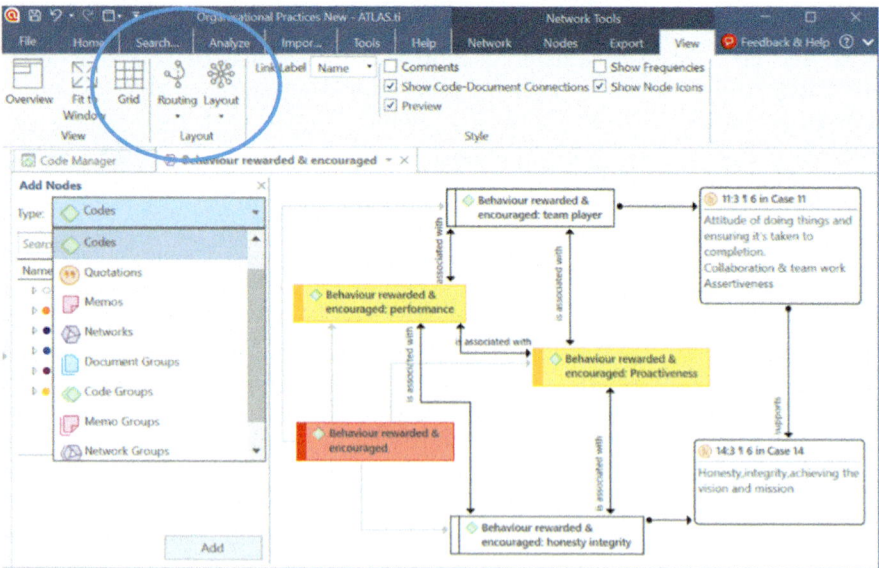

Fig. 6.1 Codes, links, relations and hyperlinks

6 USING NETWORKS AND HYPERLINKS   183

NODES AND LINKS

Figure 6.1 shows the network for the code category 'Behaviour rewarded and encouraged'. The network has two quotations and five codes, meaning that there are two quotations with five codes, with all codes indicating 'Behaviour rewarded and encouraged'. The relationships among quotations are shown by hyperlinks, and the codes are linked by code-to-code relations. Hyperlinks and code relations are used to show the linkages between quotations and codes respectively.

The network in Fig. 6.1 shows codes and quotations, showing their relationships. In the figure, the codes and quotations are the nodes. A list of nodes is shown on the left side of the figure. The link connecting codes represents the code-to-code relationship, while the link connecting quotations is called a hyperlink.

A node is the point at which several lines meet. A network is a set of nodes and links and can have multiple nodes. The links indicate the richness of the concepts, as seen in the nodes. However, a node with maximum links has the highest degree of linkages (Fig. 6.2). Perception of Qualitative Research has the highest degree of linkages.

In ATLAS.ti, code density shows the degree of the code, meaning that it shows the number of codes linked with a particular code. A code having a density of 06 means that it is linked with six codes. To find the density of the code can be seen using Code Manager. Similarly, various quotations can also be hyperlinked. In other words, quotations to quotations linkage are performed by using hyperlinks.

The connection between two nodes is shown by a directional arrow. The starting node is the source node and the node it is connected to is known as the target node. Quotation 11:3 is the starting node and quotation 14:3 is the target node (Fig. 6.1).

When a link connects two nodes without showing a relation, it becomes a non-directed link. A relation-based connection is called a first-class link, whereas a link having only a connection and no relationship between nodes is

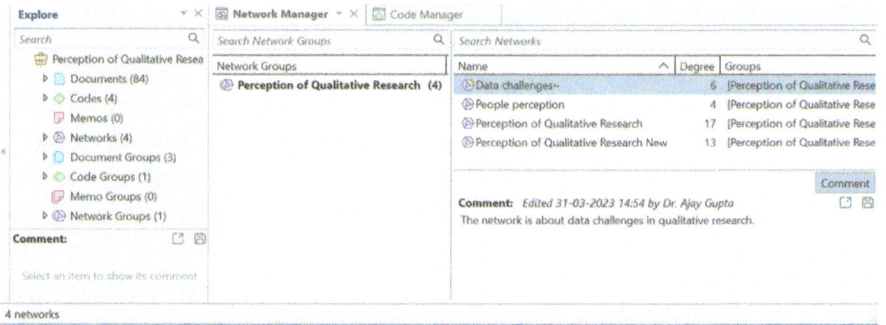

Fig. 6.2 Network group and degree

called a second-class link. Code-code and quotation-quotation links are examples of a first-class link. A code-quotation link is an example of a second-class link. The link between codes and memos and a group and its members are also examples of a second-class link. Quotation-to-quotation links are known as hyperlinks.

To make the network informative, the relationships between entities must be shown. Only then will the researcher be able to demonstrate how themes and concepts are related.

Figure 6.1 shows first class links for code-code relations and hyperlinks. Code-code relations show how codes are connected in a thematic network, whereas hyperlink relations show how quotations are connected and support the theme in a thematic network.

Second-class links do not have specific properties. They are links between code categories and subcodes, folders and categories, quotations and codes, groups, and members etc.

Figure 6.3 shows the code category—Methodological challenges about qualitative research to the sub codes 'Difficulty understanding', 'Epistemological issues, and 'Feminism'. The network shows the document group of male and female, the code category, codes, and quotations. Code 'Feminism' and 'Difficulty understanding' have code to code linkage and have first class link. Both quotations are linked using hyperlinks and have first class link. Male, Female, and code category to codes have second class link.

The network manager stores all the saved networks created by the researcher. One can create a new network or work with existing networks. One can create, edit and modify existing ones as required.

One can create a network, a new group, and a new smart group using the Network Manager. To create a new network, drag and drop the codes into the network. The number of nodes in a network are shown in the degree column in Network Manager. A network with eight nodes will show the number 08 in the degree column. One can create a network group in the same way a code group is formed. A new smart group can also be formed by combining existing network groups and using Set operators. Figure 6.2 shows a list of four networks and one network group in a project. The number of nodes in each network can be seen in the degree column.

Presenting Networks

While presenting networks, links and relationships help the researcher discuss data and findings with greater clarity and more effectively as visual representations make a better impression on audiences. Used properly, networks can convey rich information about concepts and evidence, which are strengthened by the links and relationships shown in them.

Computer-related hypertext displays present operational information in conveniently small chunks (as compared to lengthy printed texts). Additionally, they also show the linkages between various segments.

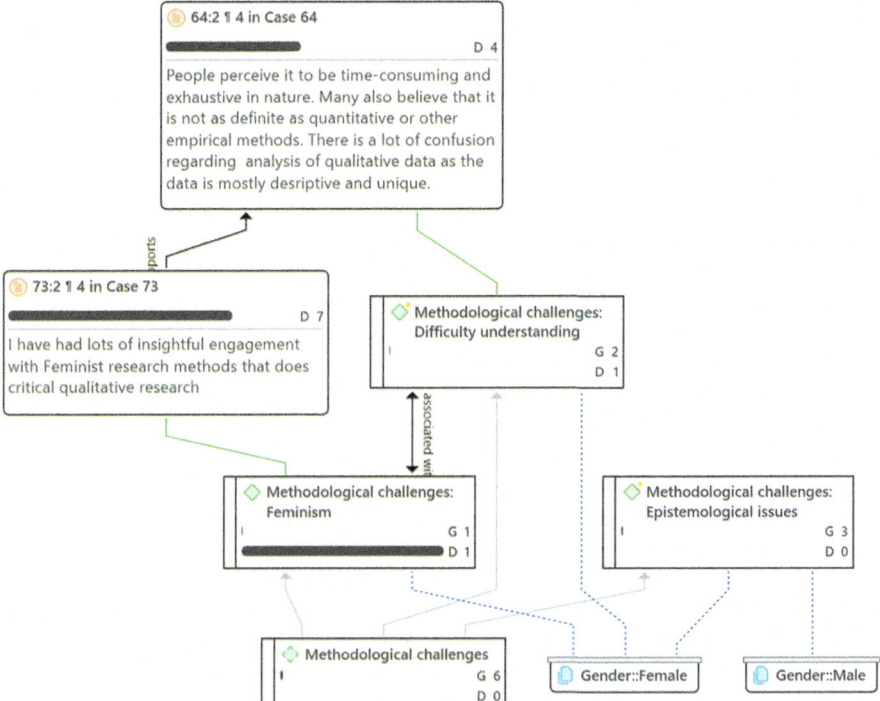

Fig. 6.3 Second class links

The World Wide Web is a popular hypermedia structure for showing text, graphical, and other multimedia information. ATLAS.ti has a similar structure with which a network with nodes, relations, and hyperlinks can present information that is directly connected with the research objectives. The example in Fig. 6.4 shows how a network can be used to present rich information.

The emerging theme in Fig. 6.4, 'favoritism practices', has code-to-code relationship with many codes—favoritism in performance appraisal, favor in promotion and transfer, HO employees get more marks, favor in promotion, favor on a regional basis, favor in transfer, management does favor, regional office people favored. All codes that have 'associated with' and 'are part of' the relationship support the theme which is the cause of Low Morale among Middle Managers. To support the theme, the researcher can import the quotations associated with these codes, which help to understand the significant statements of the respondents.

Using Networks

Networks can be edited and presented in various ways. One can create a layout for a network group in the same way it is done with code. While creating a

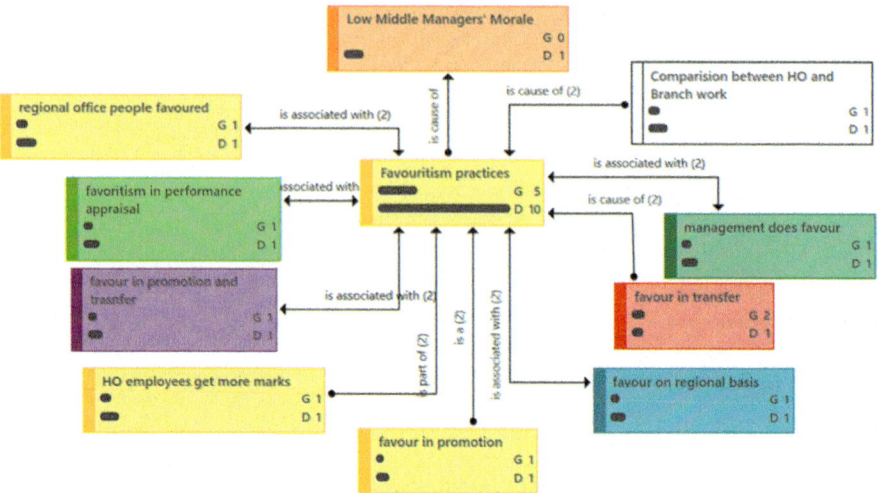

Fig. 6.4 A complex arrangement of networks, nodes, and links

network, linked codes will also be shown in the network. Networks provide the option to add codes, useful for explaining concepts.

A dummy code is not associated with a quotation (Fig. 6.5). It is used only to modify relations between codes to help understand the relationship between codes. In Fig. 6.4, code group 'Favoritism practices' includes ten connected codes. A dummy code was added in the network to show 'Low middle managers' morale', the result of 'Favoritism practice'. Since the dummy code is not attached to a quotation, the G value is shown as 0.

To create a network,

1. Open New Entities in the Home tab.
2. Open a New Network and give it a name.
3. Drag the codes to the network. Link the codes to show the relationship between them.
4. Open View and check (select) Show Frequencies. The codes used in the project are shown with a number which is the number of quotations linked with the code. The number 0 (zero) before G and D shows that there are no linked quotations, indicating that the code is a dummy.

To add a comment for the new code, click on the code, write in the Edit Comment section, and save. The comment is saved after checking the comments box.

To add nodes to a network, open Nodes and select Add Nodes. One can add a document, quotation, code, memo, network, document groups, code groups (Fig. 6.6). Two options can be seen in the same window: Add Neighbors and Add Co-occurring Codes. One can add neighbors to import

6 USING NETWORKS AND HYPERLINKS 187

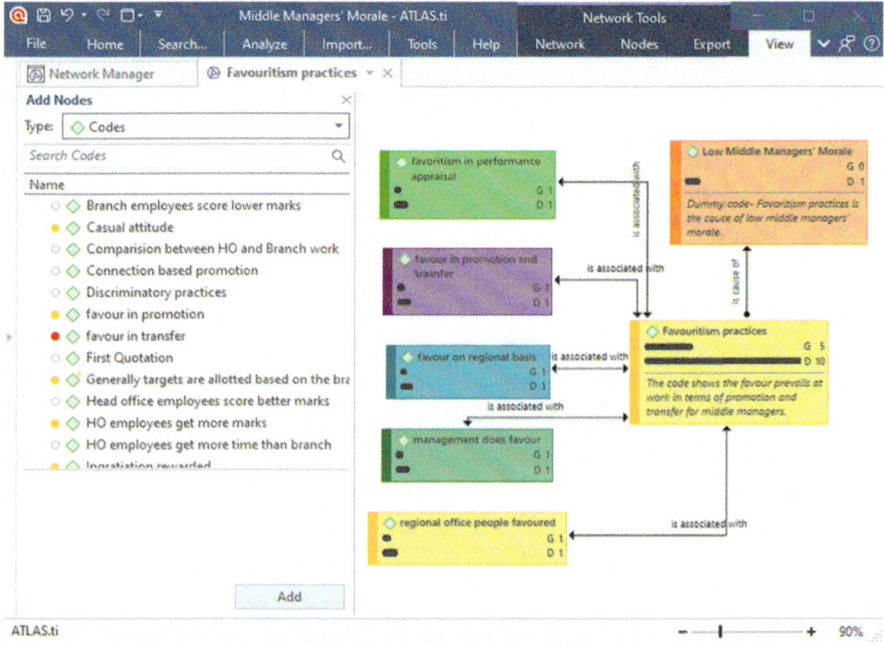

Fig. 6.5 Dummy code in the network

codes, memos, quotations, and groups, which are attached to the codes, to the network. The Add Neighbors facility is activated by clicking once on any item in the network. The user must first highlight the item in the network before adding neighbors.

To rename an entity in the network, the Rename Entity option is used after selecting the entity — code, quotation, memo, or other.

RELATIONSHIPS AND HYPERLINKS

There are seven default relationships in ATLAS.ti (Fig. 6.7). One can add, modify, delete, or rename these relationships using the Relation Manager menu. The numbers in the Usage column represent the times a particular relation was used in the project.

Relationships among codes (and other entities) are important for constructing theory. Concepts emerge during data analysis. They are important in the search for answers to research questions and fulfilling the research objectives. Linked codes, which show relationships, reflect aspects of the problem domain under investigation. On the other hand, the relationships used to link these domain concepts are part of the methodology of the study. As important epistemological tools, relationships constitute the main questions that guide the development of a model or a theory.

188   A. GUPTA

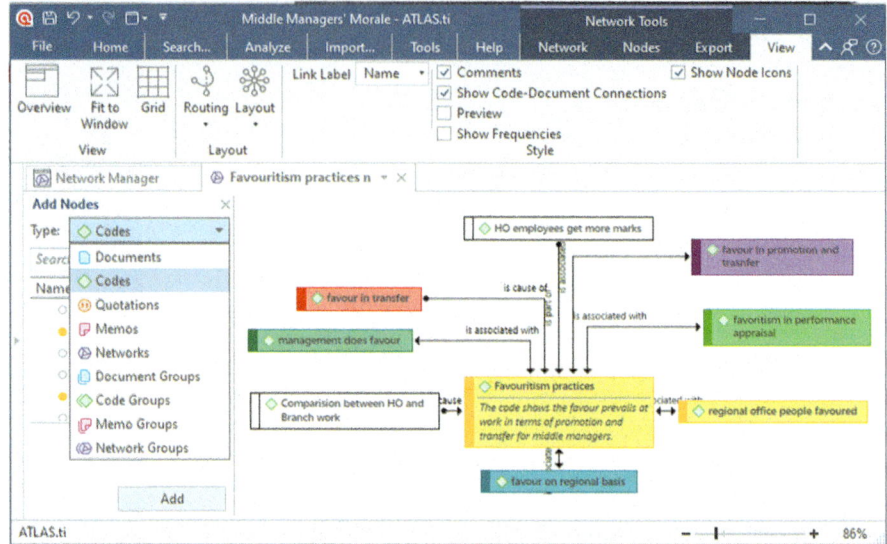

Fig. 6.6 Adding nodes to a network

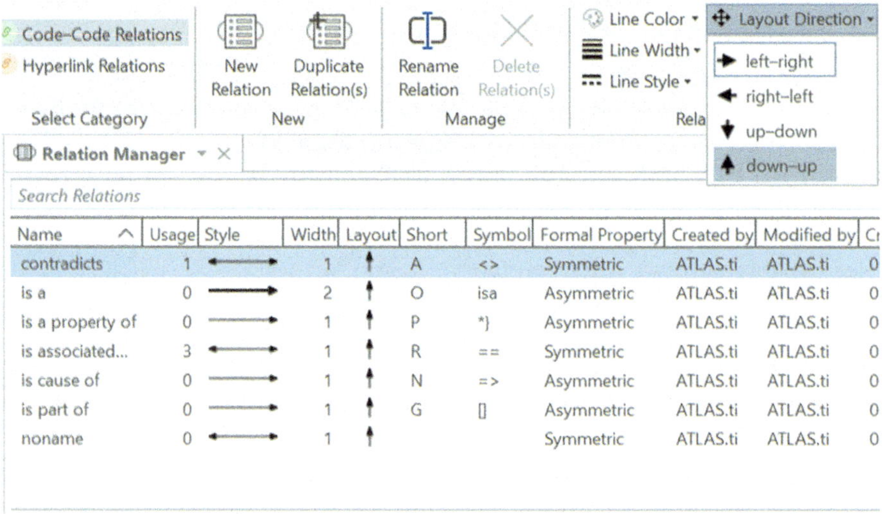

Fig. 6.7 Code to code relationships

There are three properties of relationships: symmetric, transitive, and asymmetric. A symmetric relation has a formal attribute that affects both the display and processing capabilities of a relation. All transitive and asymmetric relations are symbolized in the network editor. An arrow points toward the target code. Symmetric relations are displayed as lines without arrows. The user can represent these relations with various colours.

A relationship that has already been used in the project cannot be deleted; however, it can be modified. But while modifying a relationship, it is advisable to write a description of the change in the comment section.

Like code-to-code relationships, hyperlinks are the most suitable for connecting quotations (Figs. 6.7 and 6.8). While discussing the conclusions, the researcher may find it necessary to show what the respondents of the study had said. The responses may be found in various segments, scattered in a document or across many. Using hyperlinks, the relationships between these segments can be shown, as well as how they support the theme.

ATLAS.ti has features for creating and browsing hypertext structures. It allows for two or more quotations to be connected using *named* relationships. One can create a network of any kind of quotations, i.e., text, graphic, audio, or video, and other formats.

Hyperlinks connect segments of interest across documents. In such cases, they are known as inter-textual links. If the connections are made within the same document, they are known as intra-textual links. The hypertext network in Fig. 6.8 shows quotations with all the contents in the preview. Other node types like memos and codes can also be included in the network.

ATLAS.ti provides eight default hyperlink options. For more options, the researchers can use code-code and hyperlink relationships to create new relationships. One can also rename, delete, or duplicate relationships. The option is activated by selecting a relationship in the table.

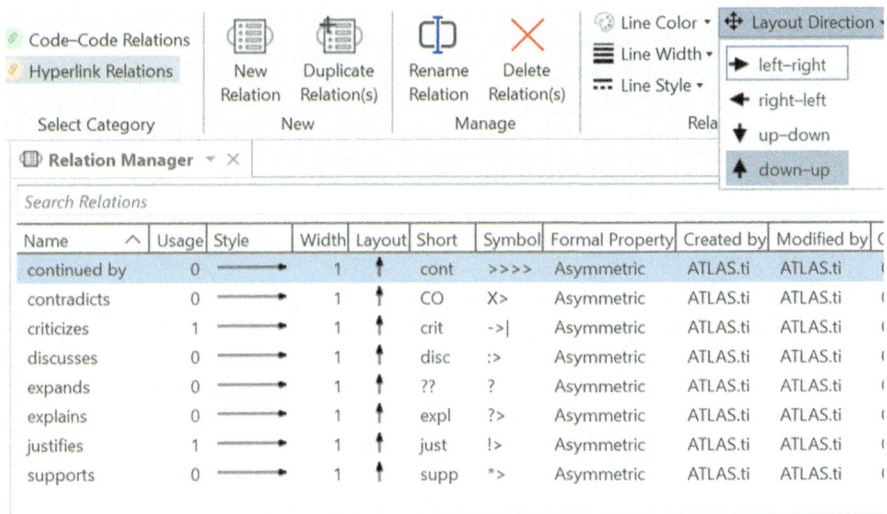

Fig. 6.8 Hyperlink relations

While forming a new relationship (or hyperlink), one can use layout direction (Figs. 6.7 and 6.8) to select a suitable layout. ATLAS.ti also allows the researcher to select the property of relationship—transitive, symmetric, or asymmetric. Line color, width, and style can be selected according to the user's preferences. Using filters, one can view the relationship in the table. By selecting 'All', all the relations and hyperlinks are displayed.

Difference Between Code–Code Relationships and Hyperlinks

Code to code relationships shows the concepts emerging from the data, whereas hyperlinks support the concepts. Just as code-code relations are shown as evidence of a pattern, the connections between quotations are shown with hyperlinks. Hyperlink relationships among quotations are evidence of the concepts emerging from the codes. Their usefulness can be understood with the following example.

In one study, employee wellbeing was the concept that emerged from the codes. To justify this finding, the researcher highlighted the segments in participant responses that were associated with employee wellbeing and explained how they were connected by using hyperlinks. A memo can be presented to strengthen the justification.

The hyperlinks between quotations can also be named. For example, if one statement in a report or interview explains another statement, the two can be hyperlinked and named "explains." Some statements may contradict or support each other. Such relationships can be shown using the default options available in ATLAS.ti.

ATLAS.ti offers a range of options for creating and traversing hypertext links. Hyperlinks can be created in Network Editor. Alternatively, hypertext links can be created "in context" or via drag and drop using Quotation Manager.

Creating Hyperlinks

Hyperlinks between quotations can be created using Quotation Manager or Network. A quotation can have several hyperlinks depending on its relationships with other quotations.

Hyperlinks using Quotation Manager

1. Select the quotation that needs to be linked with another segment.
2. Select "Link Source" or "Link Target" depending on the relationship to be shown.
3. On clicking Link Target, a screen showing various relations will appear.
4. Select the appropriate relationship for the source and target quotations.
5. One can also use drag and drop option to link quotations:

6. Drag source quotation and drop on the target. A flag showing the relationship options will appear.
7. Select the appropriate relations and click. The quotations are prefixed with a symbol (as source, target, or both) with which the relationship can be easily understood.

Hyperlinks in Network

1. Open the Network.
2. Select the quotation. A red symbol will appear.
3. Drag and link to target quotation. A flag showing the relationship options will appear.
4. Select the appropriate relations and click.
5. The relationship between source and target quotations will be shown.

Table 6.1 shows the symbols provided in ATLAS.ti. One can also see the link source and link target under quotations in the menu bar. Quotation 1:1 is the link source, while quotation 1:2 is the link target (Fig. 6.9). It is important to understand the notation. The first number indicates the number of the document in the project and the second number denotes the quotation number in the document.

One can view a quotation's relationship with other quotations in a network. In Fig. 6.10, the quotation in the middle is shown as linked to other quotations. There are several ways to create quotation-to-quotation relationships in the network. The relationships can be created at any time during data analysis.

1. Create an empty network.
2. Drag all the quotations that are part of the network.
3. Click on any quotation (the source) and drag the red dot to another quotation identified as the target. Drop the dot, and the relationship flag will display from which one can select a suitable relationship and click. A quotation can also be a target for many source quotations.

Table 6.1 Symbols and their meaning

Symbol	Description
<	The quotation is a source of a hyperlink
>	The quotation is a target of a hyperlink
~	The quotation has a comment
<>	The quotation is both the source and target link. In the quotation, at least two other quotations are linked

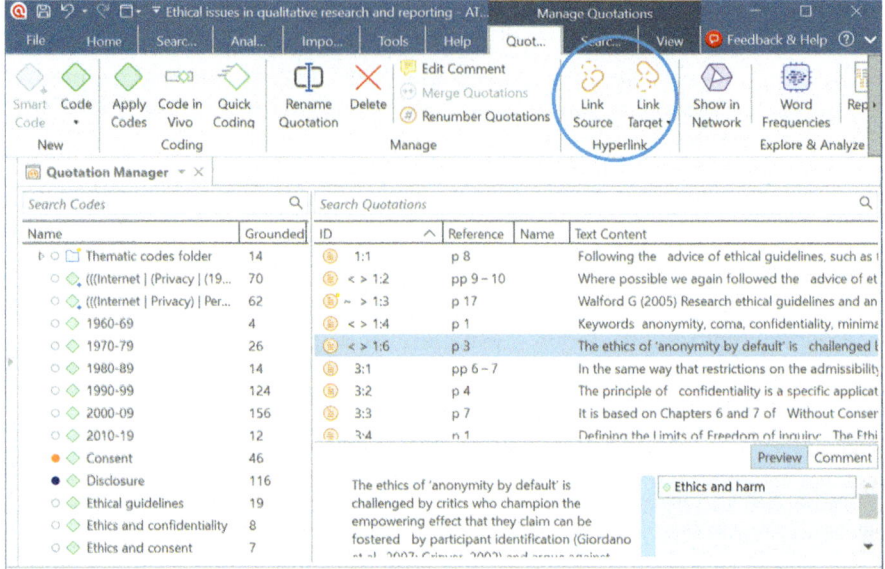

Fig. 6.9 Link source and target quotations

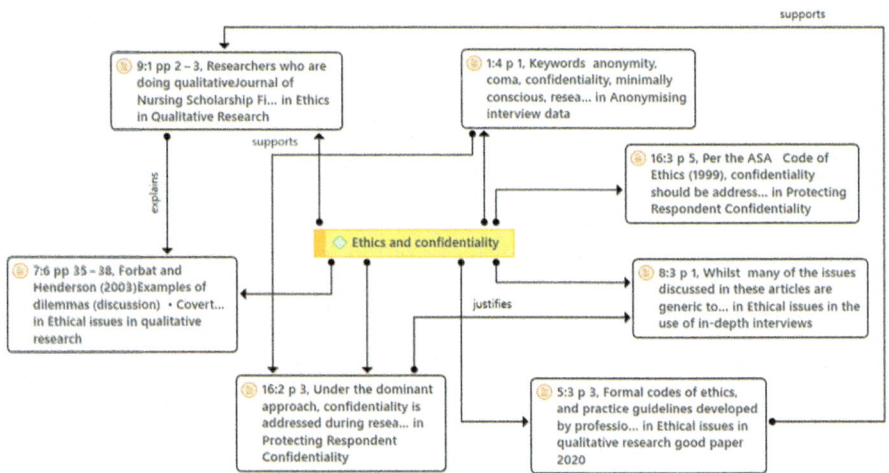

Fig. 6.10 A network showing hyperlinks relations

A quotation can be both the source and target. If a suitable relationship cannot be found in the default list, one can be created using the Relation Manager. ATLAS.ti also offers the option of showing hyperlinks using symbols. The user must go to the View section and select a suitable style for representing the network.

A star-like network can be formed by linking several target quotations with one source quotation. One can also form a chain link of source and target quotations. Here, a target quotation becomes the source quotation for another target quotation which is the source for another target, and so on.

Creating Hyperlinks with Multiple Documents

ATLAS.ti offers researchers the option of viewing multiple documents simultaneously. One can open two documents alongside each other and move from one document to the other to link segments with the appropriate relationships. This facility is particularly helpful for researchers who may want to search many documents for quotations associated with concepts that support or contradict each other. With the help of Tab and Tab Groups (explained in the next section), one can also scroll a document to the left or right, up or down. A document can also be 'floated' if needed.

A floating window will disappear in the background if one clicks outside the window. One can also arrange the windows one over the other instead of alongside each other. Right click on the button at the top left of the window and select the option "Always on Top." The floating window will only appear on the top of the project.

Tabs and Tab Groups

Tabs and tab groups are used for side-by-side display of documents, codes, quotations, memos, links, and networks. To open tabs or tab groups, click on the down arrow for the entity (documents, quotations, codes, memos, networks, and links). A flag (Fig. 6.11) will appear, and users can open a new window by selecting an option i.e., right, left, down and up.

Users have the option to open a window in the dock or floating mode. The Floating mode will open a new window, while the dock option (the default in ATLAS.ti) will open a window in the ATLAS.ti application. Using the float option in the main window, users can float the new window. With the dock option, the user can open a new window in the main window.

The Document and Entity Manager will always open in the dock mode. If one wants to open the project in the floating mode, the window will appear separate from the project. One can open many items in the floating mode. To return to the project window, one can click on the key "Dock in the Main Window", which is shown at the top of the floating window.

Tabs help the researcher to work on two or more documents simultaneously. The steps for using tabs are described below.

1. Open the two documents, either in project explorer or the navigation pane. The documents will appear in the main window and can be seen alongside each other in a different tab group.

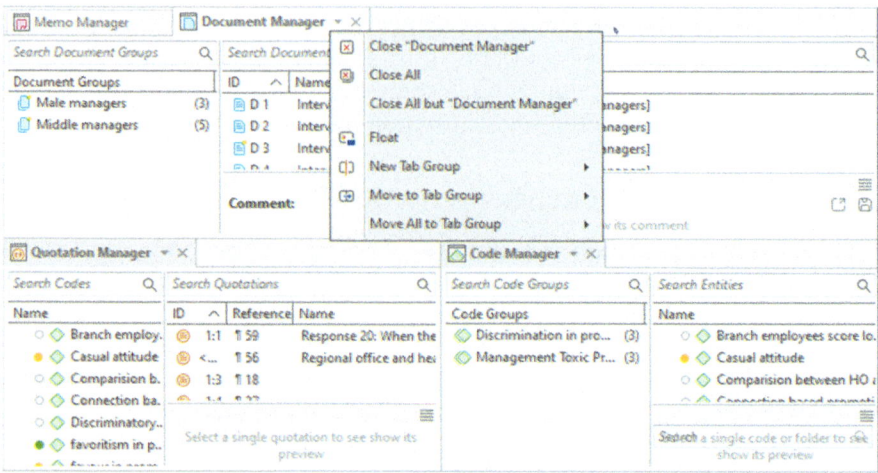

Fig. 6.11 Document view in a new tab group

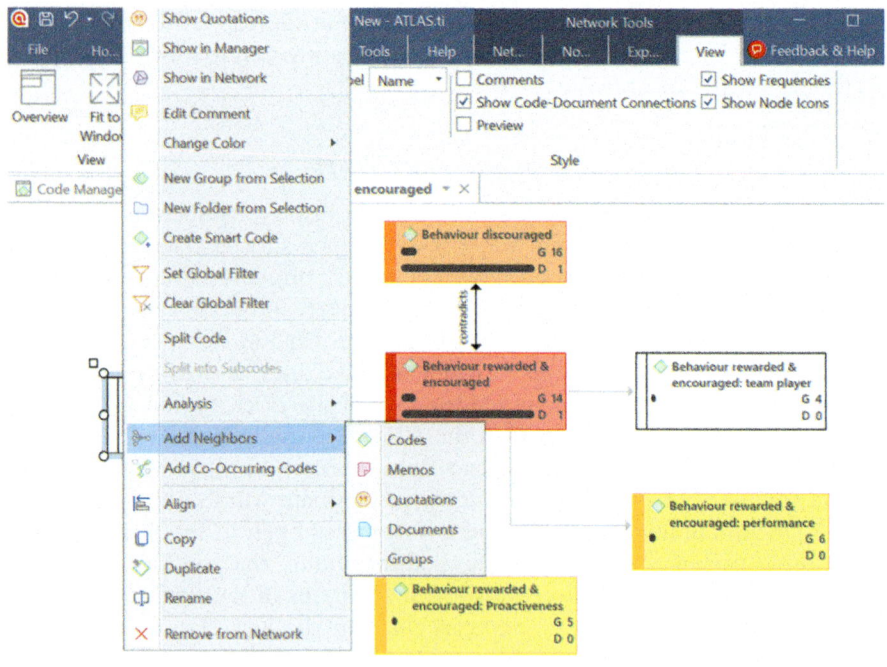

Fig. 6.12 Neighbors and co-occurring codes

2. Click on the down arrow appearing in the document name and select "New Tab Group." Four options will be displayed: Right, Left, Down, Up (Fig. 6.11).
3. Select the desired option and the window will appear at the location selected. One can move the window into a tab group anytime. One should remember that the new tab option will open when there are two or more documents in the main window. When only one document is opened, only the "float" option can be used. However, when two or more documents are opened, both "float" and "new tab group" options are available.

Figure 6.11 shows four-tab groups. The Memo and Document Manager tabs appear on the upper window, while Quotation and Code Manager tabs are displayed on the lower window.

Tab Group offers much needed flexibility to individualize workspace to suit the researcher's needs. The Tab option allows the user to view a document several items in one window, simplifying the task of examining multiple documents and their links.

Properties of Relationships

1. A symmetric relationship exists if it holds for all A and B related with R such that that A is related to B if B is related to A.
2. The transitive relationship exists if A has a relation R with B, and B has a relation R with C, then A has a relation R with C For example,

Example Luca is a student. All Students are smart. Therefore, Luca is a smart person.

A typical transitive relation is the "cause of relation." If C1 is the cause of C2 and C2 is the cause of C3, C1 is the cause of C3. Transitive relations enable semantic retrieval. There are three semantic operators in ATLAS.ti: Up, Down, Siblings. The Up operator looks at all directly linked codes and their quotations at a higher level. In other words, it searches all parents of a code. The Down operator traverses the network from higher to lower concepts, collecting all quotations. Only transitive code-code relations are processed, and all others are ignored. It is important to note that the Down operator may produce larger result sets than the Up operator. The Down operator yields more precise results than the Up operator. The Sibling operator finds all quotations connected to the selected code or other descendants of the same parent code.

One must note that many directed relations in ATLAS.ti are likely to be asymmetric. If A leads to B, it does not automatically lead to/result in/is part of C. Therefore, it is not transitive.

Neighbors and Co-occurring Codes

Neighbors and co-occurring codes are important to networks. One can import all the linked information connected with an entity. Every item that is directly attached to the entity will be shown when a network is created for a code, code group, memo, or documents. But although every linked entity may be linked with another entity, it will not be automatically shown in the network. Therefore, the "Add Neighbors" option must be used to import such entities.

One can import either all attached items, or selected items. For example, a code may be connected to a memo, code group, smart code, or smart group. But the network for the code will not show the relations until neighbors or co-occurring codes are added.

Direct import of neighbors allows the researcher to construct a connected network step-by-step. In a connected graph, there is always a direct or indirect path between two nodes. One can import neighbors for the nodes by selecting one or more nodes.

1. Select the node and right-click. A flag will appear.
2. Select 'Add Neighbors'.
3. Several options will be shown: all Common Codes, Memos, Quotations, Documents, and Groups.
4. Select an option.

The code co-occurrence function is useful for showing the relationships among codes. Code co-occurrence is possible if a code is used to code a quotation that is embedded in, overlapping with other quotation, or if two or more codes are applied to the same quotation. The Query Tool generates quotations for explicitly specified codes.

To import co-occurring codes, one must select one or more codes for importing the neighbors, and with a right click, select Add Co-occurring Codes. Alternatively, one can select Nodes/Add Co-occurring Codes from the ribbon. All the co-occurring codes will appear in the network. A global filter can then be used for generating a report on co-occurring codes.

CASE ANALYSIS USING NETWORKS

One can analyze concepts or a case using networks. ATLAS.ti provides researchers with the option of importing selected items from a specific document, or document groups, using a global filter. Without a global filter, there is a high chance that irrelevant items will also be imported. For example, in the study of employee morale, the researcher may be interested in knowing the reasons for employee motivation in a specific document or document groups. The researcher may also be interested in knowing the reasons for lack of employee motivation in a document or document group. In each case, setting the global filter is necessary for generating a precise report.

To global filters, we will assume that codes for employee motivation and demotivation (lack of motivation) were imported from a document. By setting a global filter, quotations and memos associated with the codes can be imported. This option is useful for presenting evidence that will support the interpretations of the findings in the study.

One can compare cases by reading the quotations in each category. It is possible to compare within a case (intra-case) or across cases (inter-case). This means that one can analyze data and use the results to support a claim in one category with quotations in the same category. The researcher can show how different quotations for the code 'the reason for employee motivation' explain, support, and strengthen the interpretations. Similarly, one can show how quotations for 'the reason for employee demotivation' explain lack of employee motivation.

One can also compare a quotation, or group of quotations, for the codes 'employee motivation' and 'employee demotivation' to explore the concepts. Such comparisons are useful for discovering concepts and ideas that may help to understand the triggers of the problem being studied, and for recommending appropriate remedial actions.

Usefulness of Local and Global Filters

Filters help to retrieve information specific to an entity. ATLAS.ti has two types of filters: local and global. Local filters are applied to each entity., whereas the global filter can be applied to the entire project. By using filters, the researcher can view only the selected entity and retrieve the information needed from that entity.

The Global Filter in ATLAS.ti is a powerful and widely used data analysis tool. It is applied to the whole project to refine searches. The results depend on whether the filter is set for a document or a document group.

Creating a cross-tabulation for action (research questions) and outcomes (responses) will produce results according to the global filter. For example, in the document group, 'managers', 'middle managers-male', is set as the global filter the table results will show the actions related to which outcome for all-male middle managers. The filter excludes other entities from appearing in the search results and only the selected entity and its associated components will be shown.

In Figs. 6.13 and 6.14, a global filter was created for documents, code, memo, and network groups. In the document group, a global filter was set for the document group 'male managers. Similarly, in the code group, 'management toxic practices' was set as the global filter. In the memo and network groups, the global filters were, respectively 'unhealthy cultural practices' 'unhealthy culture and practices'. The report generated after applying these filters will be limited to the search results for these filters only. For example, in Fig. 6.13, the report is for the document group 'male managers'.

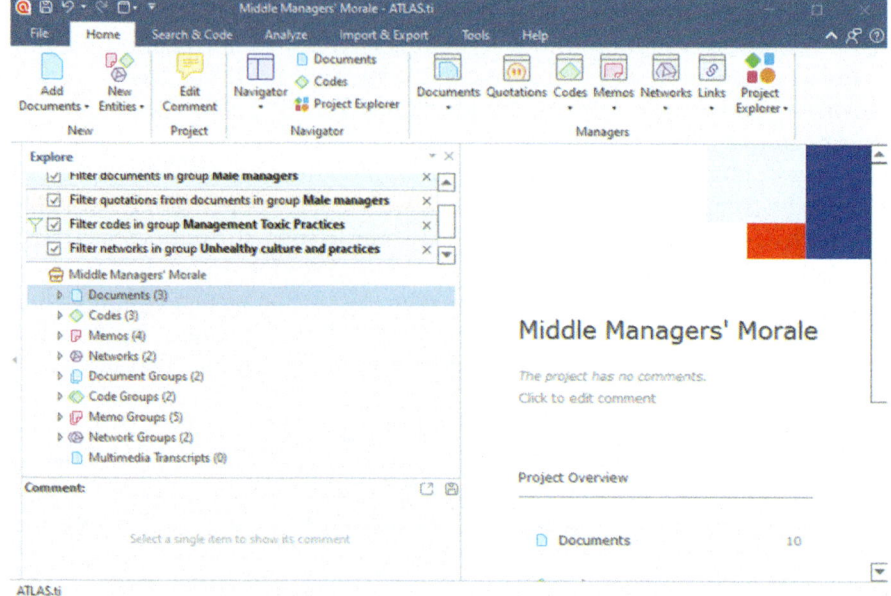

Fig. 6.13 Setting a global filter for memo group

Similarly, the output reports after setting global filters for code, memo and network groups will be restricted to the selection only.

Global filters are set using Project Explorer in the Manager option. One selects either the document, code, memo, or network for which the global filter is to be created. To set a global filter,

1. Right click and open Manager.
2. Select the groups of interest (document, code, memo).
3. Right click and set global filter.

The global filter also functions like a smart group in some ways. For example, a code document table (Chap. 7) allows one to combine two variables without the need to create a smart group. If one wants to create a code-document table for action codes (research questions) for the document groups 'Motivated Employees' and "Demotivated Employees," the results will show quotations for the selected options only.

One can create a network by applying global filters. In this case, only the entity for which the filter is applied will show from which the researcher can generate a report and retrieve information. All other entities will be excluded. Alternatively, if one selects the Import Neighbor or Import Co-occurring Options, only entities that pass the filter criteria are imported. The global filter

6 USING NETWORKS AND HYPERLINKS 199

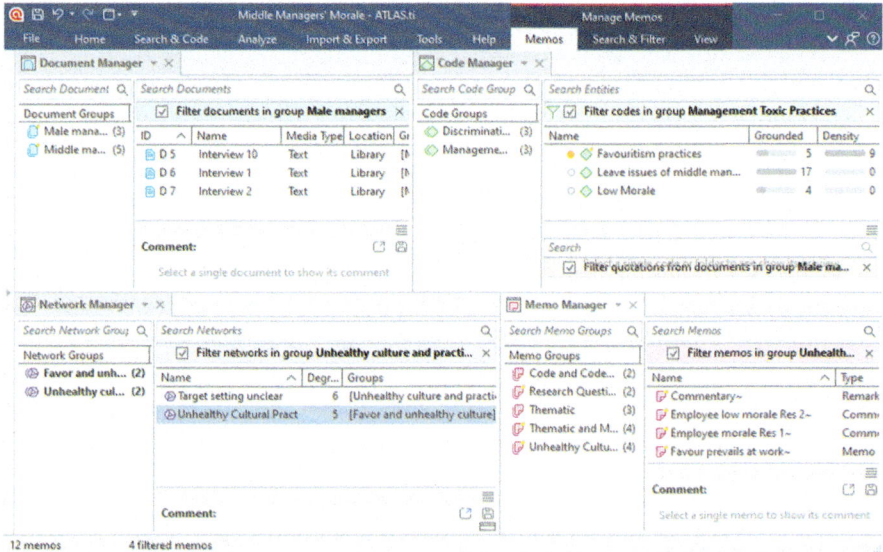

Fig. 6.14 Global filters for document, code, network and memo group

helps the researcher to focus only on certain aspects (variables or concepts) of the study and investigate them in greater depth.

One can also create document groups of only a subset of the documents for analysis. Global filters can be set up only for groups. They cannot be applied to individual documents, codes, or memos. If one wants to set up a global filter for an individual document or code, a group must be formed first. This is the only workaround available.

Setting up Global Filters

The global filter is set up in Project Explorer. At the left of each Manager, one can right-click on a group and select the option "Set Global Filter." Alternatively, one can also "Set Global Filter" by clicking once on the document group on the left and then clicking on the Set Global Filter in the "Search and Filter" tab in the main menu.

1. Open the entity manager, for which a global filter must be set
2. Select the document, code, or memo group.
3. Create a global filter, either with right-click or by using the "Set Global Filter" option in the "Search & Filter" tab.

In Figs. 6.13 and 6.14, four entity managers were used to set up a global filter. The example shown is for setting a global filter in the memo group. One can set a global filter for other entities using their respective entity managers.

To remove a global filter, click on the Remove Global Filter option below Set Global Filter. One can also remove the global filter with a right click on the group which was set as global filter (Fig. 6.15).

For each entity, global filters are shown in a different color. Global document filters are shown in blue, global quotations in orange, global code filters are in green, global memo filter in magenta, and global network filters in purple.

It must be remembered that document group filters also affect quotations. This means that quotations will be retrieved only for the selected document group. If one opens the quotation manager and a document group is set as a global filter, an orange global filter bar will be seen at the top of the list of quotations.

In Fig. 6.16, a global filter was set for the document group, which can be seen at the top of the quotation list. One can also see the global filter in the left margin. The network for an entity created by using the global filter will show the global filter set for the entity at the top. If the global filter is set for memos and codes, the indication will appear in the network on the top. One can also use a global filter in the code-occurrence table to help refine the search and output option for a specific code, concepts, or the research questions.

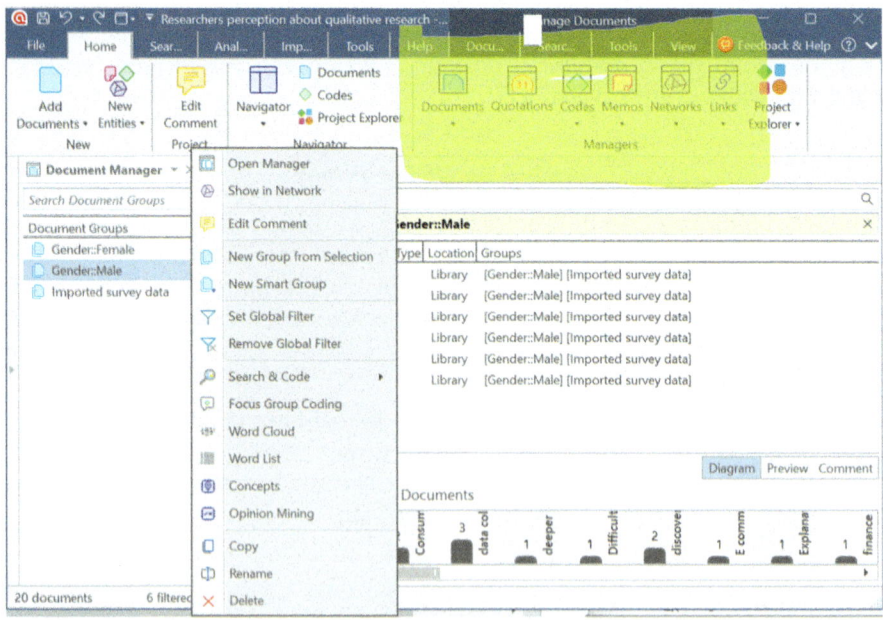

Fig. 6.15 Entity managers

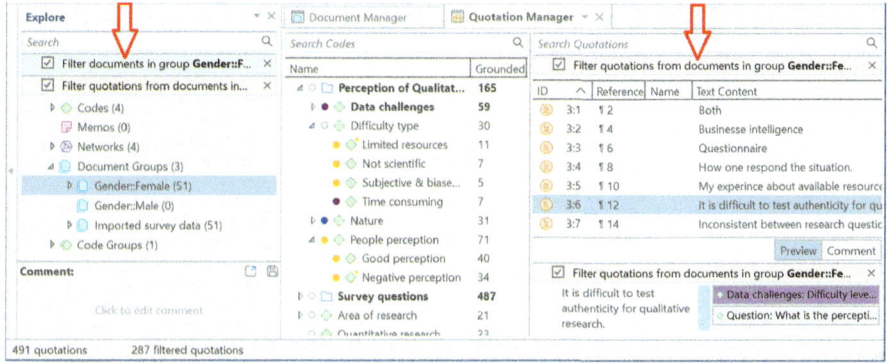

Fig. 6.16 Global filter in document group

Recommended Readings

Bhowmick, T. (2006). Building an exploratory visual analysis tool for qualitative researchers. In *Proceedings of AutoCarto, Vancouver, WA*.

Bryman, A., & Burgess, R. G. (Eds.). (1994). *Analyzing qualitative data* (Vol. 11). Routledge.

Friese, S. (2019). *Qualitative data analysis with ATLAS.ti*. Sage.

Friese, S., Soratto, J., & Pires, D. (2018). *Carrying out a computer-aided thematic content analysis with ATLAS.ti*.

Grbich, C. (2012). *Qualitative data analysis: An introduction*. Sage.

Hollstein, B. (2011). Qualitative approaches. In *The SAGE handbook of social network analysis* (pp. 404–416).

Kelle, U., & Bird, K. (Eds.). (1995). In *Computer-aided qualitative data analysis: Theory, methods and practice*. Sage.

Luther, A. (2017). The entity mapper: A data visualization tool for qualitative research methods. *Leonardo, 50*(3), 268–271.

Miles, M. B., & Huberman, A. M. (1994). *Qualitative data analysis: An expanded sourcebook*. Sage.

Richards, T. J., & Richards, L. (1994). Using computers in qualitative research. In *Handbook of qualitative research* (Vol. 2, No. 1, pp. 445–462).

CHAPTER 7

Discovering Themes in ATLAS.ti

Abstract Themes are discovered during analysis of qualitative data. In ATLAS.ti, as this chapter explains, code co-occurrence, code-document, and Query Tools are important functions in qualitative data analysis. They use set and proximity operators, and scope tools to sort and segregate data and generate reports on the parameters specified by the researcher. In this chapter, the author explains how code coefficients, Sankey diagrams, binarized and normalized data are used to generate customized and thematic reports. Examples are discussed with screenshots, tables, diagrams, frameworks, and themes.

Data Analysis

The next process in ATLAS.ti, after coding, memo writing, code categorization, adding comments, and grouping of codes and memos, is data analysis. In data analysis, depending on need, the researcher forms code networks, groups of codes and memos based on shared characteristics, smart codes, smart groups (discussed in Chap. 8) or document groups, and then tries to identify the relationships among them. The importance of these relationships was explained in Chap. 6.

Figure 7.1 shows the various data analysis options available in ATLAS.ti 22 and 23: Query Tool, Co-occurrence Explorer, Code-occurrence Table, and Code-Document. The tools are useful for retrieving reports.

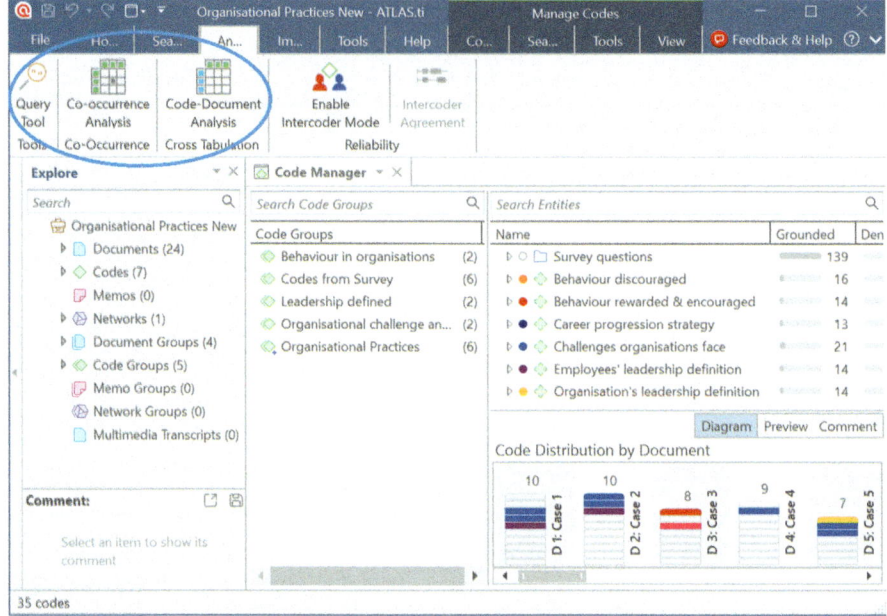

Fig. 7.1 Query writing and analysis tools in ATLAS.ti 22 & 23

Data Analysis Using Query Tool

The Query Tool option enables the researcher to generate project reports in various forms using the set and proximity operators. The reports can be created in text, Excel and network formats for documents, documents groups, codes, code groups, smart codes, and smart groups.

The query tool is used for retrieving quotations, memos, code, code groups, and memo groups, which are connected with a code or any combination of code, code groups, smart code, and smart groups. It can be used with or without operators. In Fig. 7.2, the researcher has generated a report of the codes "Subjective target setting process" and "Connection based promotion" using the OR operator. Then, with the help of Edit Scope, the researcher generated a report for the category 'Middle Managers Bank B and Middle Managers Bank A'. If needed, the report can also be generated using set or proximity operators.

> **Important** Smart Codes and smart groups are dynamic. Any change in code or code groups will change the nature of the smart code or smart group.

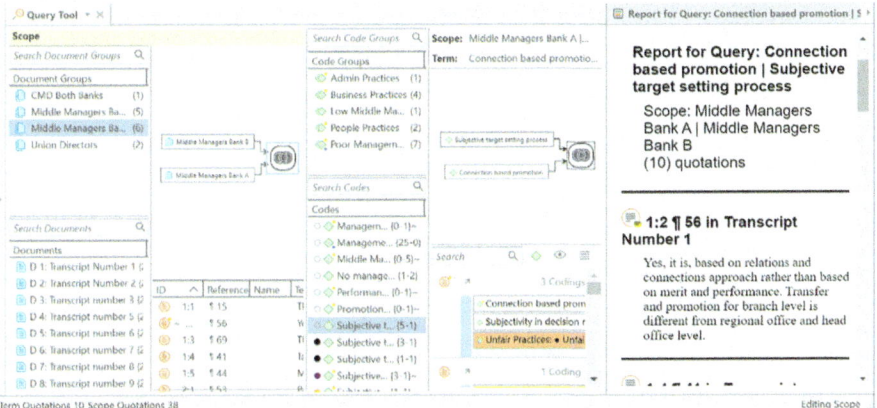

Fig. 7.2 Query tool and options in ATLAS.ti 22 & 23

The Query Tool option helps the user to create smart codes (discussed in Chap. 8). Smart codes are based on individual codes and created from queries stored at an aggregate level. Smart codes and smart groups are created from the codes or code groups they contain. In the sample project under discussion, one can also create a smart group of middle managers, top management, and the union. Since the reports are stored queries, they can be created by clicking on a smart code or smart group.

The arrows in Fig. 7.2 show the various tools that the researcher can use in their analysis. The arrows also show how reports can be generated for codes, smart codes, code groups, smart groups using set and proximity operators to define the scope. The scope option refines the reports to describe a specific group of respondents based on selected parameters.

In Fig. 7.2, the report for the codes 'Connection based promotion' and 'Subjective target setting process' was generated using the OR operator for the scope 'Male managers and Middle managers' category. The Edit Scope option was used to set the scope. The OR (|) option is used in ATLAS.ti to add the quotations associated with both the selected codes and respondent categories. The resulting report has retrieved 10 quotations, which are the key quotations (or informative segment) that support the theme—connection based promotion and subjective target setting process—identified in the responses of middle managers bank A and middle managers bank B.

After selecting the code, or code group, the Edit Scope option is used to select the specific document group. One can also apply a global filter to document groups to make the search more focused. If a document group is set as a global filter, the codes-document or code-co-occurrence table will be calculated based on the data in the filter and not across the project. For example, in this project, a global filter can be set up for male respondents in the document group. Then, the report will be generated only for these items (Fig. 7.3).

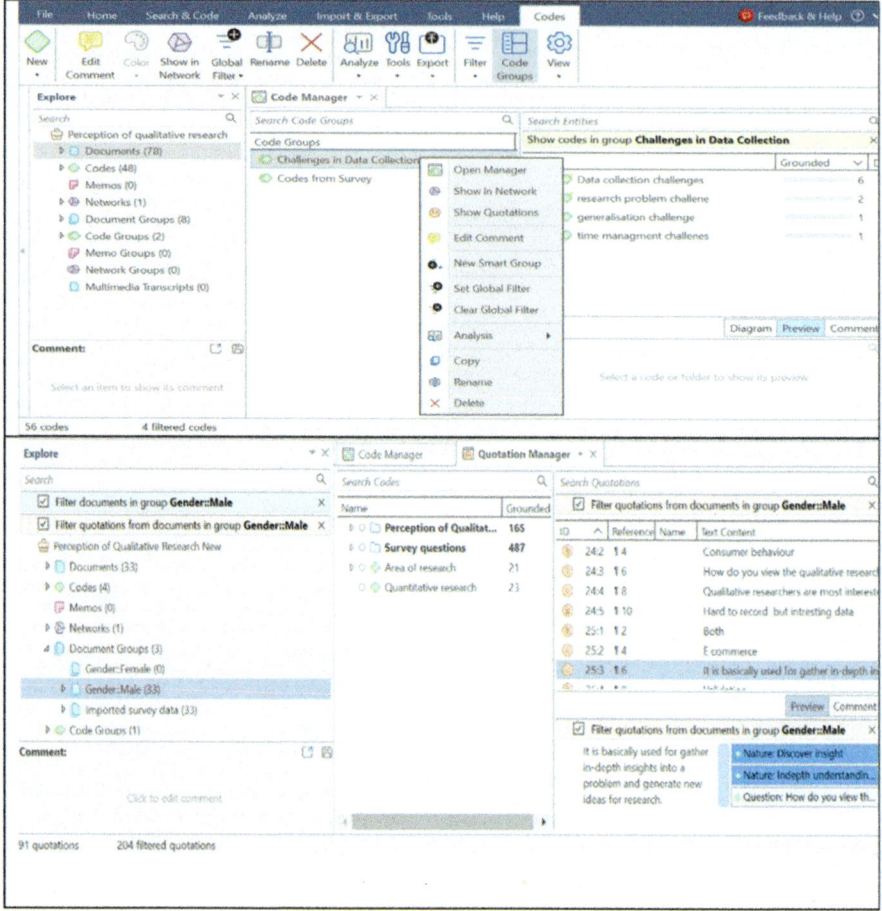

Fig. 7.3 Global filter in ATLAS.ti 22 & 23

Every report will be specific to the filter selected and the entity to which it is applied: code groups, code, smart code, smart groups or code categories.

Code Co-occurrence Analysis

Code co-occurrence is used to find the number of quotations co-occurring between codes. In other words, it is the number of quotations shared by the codes. The Code co-occurrence function searches for codes that have been applied, either to the same quotation or to overlapping quotations. Code co-occurrence shows the relationship between codes and indicates its strength.

In ATLAS.ti Code co-occurrence is presented in tabular form. The Co-occurrence table shows the frequency of co-occurrence in a matrix, similar to a correlation matrix. The Code Co-occurrence tool allows the researcher to ask various types of questions. For example, one can instruct ATLAS.ti to show all codes that co-occur across all primary documents. The result will be presented as a cross-tabulation of all codes.

The Co-occurrence table shows the strength of relationship between codes, which is represented by the c-coefficient whose value is shown within the brackets. After generating a code co-occurrence table, the user can use Refresh, Color (blue, red, green and no color) and other buttons in the menu to present the table in the desired format. The table cells can be shaded in the color (or colors) chosen by the researcher. The lighter the shade, the lower the value in the cell. In Fig. 7.4, higher values are represented by darker shades.

The user can check or uncheck the boxes accordingly as needed. The quotations box shows the number of total quotations connected with the two linked codes. When the Cluster Quotations option is checked, the quotations common to the two codes will be shown. One can also change the appearance of the code co-occurrence table by using Rotate/rows ⇒ columns.

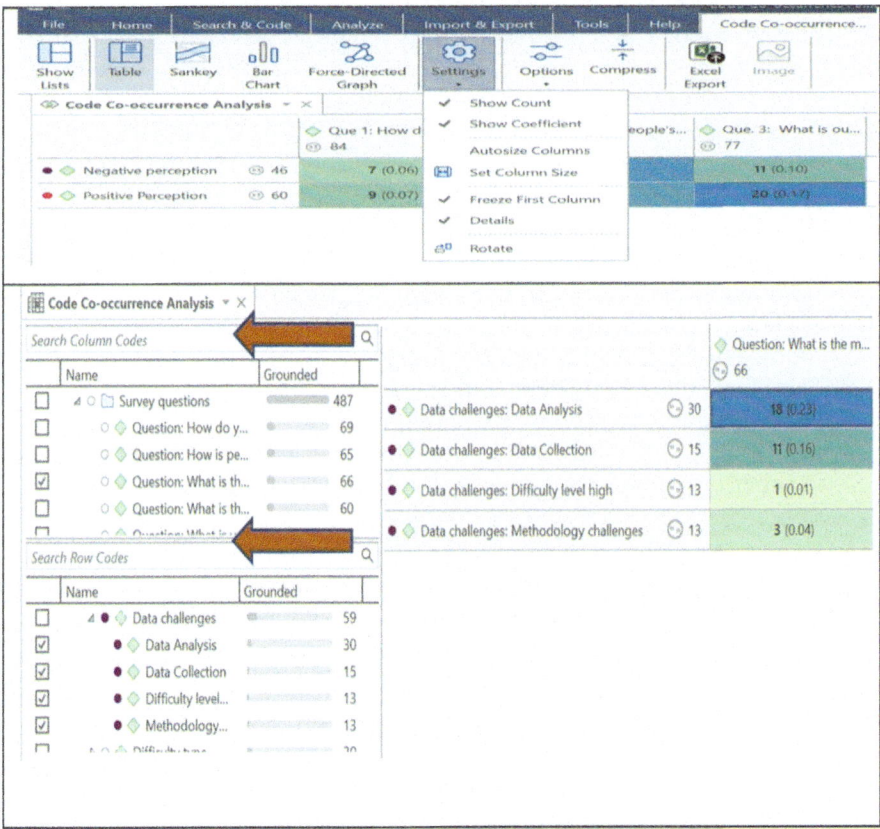

Fig. 7.4 Code co-occurrence in the query tool

The c-coefficient can be meaningfully interpreted only with large data sets or when there are many quotations in a document, such as fine-grained video data. All codes appear in two sections when a code co-occurrence table is created. The user must check the codes in both the sections for which a report is required. In other words, codes appearing in columns and rows should be checked for which report is needed.

After checking the codes, the code-to-code relations will show in a cross tabular form. Using different options, one can find the number of quotations and the c-coefficient for the selected codes.

C-coefficient

The c-coefficient indicates the strength of the relationship between two codes. It is similar to the correlation coefficient. The c-coefficient option can be activated in the ribbon of the Code Co-occurrence table. Calculation of the c-coefficient is based on quantitative content analysis. The value of c-coefficient lies between 0 (which indicates that codes do not co-occur) and 1 (when the two codes co-occur wherever they are used). The c-coefficient is calculated in the following way:

1. C-coefficient = $n_{12} / (n_1 + n_2 - n_{12})$
2. Where n_{12} is the number of quotations that co-occur between two codes
3. n_1 is the number of quotations in one code (groundedness)
4. n_2 is the number of quotations in the second code (groundedness)
5. In Fig. 7.4, there are 66 quotations for "question: what is the most challenging part of qualitative research", while there are 30 for "data challenges: data analysis". Eighteen quotations co-occur between the two codes. For calculations, the quotations are clustered.
6. Therefore, the c-coefficient can be calculated as under
7. C-coefficient = $18/(66 + 30 - 18)$
8. $18/78 = 0.2307$ which is significant as the value is closer to 1 compared to other c values in the tables.

Sankey Diagram of Code Co-occurrence

Sankey diagrams are a visualization tool that shows the strength of relationship between codes /themes and research questions. They help to visualize the flow from one value to another. They were named after the Irish Engineer, Captain Matthew H.R. Sankey (1853–1925), who introduced this concept in 1898 to show the energy efficiency of the various parts of steam engines. The French engineer, Charles Joseph Minard, had used a similar concept to visualize Napoleon's Russian Campaign of 1812. Sankey diagrams began to be used widely in the early twentieth century, especially in Germany as the country tried to maximize efficiency in the use of material and energy, both of which were in short supply due to the reparations the country was obliged to make after World War I.

The width of a path in a Sankey diagram in ATLAS.ti is proportional to the code coefficient value. Codes/themes are appearing on the left side in the row and research questions appear in the column. Code coefficients appear in the cells.

A Sankey diagram shows flow of resources between at least two nodes. In ATLAS.ti, a Sankey gives visual information about the structure and distribution of concepts and phenomena. The width of the arrows is proportional to the value of the node.

In Fig. 7.5, quotations co-occurring with the research question is shown (Que 2: How do people manage their career progression.) and code 'training, mentoring, learning'. In other words, out of 23 quotations linked to question 2, and 8 quotations linked to the code 'training, mentoring, learning', 8 quotations were associated with both codes. This means that there are 8 quotations co-occurring between research question 2 and training, mentoring, learning.

In the code co-occurrence table shown in Fig. 7.6, themes, as indicated by the answers to the question "what behaviour organisations reward" and 'what behaviour organisations discourage", are represented in a Sankey diagram. In this project, the respondents in the study were managers and senior managers working across disciplines in several organisations. Responses were coded and a code-occurrence table was created.

In Fig. 7.6, the thickness of each component of the challenges in the Sankey diagram indicates the relative importance of the issue. It can be seen that thickness of themes shows the significance of a theme. By clicking on the components, the quotations will be displayed on the right. Alternatively, the user can generate customized reports by using the Query Tool.

The Sankey diagram is the representation of code co-occurrence/code-document table. The user first creates the code co-occurrence/code-document table and then selects the Sankey diagram to visualize the significance of themes based on the c-value. In Fig. 7.6, the Sankey diagram shows the research questions on the right and emerging codes on the left. Used in this manner, the Sankey diagram can help researchers to identify key patterns emerging from data analysis. In the code co-occurrence table, the value of c-coefficients indicates the significance of these patterns.

Applications of C-coefficient

The two Co-occurrence tools are of immense analytical value. They are used to measure the occurrence of quotations for selected codes and code groups. As mentioned earlier, code co-occurrence is used to measure occurrence of quotations between two or more codes. A Code document table is used to study the patterns in codes/code groups across, for example, demographic,

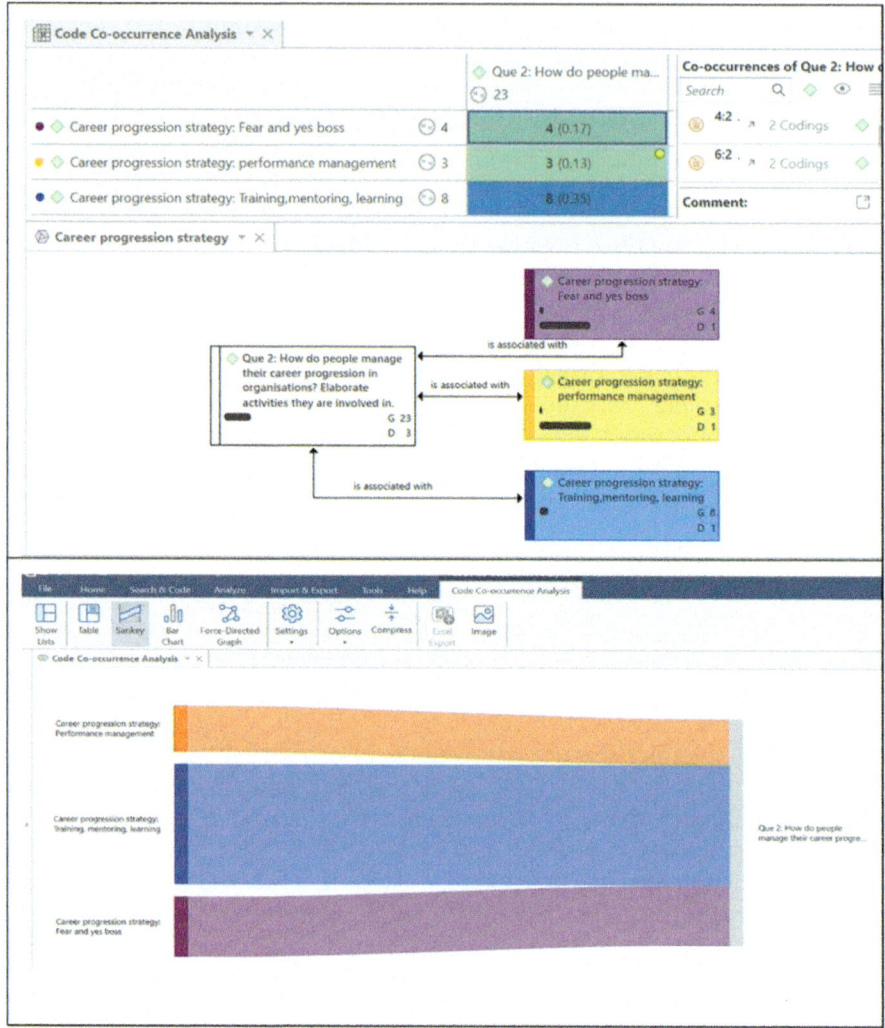

Fig. 7.5 Code co-occurrence table and Sankey diagram

socio-economic, socio-cultural, and contextual parameters. ATLAS.ti calculates the values of code coefficients based on co-occurrence of quotations between selected codes.

However, the user should also be aware that code co-occurrence table cannot be used with all types of data. For example, the C-coefficient is likely to have limited value in small data sets, such as the ones obtained from a ***study involving 10–20 respondents. In such cases, the frequency count/occurrence of quotations may be more helpful in providing ideas and insights for analysis.***

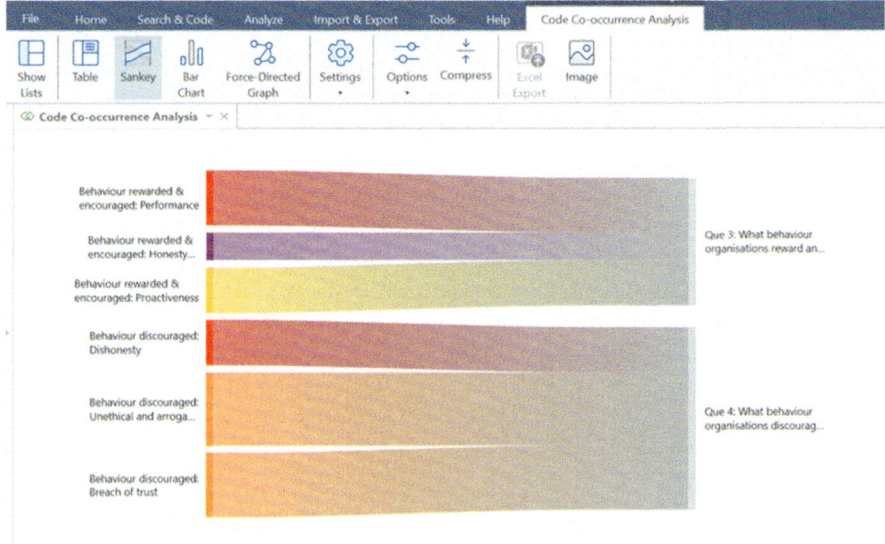

Fig. 7.6 Research questions and Sankey diagram

Boolean and Proximity operators should be used along with the Edit Scope option to generate reports for code groups, smart codes, and smart groups.

C-coefficients are more useful when the size of data is large, as in cases of large samples where the number of open-ended responses is high.

It is necessary to understand how co-occurrence is measured because of the mechanical and semantic issues involved in their meaningful interpretation. It is possible that many codes co-occur with several other codes. In such cases, researchers should interpret their significance using the c-values of every co-occurrence. These higher order co-occurrences need more elaborate methods of data analysis and interpretation.

Important C-coefficient indicates the strength of the relation between two codes. C-coefficient helps to identify the variables from the research and is used in qualitative data analysis. The value of C-coefficient lies between 0 to 1. C-value towards 1 is significant and towards 0 is not significant. If one uses the c-index, attention must be paid to the additional colored hints. Code coefficient might appear like Pearson Correlation Coefficient to quantitative researchers and readers. Therefore, it is important to distinguish between code coefficients in qualitative research and Pearson correlation in quantitative research.

Frequency Count and Listed Quotations

Code co-occurrence is the number of quotations shared by two codes, both codes appearing in the code co-occurrence table. Figure 7.7 shows that code co-occurrence value between the codes 'Disclosure' and 'Personal' is 20. This means that there are twenty quotations with two codes.

Co-occurrence frequency does not count single quotations. It only counts co-occurrence events. If two codes are linked to one quotation, it would count as a single co-occurrence. The drop-down list will display an ordered listing of all quotations for all co-occurrence events for the pair of codes. In Fig. 7.7, the number 20 is the number of quotations coded as "Disclosure" *and* "Personal." Individually, the number of quotations for 'Disclosure'

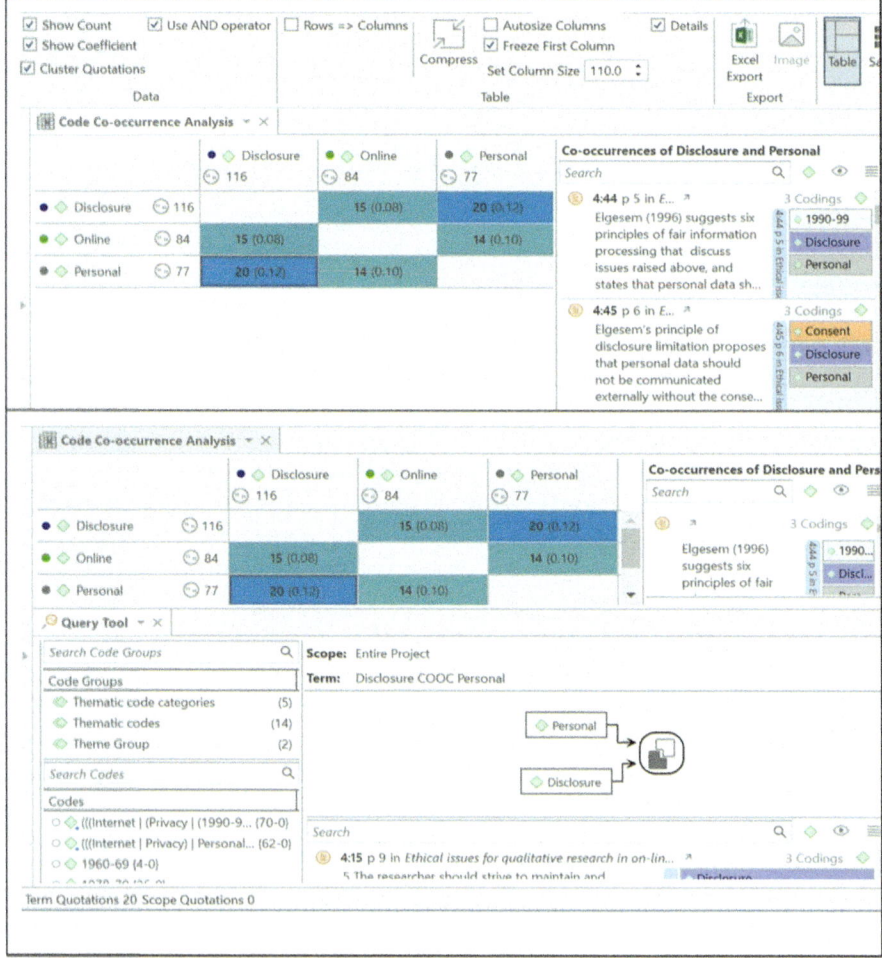

Fig. 7.7 Code co-occurrence table and query tool

is 116, while that for "Personal" is 77. By using the set operator OR to obtain co-occurrence, 20 co-occurring quotations was the result.

Determining code co-occurrence using the query tool will yield the same result. While checking the cluster quotations, only the embedded quotations are counted as a single hit by ATLAS.ti. Therefore, there are twenty quotations in code co-occurrence as shown in Fig. 7.7. By unchecking cluster quotations, this number could be higher as it counts every quotation.

The code co-occurrence table serves an important function by showing the linkages between codes and then interpreting them based on their c-coefficient values. The value of a code co-efficient will lie between zero (0) and one (1). A value closer to one indicates a stronger link than a value closer to zero. Thus, in Fig. 7.7, a c-value of 0.12 indicates a stronger relationship as compared to the one with a c-value of 0.08. In this example, the number of quotations does not decide the strength of relations between codes. The quotations are used to justify the relationships whose relative strength is based on the c-values.

The co-occurrence report generated in Excel can be formatted or analyzed according to the user's needs. One can export the table to Excel and save using the Excel report option. The Excel table will be in exactly the same format as the code co-occurrence table displayed in ATLAS.ti. One can also generate a report of c values using the Query Tool and applying various operators.

Three color indicators in code co-occurrence analysis table and their importance have been explained in Table 7.1.

Important: Yellow dots draw the attention of the researcher to the co-occurrences despite a low coefficient (Table 7.1). For example, if 20 quotations are coded as A, and 10 quotations with B, and there are 5 co-occurrences, then code coefficient is:

$$C = 5/(20 + 10 - 5) = 5/25 = 0.20$$

It shows that code B appears in 50% of all its applications with code A.

The red dot will appear in case of overlapping quotations. For example, the first quotation is coded with code A and B. An overlapping second quotation is coded with code B. Then, the code coefficient is:

$$C = 2/(1 + 2 - 2) = 2.$$

Table 7.1 Colors and descriptions

Hint color	Meaning	Description
Yellow	Distortion due to unequal quotation frequencies	Unequal quotation frequencies—the ratio between the frequencies of the code and row code exceeds the threshold of 5
Red	Out of range	C-index is outside the 0–1 range
Orange	Out of range	The orange dot is a combination of red and yellow conditions

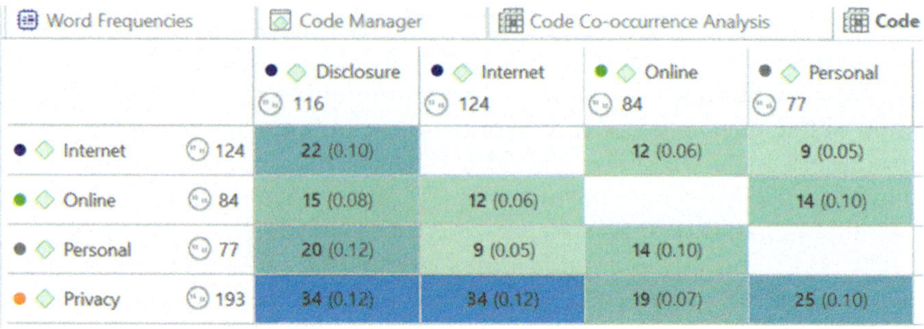

	Word Frequencies	Code Manager	Code Co-occurrence Analysis	Code
	◇ Disclosure 116	◇ Internet 124	◇ Online 84	◇ Personal 77
◇ Internet 124	22 (0.10)		12 (0.06)	9 (0.05)
◇ Online 84	15 (0.08)	12 (0.06)		14 (0.10)
◇ Personal 77	20 (0.12)	9 (0.05)	14 (0.10)	
◇ Privacy 193	34 (0.12)	34 (0.12)	19 (0.07)	25 (0.10)

Fig. 7.8 Code co-occurrence table with c-value

This value indicates the presence of redundant codes and hence the need to clean them.

Operators and Code Co-occurrence

Queries use operators to refine reports. In code co-occurrence tables, the code co-efficient measures the strength of a relationship between concepts. The code co-occurrence table generates more refined results than do queries which use operators. It is important to understand the differences in the functioning of the two. While the operators (Boolean/proximity) in queries help by acting as filters and generate quotations between codes, the tabulated reports are based on co-occurrence and code coefficient values. As examples, two reports of the same codes, one generated with operators and the other with code co-occurrence table, are shown in Figs. 7.8 and 7.9.

The report shown in Fig. 7.8 was generated using a code co-occurrence table, while the report in Fig. 7.9 used operators for the same codes. It can be seen that there are only minor differences between the two. The number of quotations in Fig. 7.9 is 62 (shown by a red arrow), while the number of quotations in Fig. 7.8 is 91. In Fig. 7.8, the code 'disclosure' co-occurs with the individual codes shown in the table. In Fig. 7.9, the co-occurrence of code 'disclosure' is shown with *all* the codes together. In such cases, the number of co-occurrences may get reduced because code 'disclosure' may be common in many codes.

Figure 7.10 is a text report generated from Fig. 7.9. The number of quotations for the selected query can be seen in Fig. 7.9.

Code-Document Table

Unlike code co-occurrence which shows code-to-code relations, code-document table shows code-document relationships (Figs. 7.11 and 7.12). A code-document table is used to find relationships of codes, code groups,

7 DISCOVERING THEMES IN ATLAS.TI   215

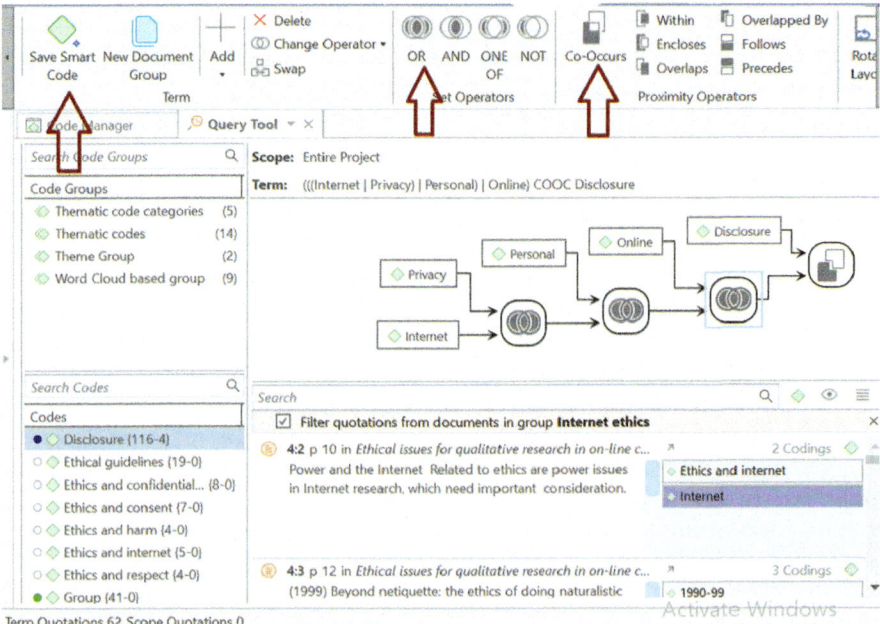

Fig. 7.9 Set operators and co-occurs

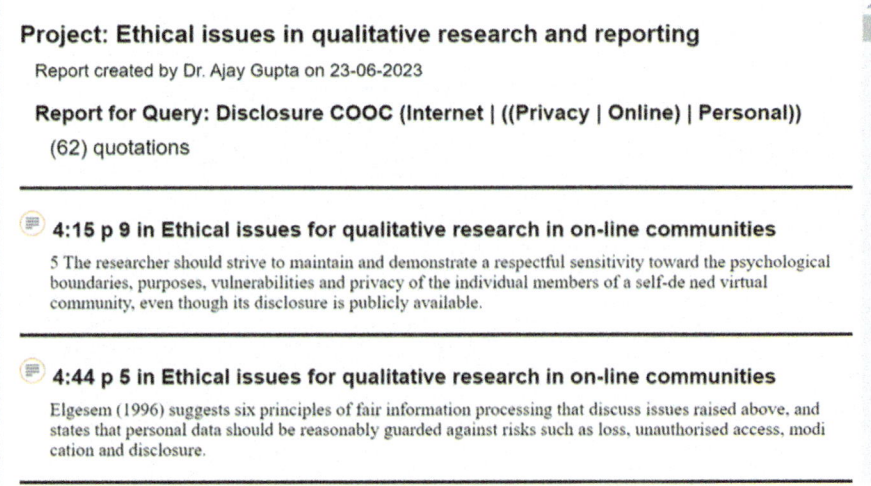

Fig. 7.10 Results using or and co-occurs

smart codes, smart groups, with documents and document groups. A code-document table gives a quotation count for each code or code group per document or document group for the coded segments.

One application of this feature is for making comparisons for a particular category of codes across different document groups. Some research questions in a study can be investigated with code-document tables. For example, in a research project, management students were asked open-ended questions to understand the 'threats to zoom classes' which were popular during the pandemic. Three themes emerged from the responses. They are shown, segregated according to gender, on the left side of the code-document table (Fig. 7.11). The table shows the row totals, column totals, absolute frequencies and row-relative frequencies.

Other options for exploring relationships and generating reports are also shown in Fig. 7.11. The Show Lists is used to view the table. If a code is changed, the user can update the tables by applying the Refresh option. Row totals display the row total of quotations, whereas column totals display the column total of quotations against the male and female category.

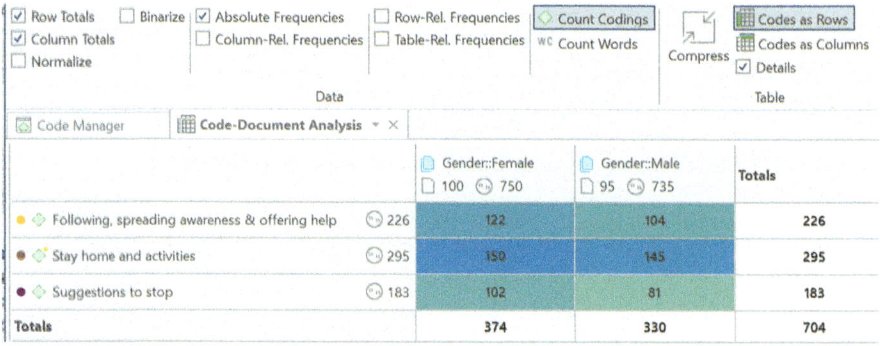

Fig. 7.11 Absolute code-document table

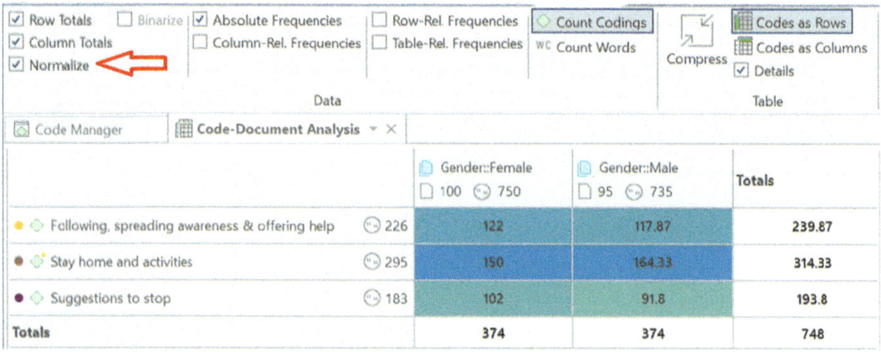

Fig. 7.12 Normalized table and row-relation frequencies

When documents or document groups vary widely in size, comparisons become difficult. In such cases, the data in them may need normalization. This means that the number of quotations per document that is selected for the co-occurrence table is adjusted. Data normalization makes the document/document groups comparable by equalizing the number of quotations under each document/document group.

Data normalization uses the document with the highest number of selected quotations as the base and suitably adjusts the numbers in the other selected documents. An example will help to understand normalization process.

Consider two documents in which one has 100 quotations and the other, 50. The number of quotations in the second document is multiplied by $100/50 = 2$. The number of quotations in the second document get adjusted accordingly, and documents become comparable. Column relation frequencies display relative frequencies based on the column quotations, whereas row relation frequencies display relative frequencies based on the row quotations. The total of each row is 100%. Table Relation frequencies display relative frequencies based on all quotations used in the table. The totals of all row and column quotations will add up to 100%.

Another method of data normalization is Binarization with which the table format addresses the occurrence or non-occurrence of a code (or codes in a code group) in a document (or document group). If they occur, a black dot is shown; if not, the cell is empty. When the tables are exported to Excel, the dots are shown as 1 s, and the empty cells are 0 s.

The user should be aware that the number of codes can be higher than the number of quotations. This is because "code co-occurrence" seems a better and more technical expression. In other words, several codes can be applied to one quotation. The Compress option is a quick way to remove all rows or columns that only show empty cells. It is the same as manually deactivating codes or documents from which results are not obtained. For this reason, a table cannot be uncompressed. Codes as Rows display all selected codes/code groups in the rows of the table. Codes as Columns display all selected codes/code groups in the columns of the table. If required, Quotation Count (shown after each code/code group or below each document/document group) can be deactivated (Fig. 7.13 and Table 7.2). In this example, an Excel report of code-document table was generated to show row, column and table relative frequencies gender-wise.

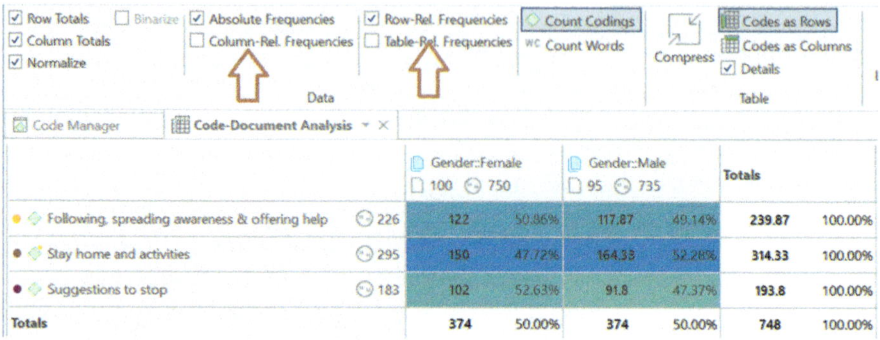

Fig. 7.13 Row-relation and table relation frequencies

Table 7.2 Excel report of absolute and relative values gender-wise

	Gender::Female Gr = 750; GS = 100		Gender::Male Gr = 735; GS = 95		Totals	
	Absolute	Row-relative (%)	Absolute	Row-relative (%)	Absolute	Row-relative (%)
• Following, spreading awareness & offering help Gr = 226	122	50.86	118	49.14	240	100.00
• Stay home and activities Gr = 295	150	47.72	164	52.28	314	100.00
• Suggestions to stop Gr = 183	102	52.63	92	47.37	194	100.00
Totals	374	50.00	374	50.00	748	100.00

BINARIZED DATA

A value of 1 or higher is indicated by a bullet, while values of 0 and less are shown by an empty cell. The Totals show bullet count. When the user checks "Binarize," either a black dot or a black dot empty cell is displayed next to the code-to-code linkage (Fig. 7.14). The Black dot indicates the relationship between the items in row and column, whereas the empty one shows that no relation exists between items in row and column. While exporting the report to excel, the result will show 1 if a relation exists, and 0 if it does not (Table 7.3).

7 DISCOVERING THEMES IN ATLAS.TI   219

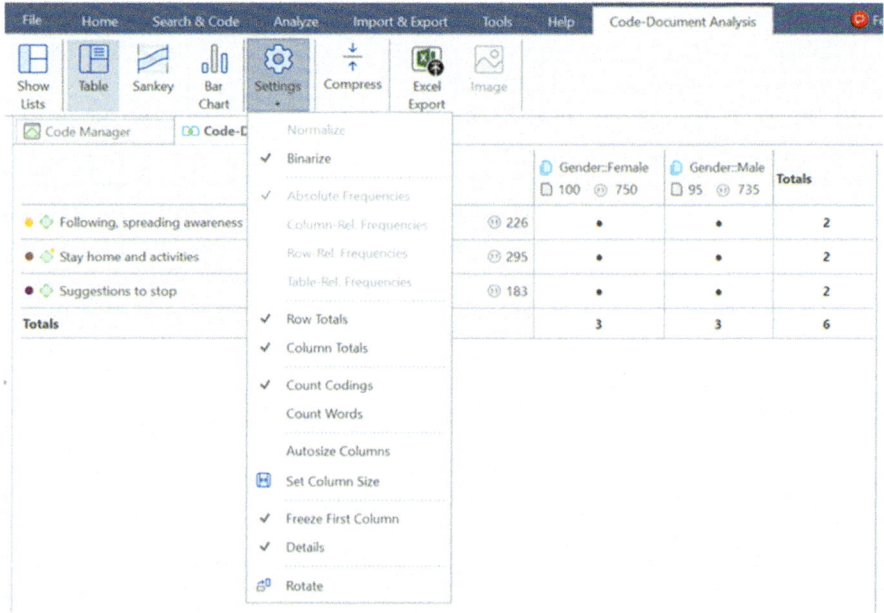

Fig. 7.14 Binarized code-document table in ATLAS.ti 23

Table 7.3 Binarized excel report

	Gender::Female Gr = 750; GS = 100	Gender::Male Gr = 735; GS = 95	Totals
• Following, spreading awareness and offering help Gr = 226	1	1	2
• Stay home and activities Gr = 295	1	1	2
• Suggestions to stop Gr = 183	1	1	2
Totals	3	3	6

NORMALIZED FREQUENCIES

In normalization, the number of quotations in all documents in the table are equalized. After normalization, every document is assumed to have the same number of quotations so that comparisons of code-relative frequencies can be made. It is good practice to normalize data before generating reports for the reasons explained below.

Figure 7.15 shows the code-document table in the absolute and normalized forms. In the upper figure, the number of quotations for three codes are shown for the male and female categories. The number of quotations for male and female is 405 and 418 respectively. The Sankey diagram is shown for the code-document table for both categories.

In the lower figure, the results were obtained after checking the Normalize option. The number of quotations for the documents is now equal. But the number of codes remains the same. Normalizing data helps in balanced interpretation of outcomes. Normalization is useful when there is an unequal number of quotations for a code across documents because of which the results of analysis may not truly reflect the characteristics of the subject of study. Normalization equalizes quotations in documents selected in code-document table after which the data becomes comparable.

Sankey Diagram for a Code-Document Table

The code document table (Fig. 7.16) was generated for the project "Blue-collar migrant workers' challenges during a pandemic in the Indian context". The respondents in the study were service providers for blue-collar migrant workers. The responses were coded, and a code-document table was created for the research questions. The resulting diagram shows the pattern of information emerging from the data.

The project's main objective was to study the challenges faced by service providers who served blue-collar migrant workers in the India during the pandemic. A table was created with codes of maximum code density. The Sankey diagram helped identify the major challenges of the service providers. The thicknesses of the components in the diagram indicate the relative significance of each challenge. Thus, food, transportation, financial assistance, and volunteers, in that order of its importance, were the challenges faced by the providers. By clicking on an individual component, the corresponding quotations will be displayed on the right. Alternatively, a report can be generated with the help of the Query Tool.

Here, the Sankey diagram was helpful to the researchers in understanding the 'intensity' of the issues identified in the analysis. As seen in Fig. 7.17, food was the major issue. Figure 7.17 also shows the absolute and normalized frequencies of the codes in ATLAS.ti 22.

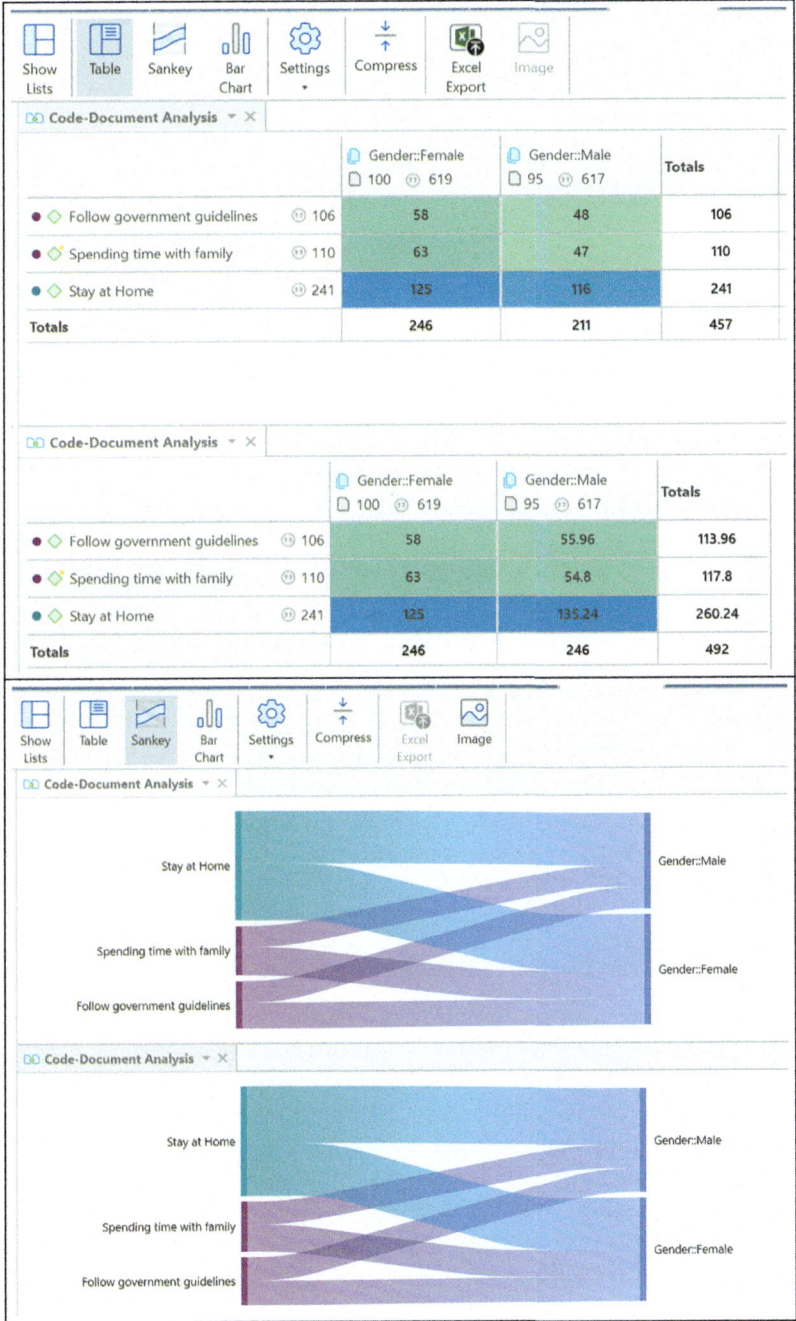

Fig. 7.15 Absolute and normalized code-document table

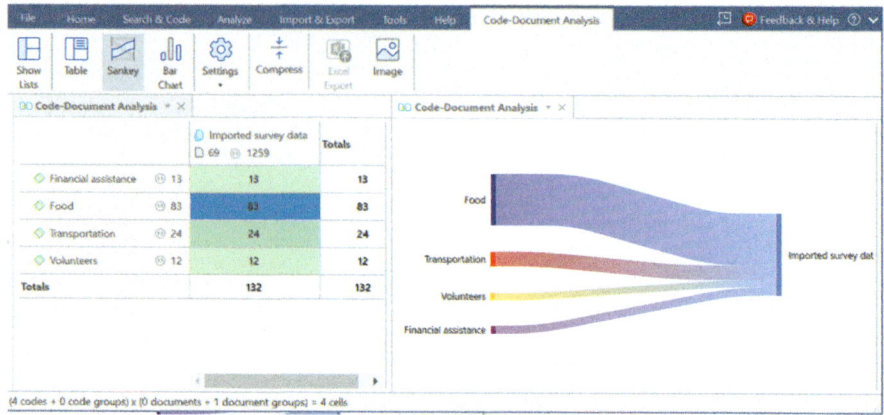

Fig. 7.16 Code document table and Sankey diagram

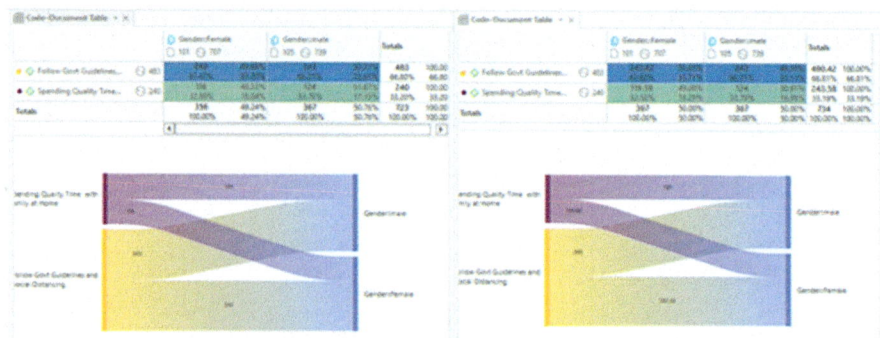

Fig. 7.17 Absolute and normalized frequencies

Showing Themes in One Report

All emerging themes can be shown in one report or network using operators and Edit Scope. Table 7.4 shows a codes-themes relationship. Code categories were created from the codes. Next, the code categories were used to form sub-themes from which themes were identified. The appropriate are used for identifying relationships between codes, code categories or sub themes. A screenshot of 11 operators is shown in that table that was used for forming smart codes and groups (Fig. 7.18). One can use smart code, smart group, and codes in any form to generate reports. The quotations, codes, comments, and memos linked with the emergent themes were imported to the network, and the text report.

With the Edit Scope option, reports can be refined for categories of respondents on demographic, socio-economic, professional and social parameters. The user can use more operators to broaden the scope.

Table 7.4 Codes and theme relations

Codes (12)	Code categories (04)	Sub-themes (02)	Theme (01)
Treating all employees equally Equity in rewarding employees Fairness in performance appraisal process	Inspirational leadership		
		Authentic leadership	
Demonstrating leadership Encouraging people to grow Leading by example	Practicing leadership		
			Exemplary Organisation
Building relations with employees Helping other employees Proving dedication	Responsible employees		
		Healthy working environment	
Conducive working environment Harmony among employees Nurturing employee relations	Vibrant culture		

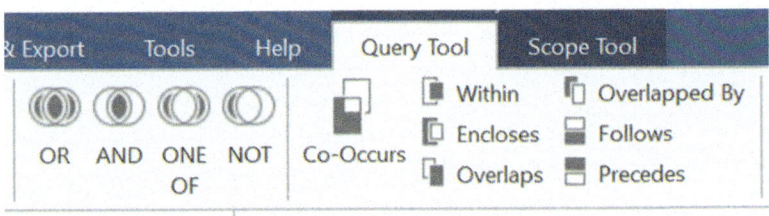

Fig. 7.18 Set and proximity operators

Recommended Readings

Atkinson, P., Coffey, A., & Delamont, S. (2003). *Key themes in qualitative research: Continuities and changes*. Rowman Altamira.

Bazeley, P. (2009). Analysing qualitative data: More than 'identifying themes.' *Malaysian Journal of Qualitative Research, 2*(2), 6–22.

Contreras, R. B. (2011, August). Examining the context in qualitative analysis: The role of the co-occurrence tool in ATLAS.ti. *ATLAS.ti Newsletter*.

DeSantis, L., & Ugarriza, D. N. (2000). The concept of theme as used in qualitative nursing research. *Western Journal of Nursing Research, 22*(3), 351–372.

Lewis, J. (2016). *Using ATLAS.ti to facilitate data analysis for a systematic review of leadership competencies in the completion of a doctoral dissertation*. Available at SSRN 2850726.

Ryan, G. W., & Bernard, H. R. (2000). *Techniques to identify themes in qualitative data*.

Schmidt, M. (2008). The Sankey diagram in energy and material flow management: Part II: Methodology and current applications. *Journal of Industrial Ecology, 12*(2), 173–185.

Wright, C., Ritter, L. J., & Wisse Gonzales, C. (2022). Cultivating a collaborative culture for ensuring sustainable development goals in higher education: An integrative case study. *Sustainability, 14*(3), 1273.

CHAPTER 8

Generating Themes from Smart Codes and Smart Groups

Abstract In ATLAS.ti, the themes of qualitative studies emerge from smart codes. This chapter discusses the principles of smart codes and smart groups, and how they differ from the conventional codes and code groups that researchers are familiar with. Smart codes are used to present themes and are constructed from existing codes using relationships. Smart codes have the characteristics of their component codes. A smart group is formed by combining two or more code groups. Text reports on smart codes and smart groups can be generated using set and proximity operators. The author then shows how smart codes and smart groups are interpreted with reference to the research questions. Screenshots and thematic visualizations of networks are used to explain the creation of themes and their presentation.

SMART CODES AND NORMAL CODES

A smart code is formed by combining two or more codes for information retrieval and theme generation. It contains the variables of the study—including, for example, demographic, socioeconomic, and contextual—and provides a query report in text, Excel, or network formats. The Code Manager and Query Tool options can be used to form a smart code.

Smart codes use relationships for presenting themes. They combine existing codes and enable researchers to generate various reports with reference to the research questions. While a code represents one concept, a smart code can represent several because it includes many normal codes. Smart codes help researchers to generate various kinds of reports. For example, in Fig. 8.1, a smart code was created from seven code categories: Leave challenges of middle managers, Connection-based promotion, Favours in promotion and transfer,

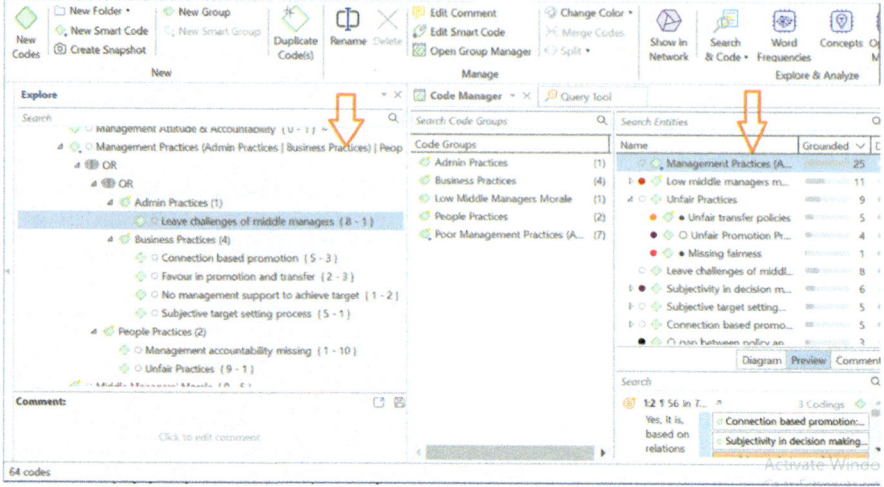

Fig. 8.1 Smart code in explorer and code manager

No management support to achieve target, Subjective target setting process, Management accountability missing, and Unfair practices.

In the example, the code "Connection-based promotion' has 3 associated codes and 5 quotations, while 'Favour in promotion and transfer' has 3 codes and 2 quotations. In ATLAS.ti, when a smart code is created with two or more codes using the "Any" operator, the quotations associated with each code are automatically assigned to it. The newly created smart code will be shown with its component codes but with a different symbol.

Smart codes can be used to generate reports using Code Manager and shown in a network. Reports can also be generated using the analysis and Query Tool. The Query Tool offers more options in using operators for creating smart codes. In ATLAS.ti 22 & 23, the Query Tool has 11 operators (set and proximity operators).

A smart code uses operators (Set or proximity) and can be used as themes. The researcher links the codes based on related attributes, causal relationships, association, and categories. Smart codes can also be grouped, based on certain characteristics, traits, and attributes. If required, code groups can be formed that contain both smart codes and normal codes.

Smart codes are created to link the attributes of two or more codes with simple as well as complex operators. For example, the codes A and B can be converted to a smart code by using "any" operator option in the Code Manager. By using "Any", the attributes associated with both codes get added, whereas by using all in the Query tool option, only the common attributes of both codes get captured. The report on the smart code will contain the attributes of both codes.

Smart codes may not be necessary when the data size is small and there may only be a few concepts for study. In such cases, it may be easier to forms linkages between and among codes, and then generate a thematic report at the code level itself.

Figure 8.2 shows code group, codes, code folder, code categories, and subcodes. There is an option to create code folder, code category, code group, and smart code in Code Manager. First, codes were created using the open coding process. Then, the relationships among the codes were identified and smart codes formed. In the network, each smart code has the name of codes in it and also of the codes associated with it. D is the density whose value shows the number of associated codes. G the groundedness, meaning the number of quotations linked with the smart code. The name of the smart code can be a default one (as shown in Fig. 8.1) or one given by the researcher.

ATLAS.ti 22 has only set and proximity operators for creating smart codes and groups. The researcher can create smart codes and groups by selecting the codes or code groups in addition to query tool option (Fig. 8.2) and then use the right click to select the appropriate option.

Reports of smart codes can be generated by applying filters in the Scope Option. In Fig. 8.3, the smart code 'Management Practices' was searched with Middle Managers Bank A using Edit Scope option. The resulting report will show Management Practices within Middle Managers Bank A. In the same Fig. 8.3, four document groups are shown from which the user can generate a

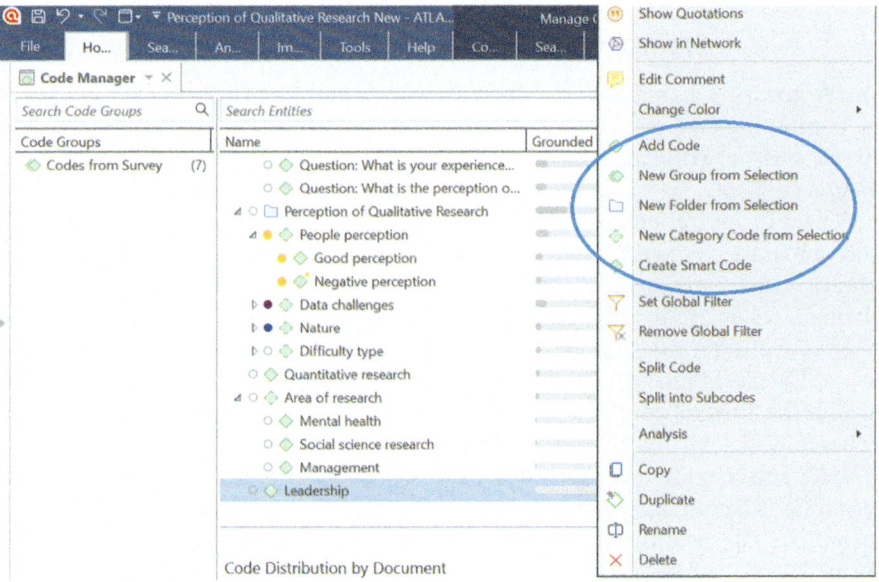

Fig. 8.2 Options for creating smart codes

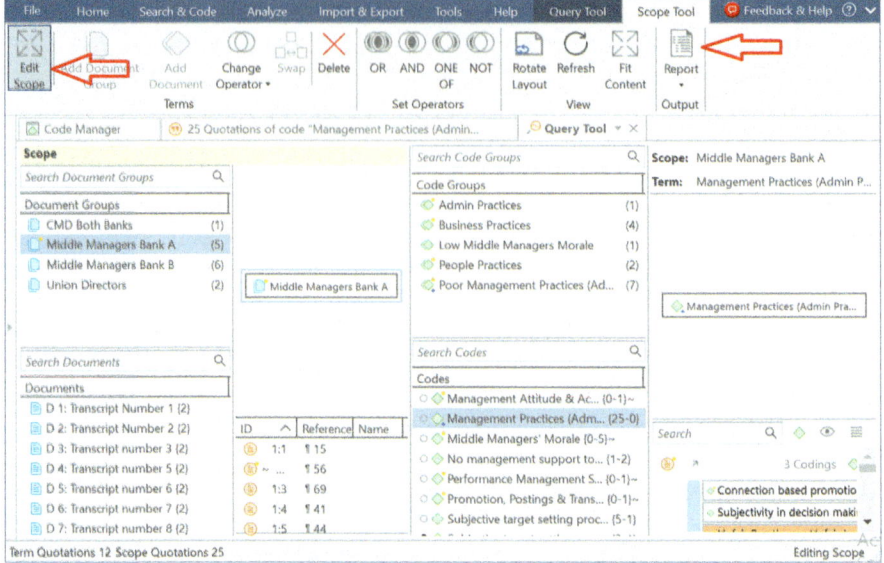

Fig. 8.3 Smart codes and edit scope

report using a filter. The set and proximity operators, with filters, if necessary, can be used to create more smart codes.

SMART GROUPS AND NORMAL GROUPS

Smart group is a theme and information retrieval code group that is formed by combining two or more code groups, to generate a queried report in text, Excel, and network formats. Smart Groups can be created in Code Manager or Query Tool.

Smart groups combine code groups to help the researcher create networks and generate reports. They contain more concepts/constructs than code groups. Smart groups are stored queries in which code groups are the core element. They are formed using set and proximity operators.

Smart groups can also be formed by linking code groups using operators in the Query Menu, or the "All or Any" option in the Code Manager menu, to link the various concepts in the groups. With smart groups, reports can be generated in text, Excel, and network forms.

Like smart codes, smart groups are stored queries based on groups. For example, if data from a study is grouped by gender, age, and location, a smart group of "All male managers under 30 years living in western region" can be formed.

Figure 8.5 shows a smart group formed from the code groups "People Practices" and "Business Practices". The 'Any' and 'All' operators were used.

The set and proximity operator options in the Query tool option was used to create smart groups (Fig. 8.4).

By selecting 'Any' both code groups are added. The 'All' option extracts only the codes associated with both code groups. The operator 'Any' is treated as OR and All is treated as AND in the Set operator.

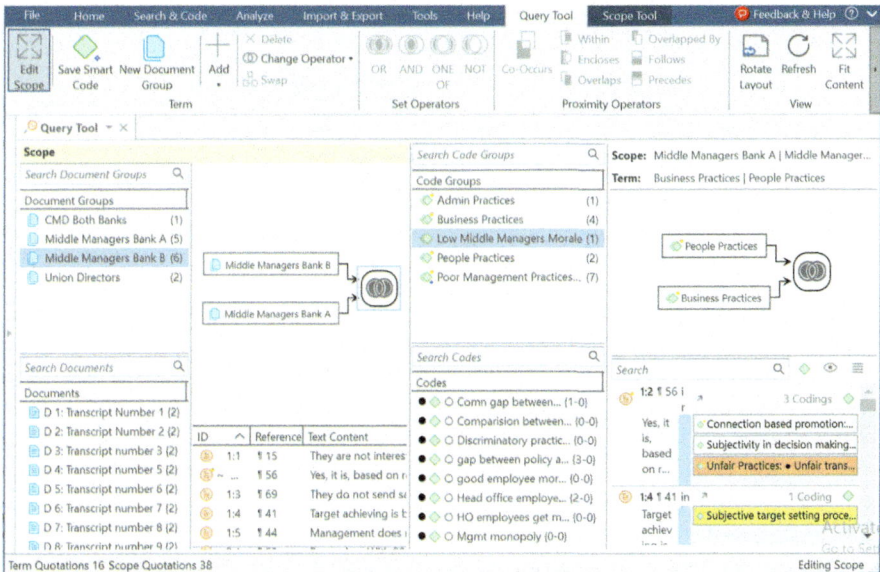

Fig. 8.4 Smart group and edit scope in query tool

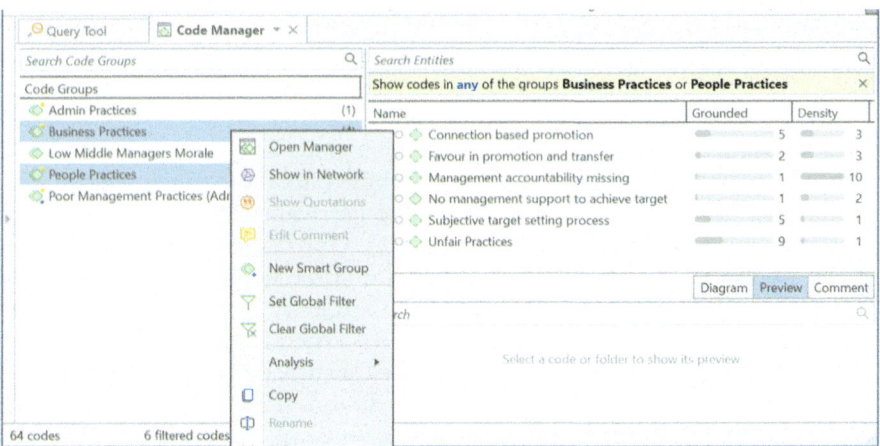

Fig. 8.5 Smart group in code manager

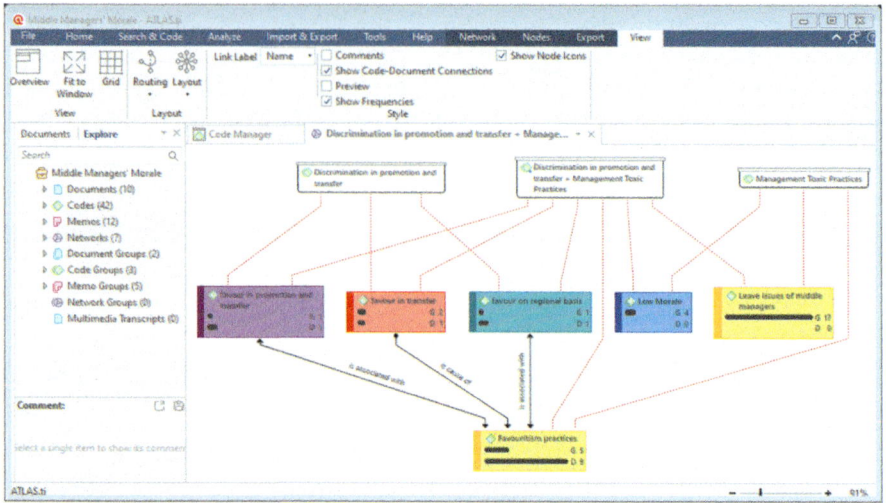

Fig. 8.6 A smart group network

Similarly, in Fig. 8.6, a smart group was created using the code groups 'Discrimination in promotion transfer', and 'Management toxic practices'. The smart group is shown between the two code groups. The text report was generated from the smart group.

Forming Smart Codes and Smart Groups

Smart codes and smart groups in ATLAS.ti are formed using Code Manager or Query Tool. While Code Manager has only a few operators, the Query tool has eleven, thus enabling the creation of a wider range of smart codes and smart groups. Smart codes are shown with their component codes in the same row.

In Code Manager there are two methods for creating a smart group. The first is by using the 'Any' operator which will include all quotations linked with each code group. The second, 'All', will include only those quotations that are common to the selected codes groups. The user can select the code groups, right-click and press the 'New Smart Group'. One can either select the 'Any' or 'All' option (Fig. 8.2). Figure 8.7 also shows the range of options—a total of eleven Set and Proximity operators—in the Query Tool for creating smart codes and smart groups (ATLAS.ti 22 & 23). When a smart group is created using the Query Tool, it is shown with its constituent codes. When a smart group is created using the Code Manager option, it is shown with the code groups. When opening a Query, the code groups will be shown on the upper pane, and codes in the lower one (Fig. 8.7).

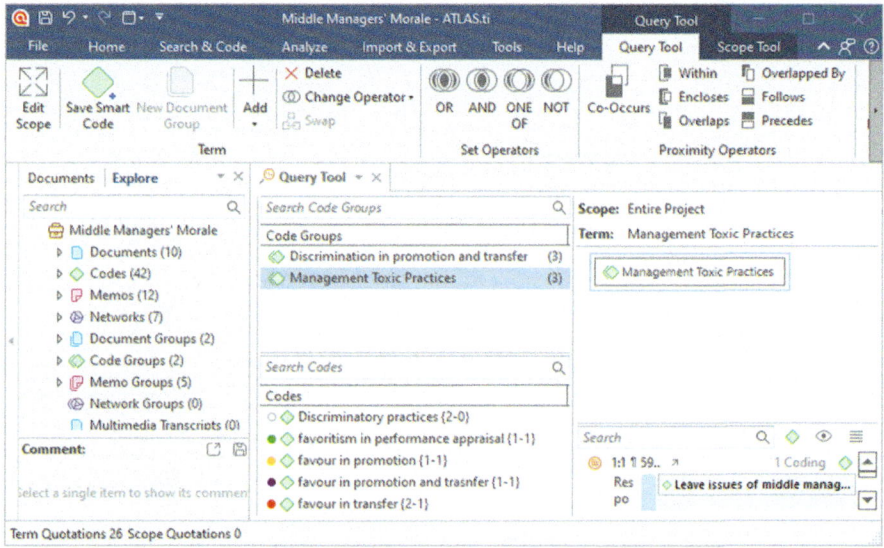

Fig. 8.7 Operators in ATLAS.ti 22

Before discussing the process for creating smart codes and smart groups, it is necessary to understand the meaning and function of each operator type in Fig. 8.7.

SET OPERATORS

Set operators are also known as Boolean operators. Using set operators, one can retrieve information applying certain selection criteria. Out of the four set operators, only the 'NOT' needs one operand. The 'OR', 'AND', and 'ONE OF', need two operands for retrieving information.

OR. The OR operators retrieve all quotations that match at least one of the arguments used in the query, such as, for example, all quotations coded with 'Data Challenges: Data Analysis' or 'Data Challenges: Data Collection'. A more complex selection may be based on a combination of queries like all quotations coded with 'Data challenges: Data Analysis' *or* coded by 'Data challenges: Data Collection' *and* 'Data challenges: Difficulty level high'. In our example, the number of quotations associated with 'Data collection: Data Analysis ' is 30, while the number of quotations associated with 'Data challenges: Data Collection' is 15 (Fig. 8.8). The OR operators combine the quotations associated with each code. Thus, the new smart code will have 40 quotations. Five quotations co-occur between two codes. In other words, five quotations have been coded by both codes. These quotations can be read in the bottom right. The quotations are identified as 7:7, 12:7, and so on. To

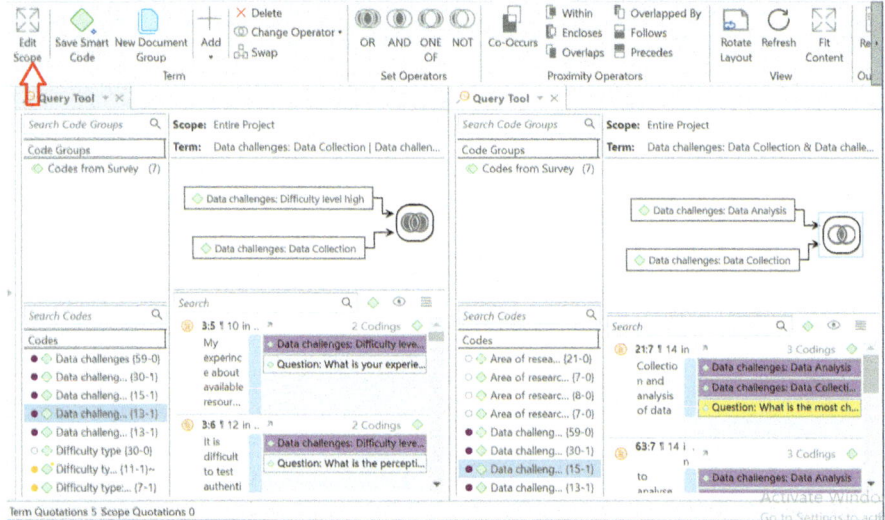

Fig. 8.8 'Or' and 'and' operators

interpret the numbering, 7:7 means quotation number 7 in document number seven, quotation number 7 in document number 12, and so on.

AND. By using the AND operator to create a smart code, the software will find the quotations that match all the conditions specified in the query. Thus, in this example, the search will include all quotations for 'Data Challenges: Data Analysis' *and* 'Data challenges: Data Collection'. The researcher must understand that the AND operator is very selective and often produces an empty result set. This is because it combines precision with low recall in its search. In this example, the result after using AND is 5, meaning that there are only five quotations coded with 'Data challenges: Data Analysis' and 'Data challenges: Data Collection'.

ONE OF. The third operator in this group is "ONE OF", meaning that exactly one of …. the conditions are met. In everyday language, this is the either/or rule. For example, applying the XOR operator means that all quotations are coded with *either* "Data challenges: Data Analysis" *or* "Data challenges: Data Collection". Thus, the result is 35 quotations, meaning that 35 quotations have been separately coded. Co-occurring quotations are excluded from each code. The first code has 30 quotations and the second has 15. Five quotations are co-occurring between two codes. The XOR operator will calculate quotations by subtracting co-occurring quotations coded with each code and then adding them.

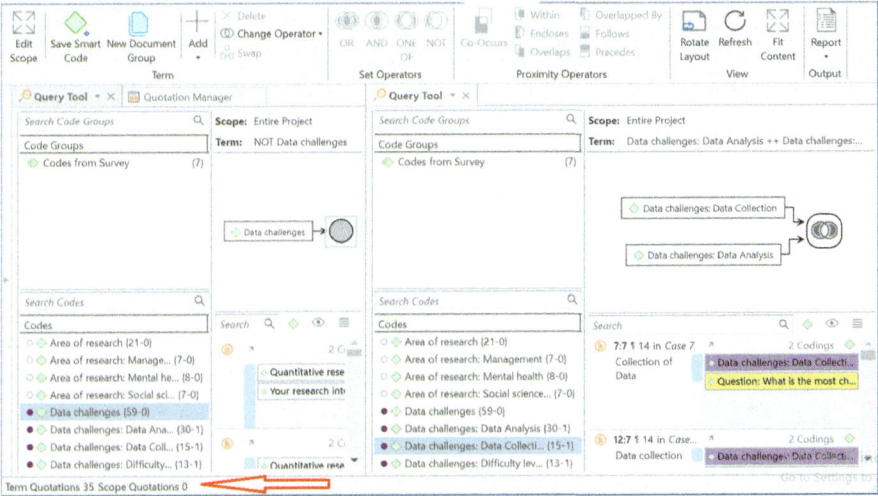

Fig. 8.9 'One of' and 'not' operators

NOT. The fourth Set operator is the NOT. NOT tests for the absence of a condition by removing the matches for the non-negated term from all quotations in this project. In Fig. 8.9, the NOT operator shows all the quotations except those coded as 'Data Challenges'.

The set operators 'ONE OF' and 'NOT' are shown in Fig. 8.9. Here, the researcher has coded 492 quotations. Data Collection has 30 quotations and Data Analysis has 15. By using the 'ONE OF' operator, the result shows quotations coded either as Data Collection or Data Analysis, but not both. In this option, co-occurring quotations will be excluded. Figure 8.10 shows the set operators and number of quotations under each operator. A smart code was created for each set operator and a folder was created for them. In the figure DA means Data Analysis and DC is for Data Collection.

Functions of Proximity Operators

Proximity operators show the closeness between or among quotations, meaning that they describe the spatial relation between quotations. They (Within, Encloses, Overlaps, overlapped by, Follows or Precedes) show how quotations are connected: Quotations can be embedded in one another; or one may follow another, and so on. Proximity operators highlight these relationships. They require two operands as their arguments and differ from the other operators in one important aspect: proximity operators are *non-commutative*. This property makes learning to use of proximity operators a little more difficult.

Non-commutativity requires a certain input *sequence* for the operands. While "A OR B" is the same as "B OR A," this does not hold for any

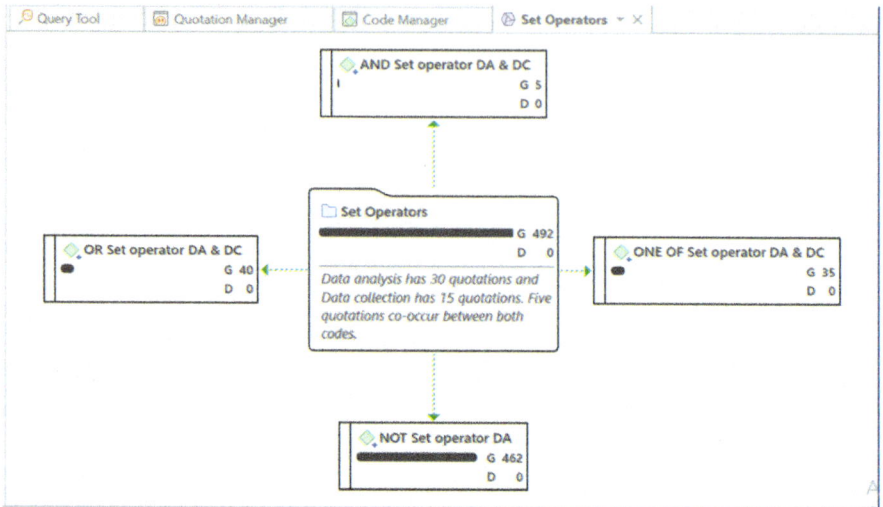

Fig. 8.10 Set operators

of the proximity operators. Thus "A FOLLOWS B" is not the same as "B FOLLOWS A."

Another important characteristic of proximity operators is the specification of the operand for which the quotations need to be retrieved. This means that researchers should know how to create smart codes using proximity operators using query tool option. For example, while "A WITHIN B" specifies the constraint, one must also specify whether the quotations must be for A or B. The code (or term) that is entered first is the one in which the user is interested. If quotations for B are required, the user must enter "B ENCLOSES A" using the query language described in Figs. 8.11 and 8.12. Figure 8.11 shows the transcript coded with several codes appearing Within, Overlaps, and Encloses, etc. Figure 8.12 shows a network of proximity operators, associated codes and groundedness.

Figure 8.12 shows the four proximity operators: WITHIN, ENCLOSES, FOLLOWS, and OVERLAPS.

1. By applying 'Promotion not performance based' WITHIN 'Low Morale' two quotations are obtained.
2. Applying 'Discrimination in promotion' to OVERLAP 'Experience rewarded over merit' yielded one quotation.
3. Low Morale ENCLOSES in 'No HRM practices' also yielded one quotation.
4. When 'Promotion not performance based' FOLLOWS 'No HRM practices' the result is one quotation.

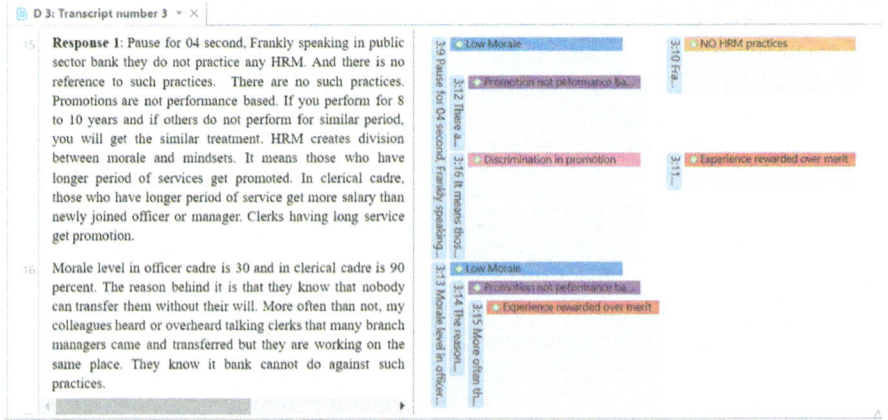

Fig. 8.11 Query tool and proximity operators

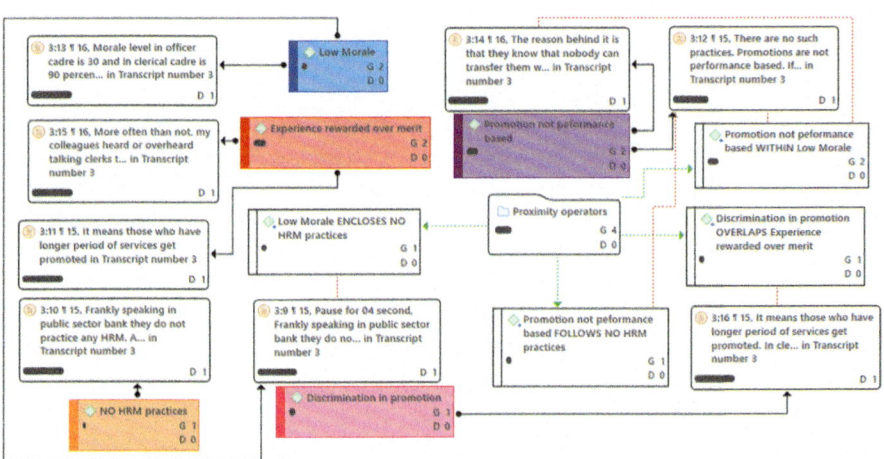

Fig. 8.12 A network of query tool and proximity operators

If the user specifies A ENCLOSES B, all quotations coded with A *and* containing quotations coded with B are retrieved. This means that all quotations that are enclosed by other quotations—for example, A being enclosed by B (or A within B)—are retrieved by the operator. In other words, all quotations coded with A that are contained *within the data segments* coded with B are retrieved.

Co-occurs is basically a shortcut for combining the AND operator and all basic proximity operators *except* FOLLOWS and PRECEDES. When A Co-Occurs with B is applied, all quotations that co-occur with B in any way are retrieved:

1. **A within B** retrieves all quotations coded with A that are contained in data segments coded with B.
2. **A encloses B** retrieves all quotations coded with A that contain quotations coded with B. This is a non-commutative operator because the sequence of arguments matters.
3. **A overlaps B** retrieves all quotations coded with A that overlap with B. This is also a non-commutative operator because of the arguments' sequence.
4. **A overlapped by B** retrieves all quotations coded A and overlapped by quotations coded B. This is a non-commutative operator because the sequence of arguments matters.
5. **A follows B** retrieves all quotations coded A that follow quotations coded B. This is a non-commutative operator because of the arguments' sequence.
6. **A precedes B** retrieves all quotations coded with A followed by quotation coded with B. This is a non-commutative operator because of the arguments' sequence.

In addition to these operators, one can also refine the search by using Edit Scope on the top pane. For example, if the researcher wants to generate a report on 'Data analysis' overlapped by 'Data collection' in the document group 'Male researchers', 'Male researcher' is selected in Edit Scope. Various document groups can be created based on various parameters like demographic, traits, experiences, social status, gender, etc.

Reports can be generated with or without using operators. For example, one can refine the search for the code 'Data collection' in 'Quantitative researchers' to retrieve only those quotations associated with 'Quantitative researchers' and 'Qualitative researchers'. Then, only the responses linked with "Data collection" will be shown.

With the help of Query section, the researcher can use code, code groups, smart codes, and smart groups to generate reports using operators or without them.

Steps for Generating Reports Using Set Operators

1. Go to the Home tab and open the 'Analysis' option.
2. Open Query. Code groups and smart groups will be displayed on the upper pane, and codes and smart codes in the lower one.
3. Select the code, code groups, smart code, or smart group for which a report is required. Two or more operands in Set Operators can be used.
4. Double click the first code/code group/smart code/smart group. It will be displayed on the right of code or smart groups.
5. Double click on the selected set operator.
6. The code along with the selected operator will appear on the right of the selected code.

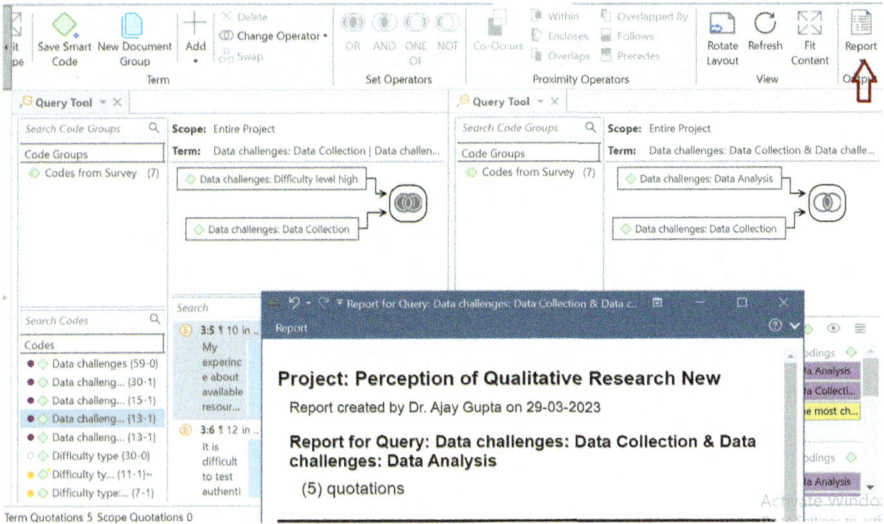

Fig. 8.13 Report using set operators

7. Then, select the next code and double click.
8. Click on report (top right corner). The report will be generated.
9. For generating a report within a specific document group, use the Edit Scope on top left. The resulting report is shown in Fig. 8.13.

If a report is needed for the code groups, 'Admin practices', 'Business practices' and 'People practices' using set operators (OR AND & ONE OF), one must double click on the selected set operator and link the code groups (Fig. 8.3).

By clicking on Edit Scope, documents and document groups will be shown. The user can then select the document or document groups for a report is required (Fig. 8.3). In the example shown in the figure, there are four document groups—CMD both banks, Middle Managers Bank A, Middle Managers Bank B, and Union Directors. There is also a list of documents. One can select any document or document group, or combinations using the operators for generating reports.

GENERATING REPORTS WITH PROXIMITY OPERATORS

1. Go to Home tab and open Analysis.
2. Open Query. The code groups and smart groups will be shown on the upper pane, and codes and smart codes in the lower one.
3. Select the code, code groups, smart code, or smart group for which a report is required. Two or more operands in proximity operators can be used.

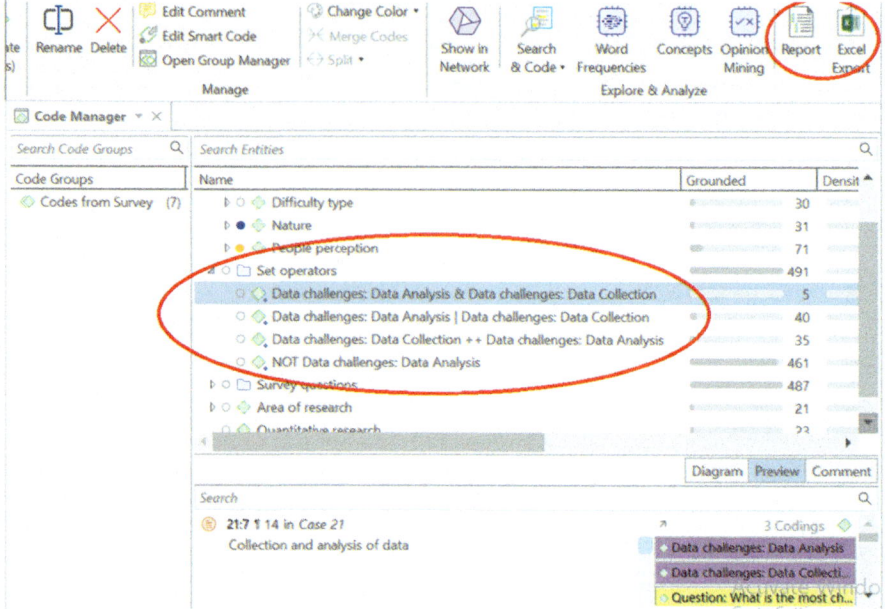

Fig. 8.14 Text report using set operator in code manager

4. Double click on the first code/code group/smart code/smart group. The selected item will appear on the right of code or smart groups.
5. Double click on the selected proximity operator. It will be shown on the right with the selected item.
6. Select the second code and double click. It will appear on the right side.
7. Click on Report at the top right corner and generate report.
8. Use Edit Scope on extreme left of the top menu bar to generate a report.
9. The report will be displayed as shown in Fig. 8.15.

Important Users can generate reports for the selected codes at document level. The Edit Scope function is used to refine the search based on various parameters (gender, demographics, social status and category of respondents, etc.) In Fig. 8.15, the researcher used the Encloses, Overlaps, Follows and Within operators to generate a report for smart codes. With the Edit Scope option, the search report was refined to include a specific document or document group.

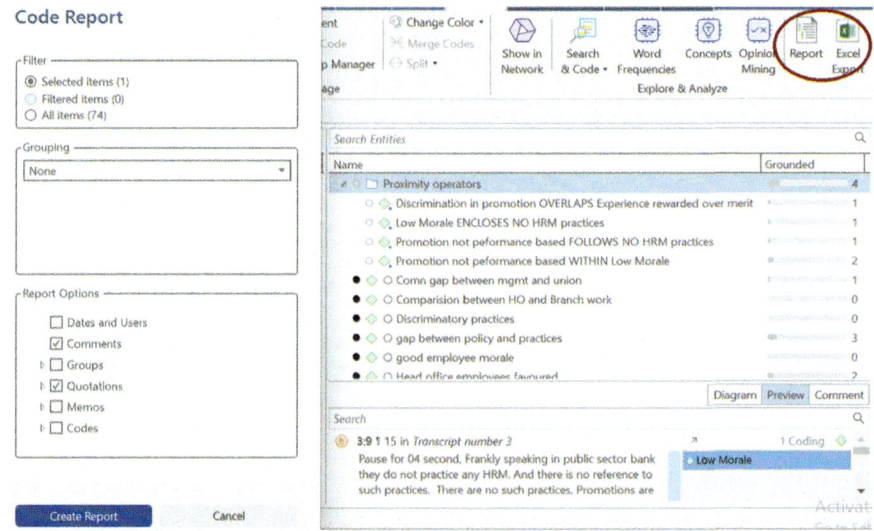

Fig. 8.15 A text report using proximity operator

Smart Codes and Their Interpretation

Smart codes store complex queries about the concept (or concepts) under study. Smart codes are dynamic, their attributes changing if the codes associated with them change. Any change, such as modification in the name of the code or comments, a change in links between memos, codes, quotations, or their mergers and splits, will automatically result in changes in the smart codes. In Fig. 8.16, the smart code used the OR operators in the query tool. It includes all the quotations linked with the codes. The default display of the smart code is Fig. 8.16. However, the display can be modified by the researcher if required.

The OR operator used for creating the smart code from the seven codes: 'Leave challenges of middle managers', 'Connection based promotion', 'Favour in promotion and transfer', 'No management support to achieve target', 'Subjective target setting process', 'Management accountability missing', and 'Unfair practices'. These codes belonged to three code groups—'Admin practices', 'Business practices' and 'People practices'. The smart code has 25 quotations, and its default name is Management Practices (Admin Practices) | Business Practices | People Practices.

Figure 8.16 shows the smart code from which a report was generated for 'Middle Managers Bank B' and 'Middle Managers Bank A' using the OR operator. The selection provided reports on management practices for 'Middle Managers Bank A' and 'Middle Managers Bank B'. Various other reports are possible with smart codes. For more complex reports, the Edit Scope option is used.

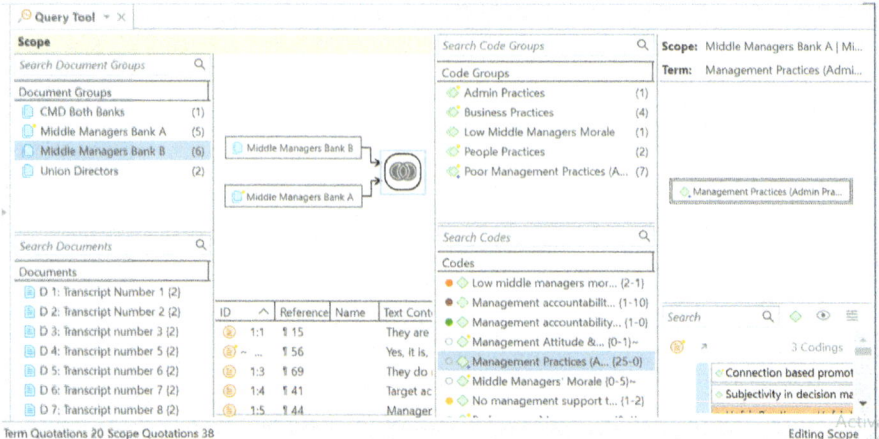

Fig. 8.16 Smart codes using edit scope

Smart codes are thematic. They contain large amounts of information relevant to the research objectives. The researcher can create multiple reports in various forms: networks, Excel, Text, etc. Existing smart codes can be used to form other smart codes, as well as for forming a network (Fig. 8.16).

Interpreting Smart Groups

Like smart codes, smart groups also store queries from which a report can be generated in any form—Text, network, and Excel. For example, one can create a smart group from code groups using operators. Figure 8.17 shows a typical smart group report about a smart group named 'Poor Management Practices (Admin practices + Business practices + People practices)'. Here, the OR operator was used to create the smart group. The report was for a smart group in the "Middle Managers Bank A" category. It contains all information for the selected smart group formed by using the Edit Scope option. Reports for smart groups will include all information (i.e., quotations and memos) from which the researcher can select and refine according to need.

A network of smart groups can be created using either the Code Manager or Code-Group Manager option. Once the network is formed, the user can use the 'Add Neighbor' option and other sub-options to import code groups, codes, memos, quotations, etc. With the View option, the user can see the comments and quotations. One can also use code-to-code relations and hyperlinks to make the network more meaningful and relevant.

Figure 8.18 shows smart groups, code groups, codes, code-to-code relations. The researcher can also generate a text or Excel report to interpret the emerging themes. This method allows for easier and more insightful interpretations of the results.

8 GENERATING THEMES FROM SMART CODES AND SMART ... 241

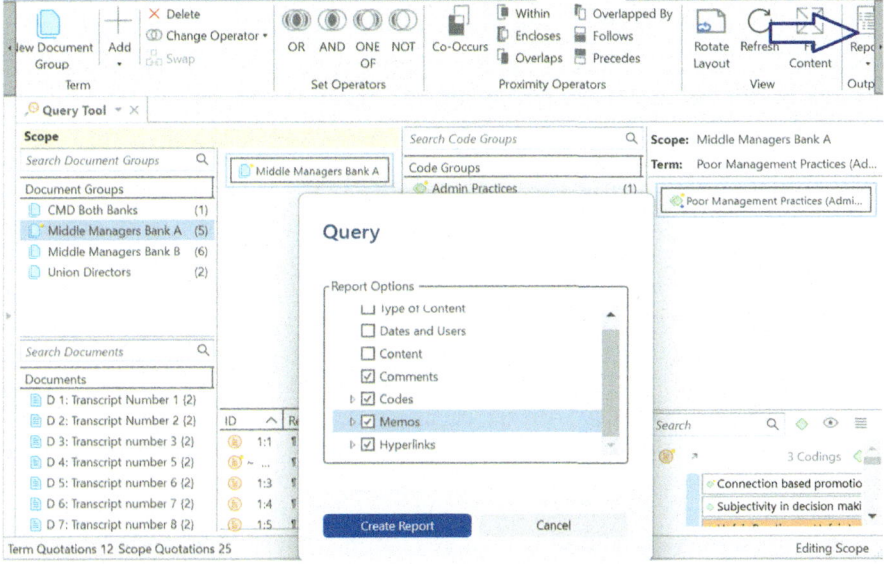

Fig. 8.17 Smart group using edit scope

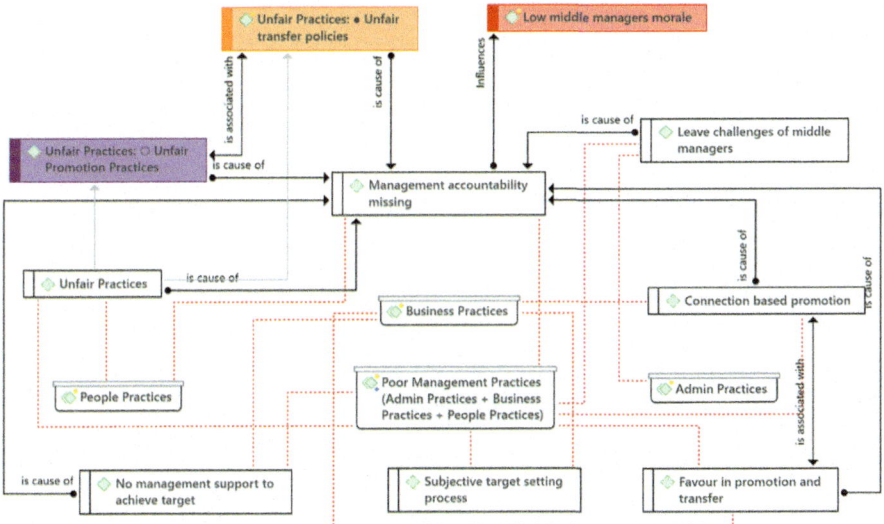

Fig. 8.18 A smart group and themes network

Thematic Visualization of Networks

Networks are a useful visual medium to present themes. They are helpful for making easier the explanation of results of analysis, and their interpretation. Networks contain all information relevant to, or linked with, the theme. They help the audience (the readers) to form images of concepts and their associations. Thematic networks can be created in ATLAS.ti using smart codes or smart group, codes or code group, depending on the size of the data. Figure 8.18 shows a theme identified with the help of a smart group. The theme is related to the causes of low middle managers' morale. All data and information supporting the concept are shown in the network.

Codes, code groups, smart groups, memos, quotations, and comments are all parts of a network. The network view allows researchers to see the nodes and interpret them suitably with reference to the research questions and objectives. In Fig. 8.18, the theme is a smart group formed with three code groups, each consisting of several codes. Each code is represented by a quotation (a key segment of response), and each quotation has a linked memo that contains the researcher's observations during data collection, which could not be captured in the text form.

In the example shown in Figs. 8.18 and 8.19, a smart group was created using OR operators that retrieved all the information associated with code groups. The smart group represents one theme: 'the sources of middle managers' low morale'. Using the "Add the Neighbors" option with a right-click allows the user to import all the information required to refine the information network. A right click on any code in the network will import information associated with the code (Fig. 8.20). By using the 'Appropriate' option, ATLAS.ti will import codes, memos, quotations, documents, and groups associated with the code.

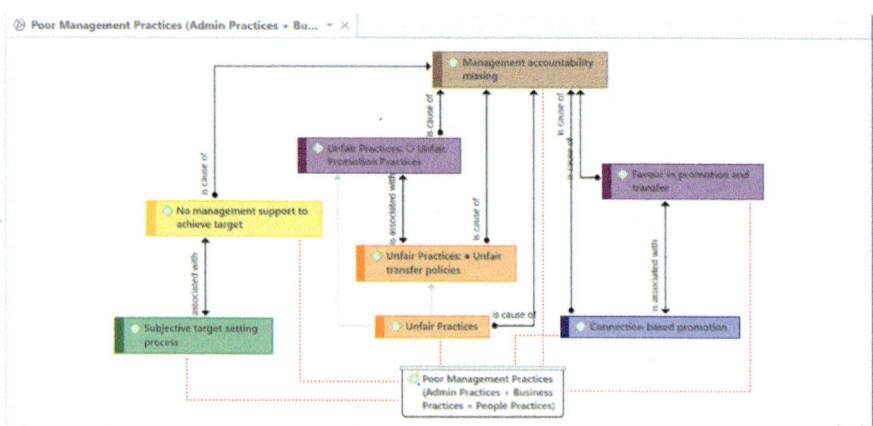

Fig. 8.19 Thematic visualization of a network

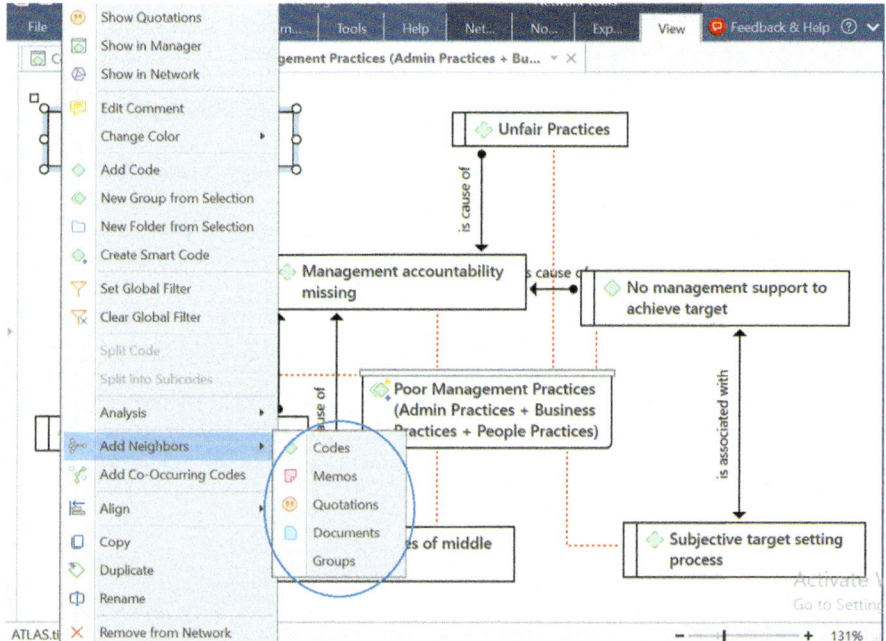

Fig. 8.20 Add neighbors option

Interpreting themes become easier once the concepts of smart codes and smart groups are understood. Therefore, it is imperative for the user to become familiar with the meanings, significance, and functions of quotations, memos, and codes. Quotations are the evidence supporting the researcher's claim about the theme. Memos are the researchers' observations and insights gathered from the field during data collection. Codes are representations of the meaningful information identified in the quotations. The linkages among quotations and codes, smart groups and code groups are necessary for supporting the researcher's claims; they are important for making the networks self-explanatory. Additionally, they are valuable for generating text reports.

Recommended Readings

Atkinson, P., Coffey, A., & Delamont, S. (2003). *Key themes in qualitative research: Continuities and changes*. Rowman Altamira.

Bazeley, P. (2009). Analysing qualitative data: More than 'identifying themes.' *Malaysian Journal of Qualitative Research, 2*(2), 6–22.

Contreras, R. B. (2011, August). Examining the context in qualitative analysis: The role of the co-occurrence tool in ATLAS.ti. In *ATLAS.ti Newsletter*.

DeSantis, L., & Ugarriza, D. N. (2000). The concept of theme as used in qualitative nursing research. *Western Journal of Nursing Research, 22*(3), 351–372.

Friese, S. (2019). *Qualitative data analysis with ATLAS.ti*. Sage.

Lewis, J. (2016). *Using ATLAS.ti to facilitate data analysis for a systematic review of leadership competencies in the completion of a doctoral dissertation*. Available at SSRN 2850726.

Ryan, G. W., & Bernard, H. R. (2000). *Techniques to identify themes in qualitative data*.

Saldaña, J. (2014). Coding and analysis strategies. In *The Oxford handbook of qualitative research* (pp. 581–605).

Schmidt, M. (2008). The Sankey diagram in energy and material flow management: Part II: Methodology and current applications. *Journal of Industrial Ecology, 12*(2), 173–185.

Smit, B. (2021). Introduction to ATLAS.ti for mixed analysis. In *The Routledge reviewer's guide to mixed methods analysis* (pp. 331–342). Routledge.

Williams, M., & Moser, T. (2019). The art of coding and thematic exploration in qualitative research. *International Management Review, 15*(1), 45–55.

Wright, C., Ritter, L. J., & Wisse Gonzales, C. (2022). Cultivating a collaborative culture for ensuring sustainable development goals in higher education: An integrative case study. *Sustainability, 14*(3), 1273.

CHAPTER 9

Group Creation for Entities and Report Generation

Abstract This chapter is about organizing data. The data collected by the researcher must be categorized or classified according to the various parameters of the study (i.e., professional, demographic, socio-economic, contextual, and cultural) with reference to the research questions. In ATLAS.ti data classification begins with grouping documents, codes, memos, and networks from which the researcher can generate various reports. With the help of examples, the author shows how to form code groups and then arrange them in a network, as well as generate reports according to need. The purpose and usefulness of memo groups are also discussed in this chapter. Following the steps similar to forming code groups, groups of memos and networks can also be created. Network groups help the researcher, as well as the audience, visualize concepts and themes and their interrelationships in network form.

DOCUMENT GROUPS

Typically, close-ended questions in a survey are aimed at gathering personal information about the respondents (age, family size, marital status, occupation/profession, income, etc.). Usually, the questions are provided with response options for the respondent to choose from. The researcher then groups the responses according to certain shared characteristics.

In ATLAS.ti, document groups reflect demographic profiles of the respondents in a way that enables the generation of reports. These document groups are clusters of documents that help the researcher to filter data. The grouping of documents is the process of combining certain data attributes for retrieval and analysis, depending on the objectives of the study. As explained earlier (see Chap. 2), the researcher must work with several types of documents: interviews, field notes, articles of relevance to the project, audio or video recorded interviews, pictures, images, photographs, etc. Documents groups can be formed for each document category.

Classification of documents is a crucial step in analysis. It helps researchers to narrow their research findings to a specific theme or concept. By classifying documents, researchers can compare findings across variables: cases, respondents, region, gender, and others. Without classification, it will be difficult to arrive at meaningful insights. Hence, researchers should aim to create as many groups as possible, which are based on various attributes, characteristics, contexts and situations. Figure 9.1 shows how documents are classified. In this example, the groups are based on the professional and demographic characteristics of the respondents in the study.

ATLAS.ti enables the user to combine documents using various operators quickly and efficiently for retrieving information (see Chap. 7 on operators). The basic aim of this chapter is to familiarize the researcher (or user) with the functioning of ATLAS.ti, beginning with the creation of a research project. The project example in the discussion is about a study of middle managers' morale in public sector banks. The study was conducted by the author who collected data from management, middle managers, union directors, and retired CMDs. Four document groups have been created based on respondents' profile. Document groups help to refine the research outcome in the specific group or in combination of many document groups.

Depending on need, reference will be made to other projects as well.

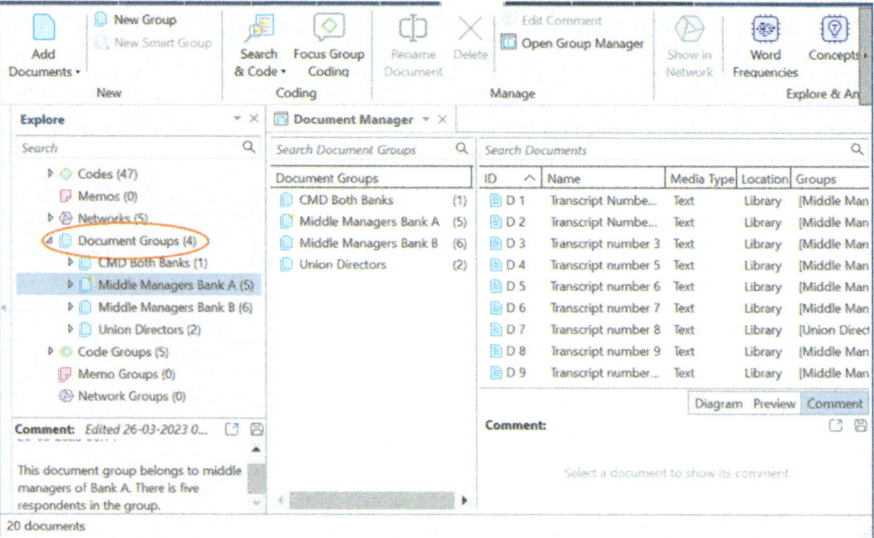

Fig. 9.1 Creating document groups

9 GROUP CREATION FOR ENTITIES AND REPORT GENERATION   247

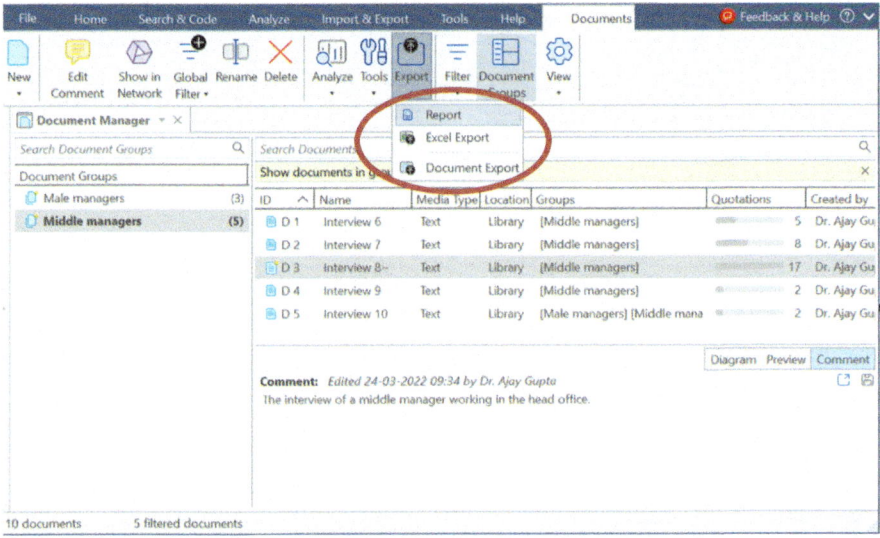

Fig. 9.2 Comment for document group

The display in Fig. 9.1 shows documents, codes, memos, networks, document groups, code groups, memo groups on the left of the screenshot. The figures in brackets are the numbers of entities (explained in Chap. 5) in each group. There are 20 documents and 4 document groups in the project. As seen in the figure, four document groups were created, the grouping based on the research objectives. The comments pane in Fig. 9.2 describes the meaning of a document. The highlighted groups refer to the middle managers, whose interview data were combined in one document group.

Grouping of documents helps the researcher generate reports for specific document groups, useful for analysing the pattern of information emerging from the particular group (i.e., middle managers). In this project, a document group is a category of respondents, which was formed based on certain characteristics defined by the researcher.

Other document groups are also listed. For each, a separate report can be generated by selecting Report.

Steps for Creating a Document Group

1. Open the research project in ATLAS.ti (Fig. 9.2).
2. Open Documents in the main menu using the down arrow.
3. Select the documents for grouping by using the control key and cursor, and open New Group (top left corner).
4. Write the name of the document group and save.

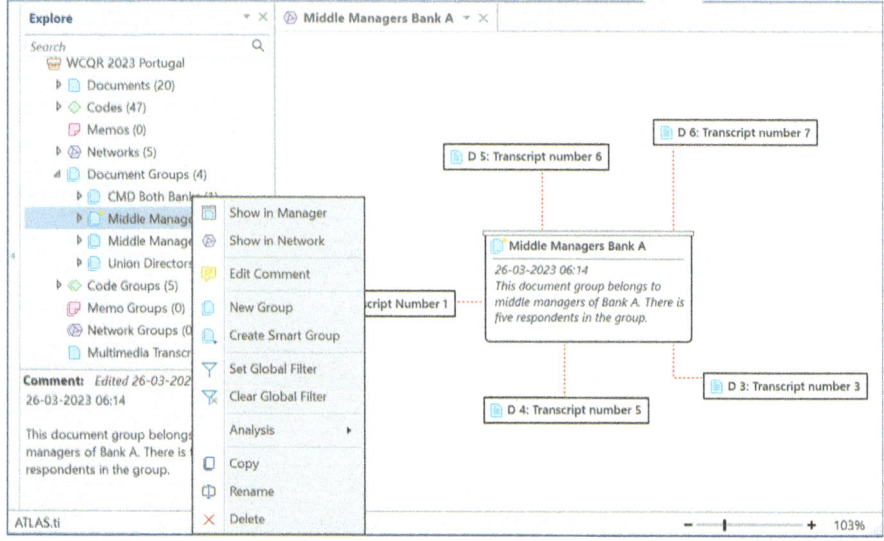

Fig. 9.3 Document group and network

Alternatively, double click on Documents in the main menu.

1. Open New Group (top left corner) and write the name of document group (shown on the left side in Document Manager).
2. Select the document for inclusion in the group. Drag and drop the document into the newly formed document group.
3. Click on Document Groups in the left margin or Document Manager (within the red shapes) on the menu bar. A window will open, showing all the document groups.
4. Select and click a document group, write a short suitable description of the document group in the comment section. Repeat for other document groups.
5. Save and exit.

Creating a Network of Document Groups

In Fig. 9.3, the five documents in the Middle Managers Bank A groups are shown in a network (see Chap. 6 for a discussion on networks).

Steps for Creating a Network

1. Select the document group to show in network form.
2. Highlight the document group in Document Manager and right click. A Context menu will appear.

3. Click on Show in Network (at the top). The network is ready.
4. Use Layout and Routing options to modify or redesign the network.
5. To add a description, right-click each document. Select Edit Comment, write description, save and exit.

Reports

Reports can be generated for documents and document groups in both text and Excel forms. Figure 9.4 shows a listing of documents in Document Manager, which was obtained using the report option. Report option in ATLAS.ti 23.2.1.26990 has been shown in Fig. 9.2.

1. Open the project and click on the Home tab.
2. Double click on Documents in the left margin or in the main menu (top). Document Groups and Documents will be shown in the window.
3. Select one or more documents for which a report is required.
4. Click on Report or Excel Report.
5. The windows for text and Excel reports are shown in Fig. 9.4.
6. Select an option.
7. Generate report or export it.
8. Save.

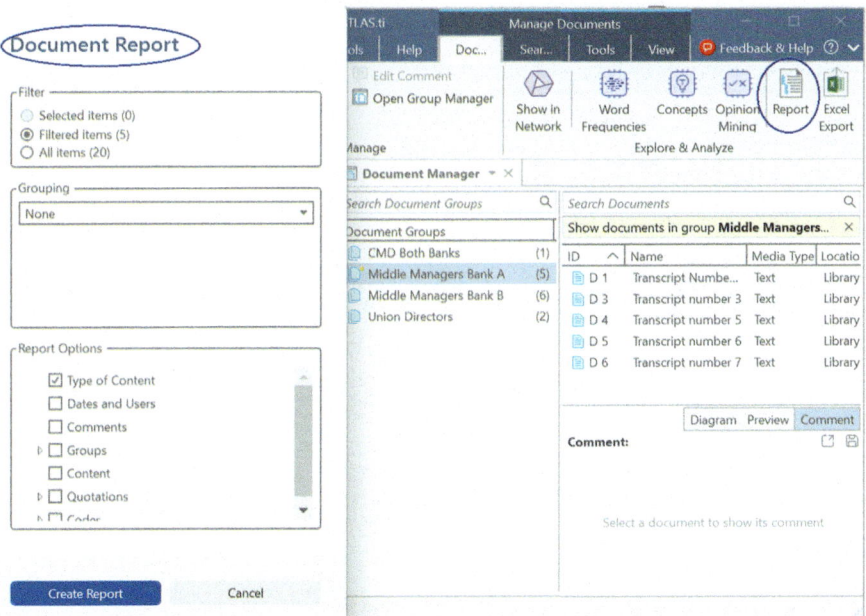

Fig. 9.4 Report for a single document

Single or Group Document

1. In Document Manager, select one or more documents.
2. Select the components of the report from the options provided in the report menu.
3. Generate report in text or Excel (Fig. 9.4).

Reports for Document Groups

1. In Document Manager, select the document group or groups.
2. Select the components of the report from the options provided in the report menu.
3. Generate textual or Excel as required (Fig. 9.4).

Reports for Quotations

The steps for generating reports for quotations are like those for reports for documents:

1. Open the project in the Home tab (Fig. 9.5).
2. Click on Quotations in the main menu. The Quotation Manager will open. All quotations will be shown on the right-hand side of the display. Quotations with comments will be shown with yellow dots and ~ symbol.

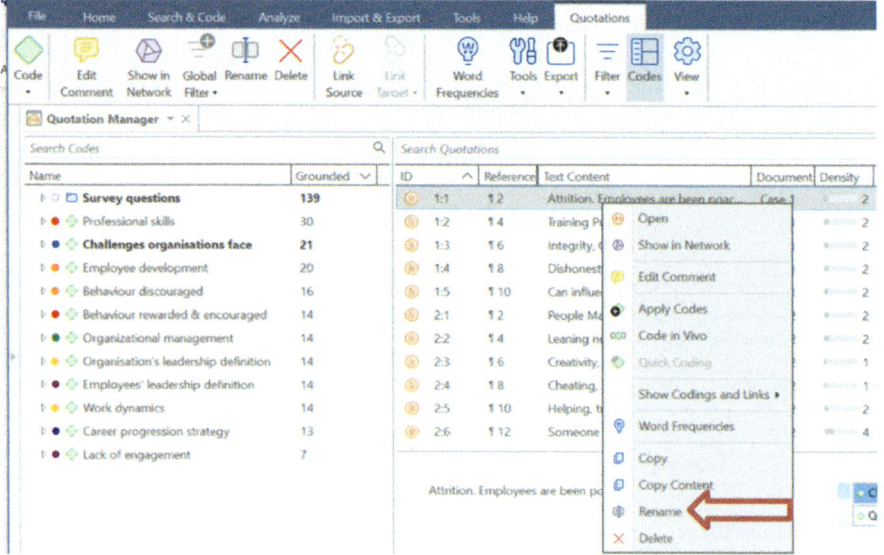

Fig. 9.5 Report for quotations and rename

3. Select one or more quotations, as needed, for the report.
4. Select the report format (text or Excel) or Export option in ATLAS.ti 23 (Fig. 9.5).

Quotations linked to codes appear on the right side. By selecting code, a report for associated quotations can be generated. All quotations linked to the code will be displayed (Fig. 9.5). The users can rename the quotations based on certain characteristics by selecting quotations and using the "Rename" option.

Depending on the need, a report can include selected, filtered, or all items. For a filtered report, a global filter must be applied. Then, the information in the report will be specific to the selected entity only. If a code is set as global filter, only information linked to that code will be shown. When no longer needed, the global filter can be removed using "Clear Global Filter" option.

CODE GROUPS

Code groups are formed after data coding. But first, one must understand the purpose and importance of code groups.

For discussion, we will assume that over 100 codes were generated for a research project. Logically, these codes must represent 100 types of meaningful information. But this is not possible or practical. Closer scrutiny of the codes will reveal that several codes reflect similar meanings, that is, the codes are homogenous. Researchers should merge them. After the process, the number of codes will get reduced. Further, several codes may present a similar concept, sentiment, or attribute that researcher may group them. Such codes can be brought together in groups known as code groups. Code groups are formed by codes sharing certain attributes. The groups thus formed are named such that they reflect the meanings of the codes associated with them.

It is also possible that some codes may have distinct characteristics and cannot be placed in a group. These codes are called outliers. Outliers are not necessarily insignificant; in fact, they may be important, especially in grounded-theory research. Outlier codes help to explore deviant cases that help for further exploration of the phenomenon.

Code groups can be created in ATLAS.ti in three ways. But first, the researcher must browse the codes and look for commonalities (or homogeneity) among the codes. After identifying the patterns, the codes can be grouped. The researcher must conceptualize the characteristics of each group and name them suitably. For example, the codes for the responses to the question about employee morale, such as those indicating that they are happy, encouraged, motivated, and committed, reflect positive feelings. Therefore, "Positive feeling" is an appropriate name for this code group. After deciding a name for the group, all the codes reflecting this characteristic are exported to

Table 9.1 Sub-codes, code categories and code groups

Code groups	Code categories	Codes (sub-codes)
Employee morale	Employee high morale	Employee high morale: satisfied
		Employees high morale: motivated
		Employees high morale: committed
	Employee low morale	Employees low morale: dissatisfied
		Employees low morale: frustrated

the code group. This process—identifying codes reflecting similar characteristics for grouping, giving the group a suitable name, and then exporting these codes—is followed for other codes.

Code Groups and Code Categories

Employee morale is a code group that includes two code categories, 'Employee High Morale' and 'Employee Low Morale' (Table 9.1). These code categories can contain many sub-codes. Code categories are created within a code group. By using code group, all information related to code categories and codes will be collected. Thus, a report for code group will contain code categories, codes and quotations associated with codes.

Figure 9.6 shows a network of the code group 'Employee Morale', which consists of two code categories: Employee High Morale and Employees Low Morale. Each category includes sub-codes that share the attributes of that code category. The sub-codes are associated with each other. The two code categories reflect opposite properties that have been shown by using code to code relationships. Networks like the one shown here are important for identifying emerging concepts/themes and their associations with other concepts.

Steps for Forming Code Groups

1. Open the Home tab.
2. Click on the codes shown on the left margin. Alternatively, click on the Code Manager icon on the top pane. The Code manager window will show all the codes.
3. Browse the codes and identify codes sharing certain attributes/similar meanings.
4. Select these codes by holding down the shift or ctrl-key.
5. Right-click. A small flag will appear. Alternatively, one may click on any code in the pane on the right side. A New Group will be shown on the top margin-left hand side.
6. Open New Group and give it a name.
7. All the selected codes will be included in this code group.

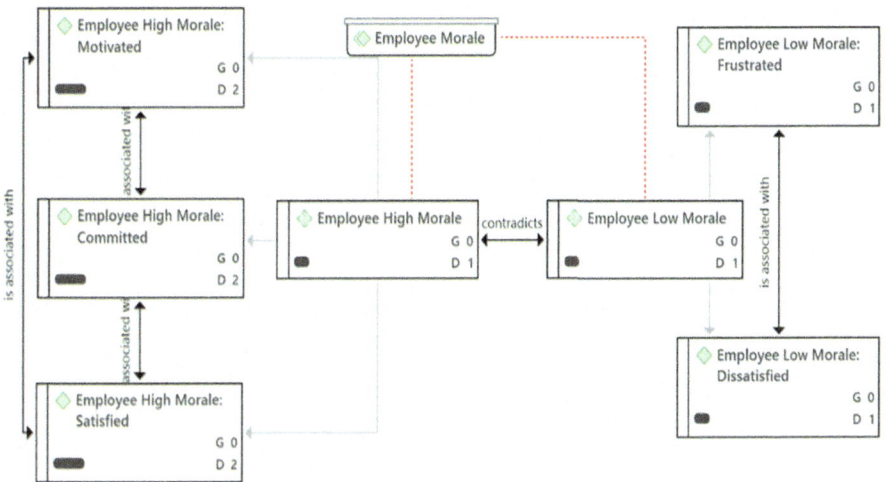

Fig. 9.6 A network of codes, categories and code group

Another method for creating a code group in ATLAS.ti:

1. Open a down arrow in code on the top pane. The Down arrow will appear in the Code Menu on the top pane. On clicking the arrow, Code Groups will appear.
2. Open Code Groups.
3. A new group will be shown on the top left, below File.
4. Name all groups.
5. Write code descriptions in the comment section and save.
6. The code groups will be shown on the upper section, and the codes on the right below (Fig. 9.8).
7. Click once on the Code Group and locate codes that need to be exported to the group.
8. Double click on each selected code. Alternatively, press ctrl or shift key. The codes will be shown below each code group.

Code groups can be formed using Code Manager or Code Group Manager (Fig. 9.7). When using the Code Manager, the researcher can select codes from the list and right click after which flag will appear with a Create Code Group option. On selecting it, a code group is created after which the researcher gives it a suitable name.

In the same Code Manager window, the researcher can select codes for inclusion in a code group, and open New Group (top left above Create Smart Group). In the Code Group Manager (Fig. 9.8), the researcher can select the codes shown on the right side of the lower pane and then selecting New Group. The code group will appear with the number of codes in brackets.

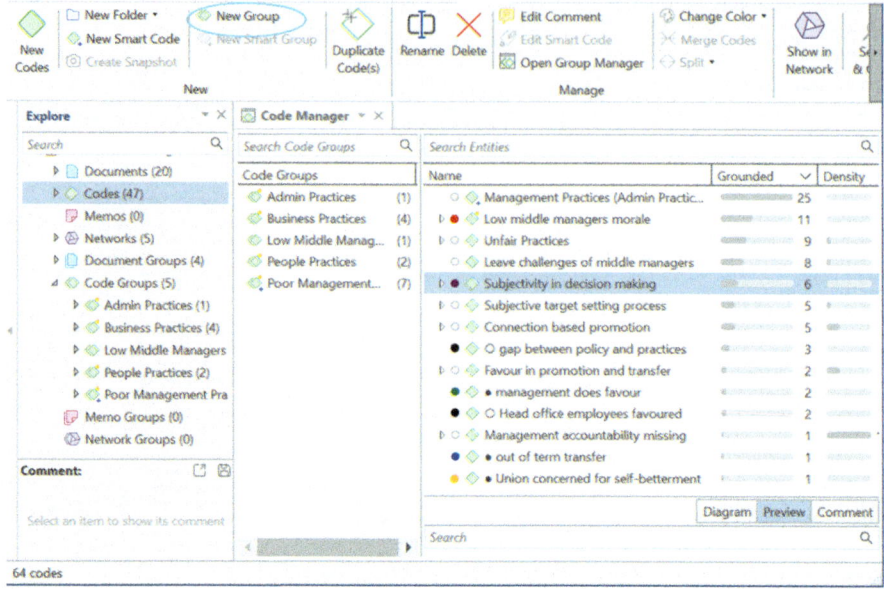

Fig. 9.7 Code group manager in ATLAS.ti

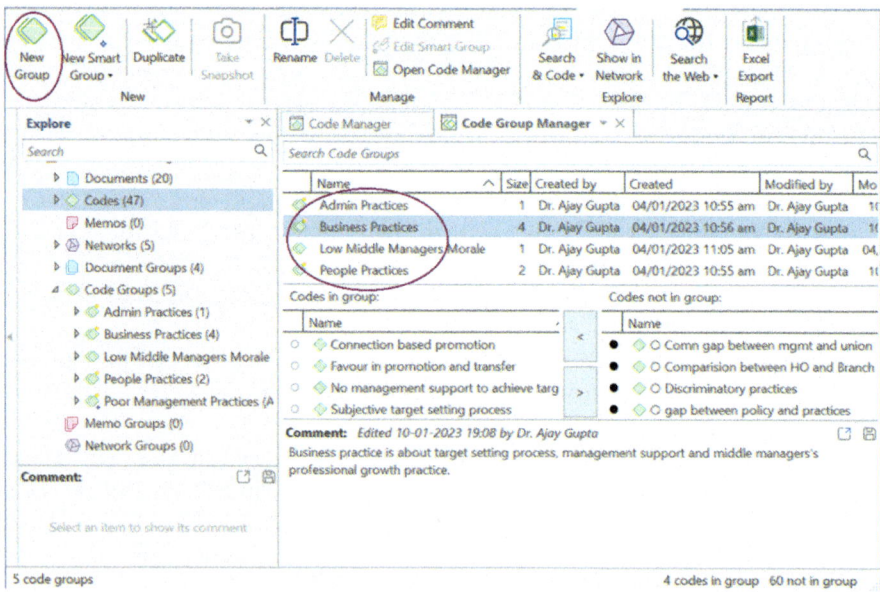

Fig. 9.8 Forming code groups

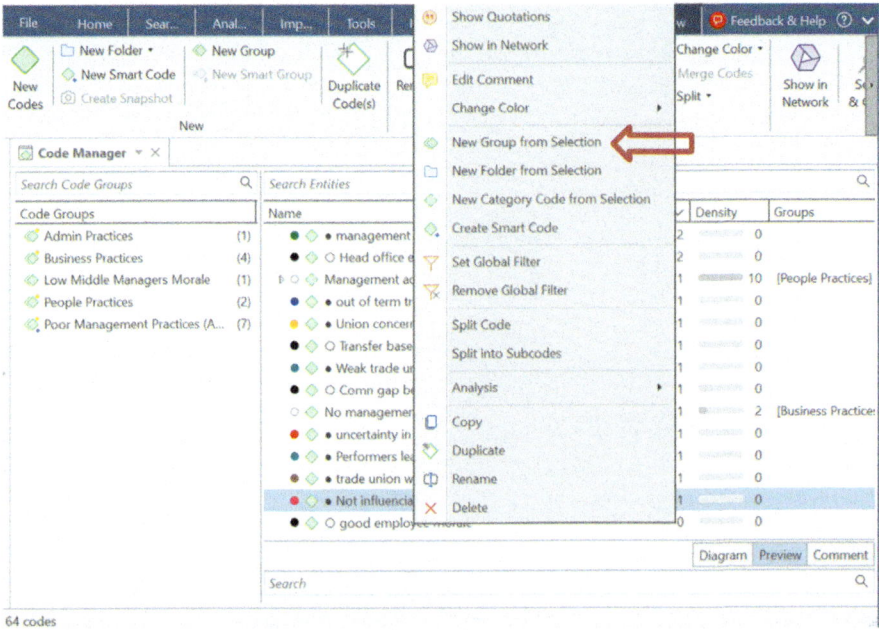

Fig. 9.9 Creating code groups

Reports can be generated in text or Excel form. Networks of code groups can also be formed.

A code group can also be formed by selecting codes in the Code Manager and then, with a right click, selecting 'New Group from Selection' in the context menu (Fig. 9.9).

After creating a code group, the user can show the links and relationships among codes with the help of a network. A graphical representation like a network helps to generate audience interest. It should give richer information about the meanings and rationale of the code groups. The steps to follow for creating a network are given below.

Creating a Network of Code Groups

1. Select any code group.
2. With a right-click, a Context menu will appear with many commands will display.
3. Select "Open Network/Show in Network" and click. The network is formed.
4. Use layout and routing to change network layout and appearance.
5. To know the definitions of codes and code groups, go to View (top pane).

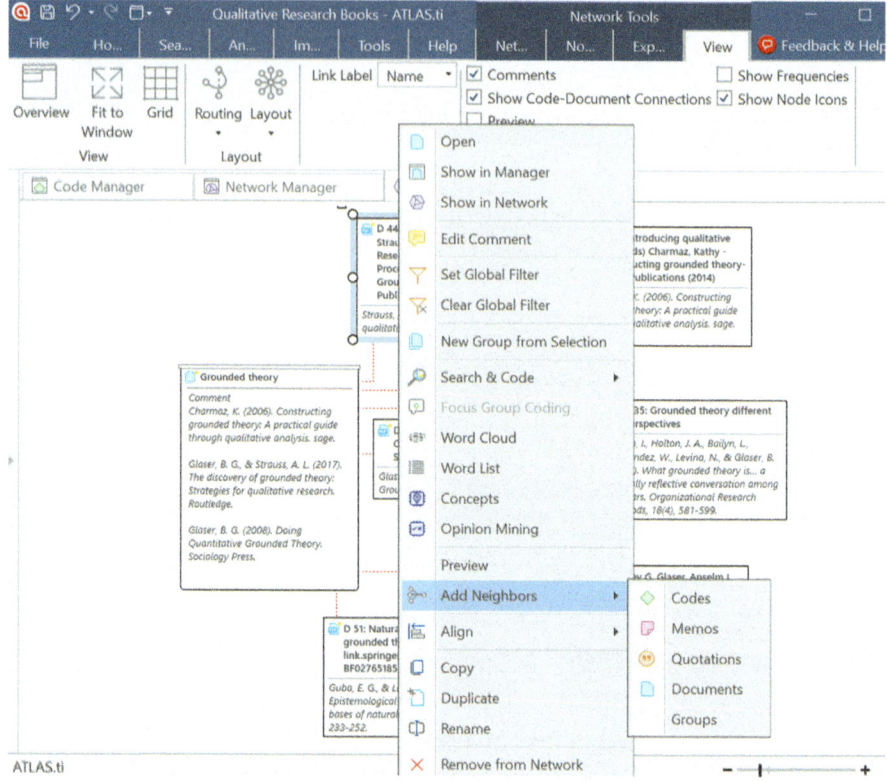

Fig. 9.10 Importing function

6. On checking (✓) comments, the definitions and descriptions of the codes and code groups will display (Fig. 9.10).

Making an Informative Network

An informative network is the one that has information to explain the concept. It includes hyperlinks, code to code relations, quotations, themes, sub-themes, research questions and memos.

1. Open the network and right click on a code to display the Context menu.
2. Select "Change Color" option (Fig. 9.9).
3. Assign a different color to each code.
4. To show the relationships among codes (Fig. 9.11).

   I. Click on any code in the network. A red dot will appear.

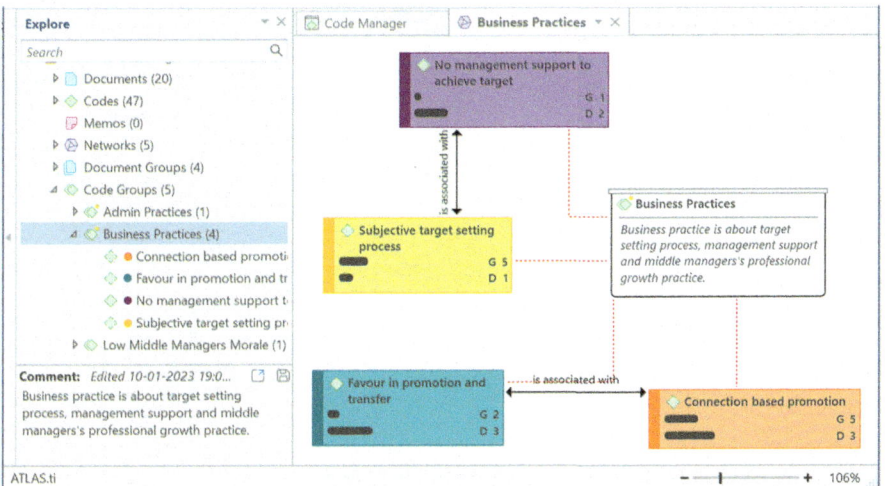

Fig. 9.11 A network of code groups

II. Drag and connect the dot with a code to link it with the first. After the connection between the codes is made, the relationship symbols will display.
III. Select the appropriate symbol and click. One can also create new relationships (see Chap. 6) between codes and quotations using code to code relations and hyperlinks.

5. Go to View to check the frequencies of codes. The letters G and D are displayed for each code. G is the number of quotations associated with the code, and D is density representing the degree of a code (number of associated codes).
6. The user can extract all quotations, memos, and codes connected with a code by using the import function/Add Neighbors option (Figs. 9.10 and 9.11). A right click on the entity will open the Context menu. Select the Add Neighbors option and import the required information for the network.

Reports on Codes

1. Double click on Codes, either in the left margin or the main menu (top, Fig. 9.12)
2. The Code Manager will be shown with the code groups on the left and a list of codes on the right.

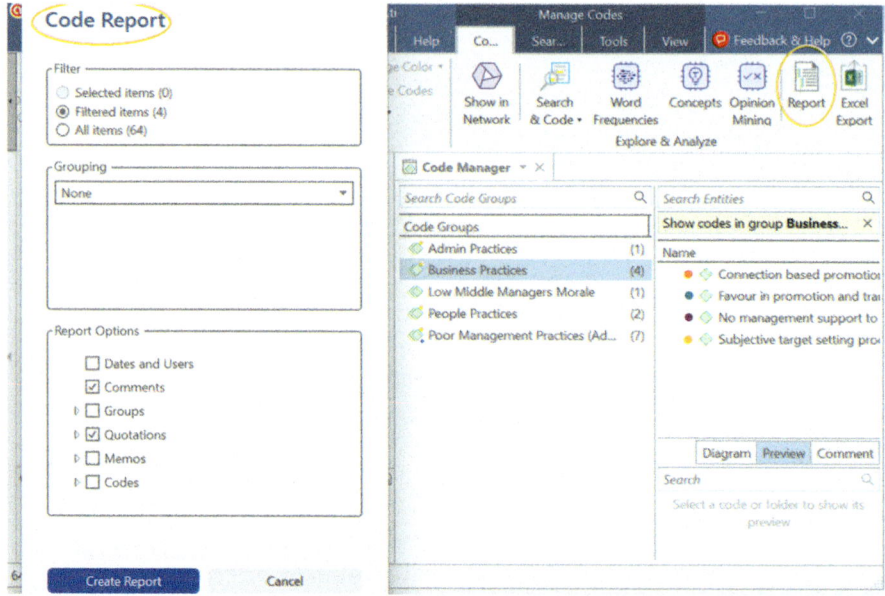

Fig. 9.12 A report for a code group report

3. Click on code group or one or more codes for which a report is needed (text or Excel). In the example in the figure, "Business Practices" is the code group. It consists of four codes (the number 4 is shown in brackets.
4. Select Report (top right corner)/Export option (Fig. 9.2 in ATLAS.ti 23).
5. Select inclusion options for the report by checking the appropriate boxes (Fig. 9.12).

Memo Groups

A memo group comprises several memos. In Fig. 9.13, the Memo groups in a project are displayed on the left. The memo icon is shown by the orange arrow. The number within the brackets show the number of memos in a memo group.

Memo groups consist of memos having certain shared characteristics and attributes (Fig. 9.13). The memo group in the example shown in the Figure, 'Thematic and Methodological', has four memos having information on theme and methodology. Comments on the memo group are shown in the lower margin.

9 GROUP CREATION FOR ENTITIES AND REPORT GENERATION 259

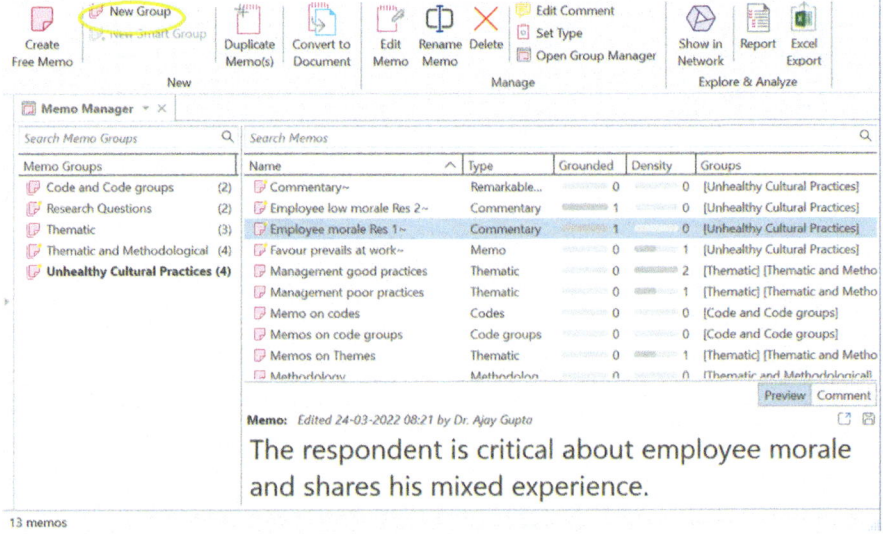

Fig. 9.13 Creating a memo group in memo manager

Creating a Memo Group

1. Open the Home tab of the project. There are three ways of creating a memo.

    (i) The first method.
    - Double click on memos on the left pane or on the memo icon on the Menu (top).
    - Click the down arrow in the memos (top pane)
    - Click on Memo Groups. Select the memos for grouping using the Ctrl or shift key. Right-click.
    - A flag will display. Select New Group. Name the group.

    (ii) The second method
    - Select the memos for grouping.
    - Click on "New Group" (top pane). Name the group.

    (iii) The third method
    - Click on Memo Groups. A "New group" option will appear on the extreme left on the top below the file (Fig. 9.13).
    - Create a new group and give it a suitable name.

2. The Memo groups will be displayed on the upper pane, while memos will appear on the bottom right side.
3. Click once on the selected memo group, double click on the memos for exporting them to the group. The memo group is ready.
4. The number of the memos in the memo group will appear as shown in Figs. 9.13 and 9.14).
5. More memo groups can be created following these steps.

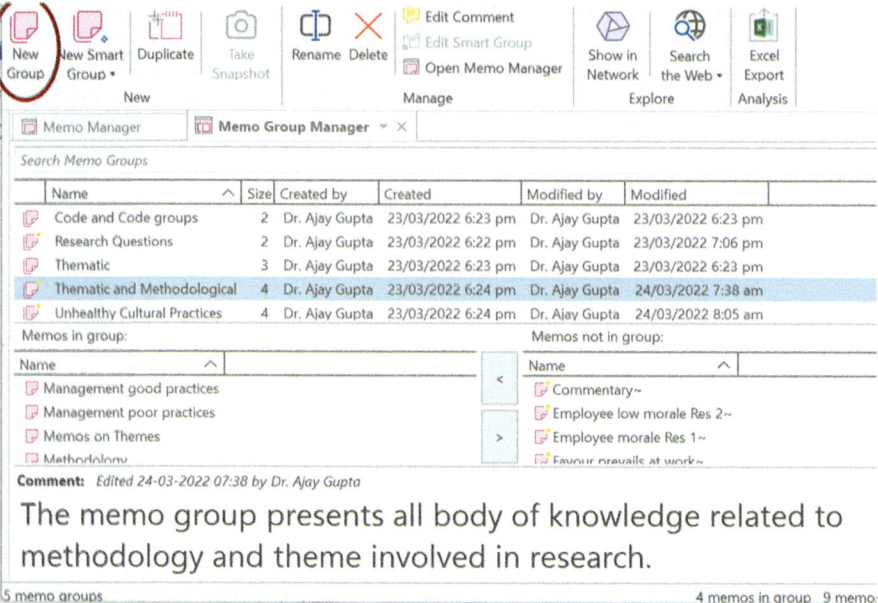

Fig. 9.14 Creating memo group in memo group manager

A memo group can also be created by selecting memo in Memo Manager, doing a right-click and selecting 'New Group from Selection' from the context menu (Fig. 9.15).

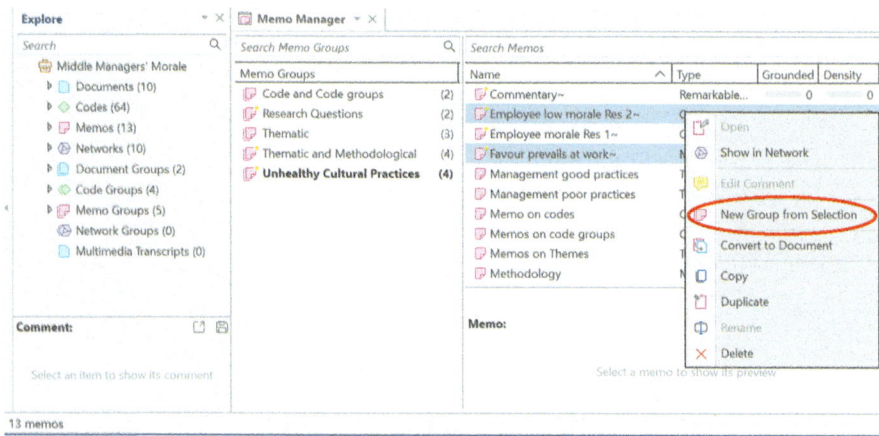

Fig. 9.15 Creating memo group in memo manager

Creating a Network of Memo Groups

Visual representations are useful for raising audience interest. After creating a memo group, the user can show the linkages and relationships among the groups with a network. The network should reflect the rich information present in the memos in the group. In a network, the user can show the definition of each memo and the memo group. The network will help the researcher show how the research questions of the study are linked with the findings, as well as how the findings are supported by the quotations. One can also create a network of all memo groups to fit in one window, and then link them to present information about the objectives of the study. The user can open a network and drag memos into the network and then link them with the research questions/objectives.

Creating a Network

1. Select the memo group that must be shown in a network. Right click.
2. A flag will be displayed showing various options.
3. Click on "Open Network/Show in Network (ATLAS.ti 23)".
4. The selected memo group will be shown in a network (Fig. 9.15).
5. Use layout and routing options to change the shape and appearance of the network.
6. Click on "View" (top pane) to display the definitions of the memos and memo groups.
7. For seeing the definitions and description of memos and memo groups, check (✓) comments.

Reports from Memos

Using the Memo Manager, one can create reports for one memo or more memos, a memo group or many memo groups. The process followed for generating reports is similar the one for codes and documents.

SINGLE MEMO

1. Open Memo Manager
2. In Memo Manager, select the memo or memos for which you a report is required.
3. Select whether the report should be in the text or Excel format.
4. Select the information for inclusion in the report by checking the appropriate boxes.

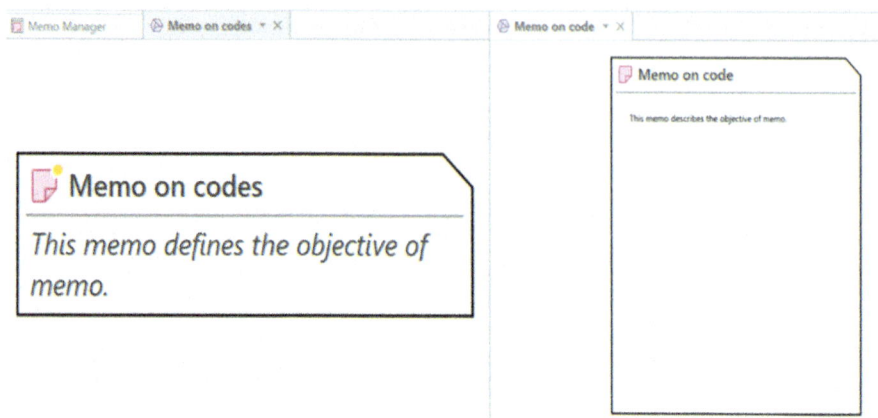

Fig. 9.16 A memo writing option

> **Important** Like code-to-code relationships, the memo has only one: Adding. One can use the drag and drop option to link one memo with another memo. If the memo type has not been set, the "Set Type" option can be used. If a blank memo has been created, one should use Edit Comments for adding thoughts, observations, and experience.

It is better to use the Edit Comments option to describe the memo than with 'New Entities' in the 'Home' menu. The reason is that when a new network is created, the comment section appears clearer and by selecting comments under 'View' section. The researcher must select 'Preview' under the 'View' option when a memo has been created using 'New Entities' under 'Home' menu. The memo description will not appear as clearly. Figure 9.16 shows the two memos, the memo on the left was written using the 'Edit Comment' option and the one on the right with the 'New Entities' option.

REPORTS FOR MEMO GROUPS

1. Open Memo Manager (left margin or main menu).
2. Select the memo group or memo groups for which a report is needed.
3. Select the type of information for inclusion in the report by checking the appropriate box in 'Memo Report' (Fig. 9.17).
4. Generate report in text or Excel format.

One can also create a network of memos and add neighbors, as it is done with other entities. Memos must be linked to other entities. The researcher should be aware that Neighbors cannot be added to free memos.

9 GROUP CREATION FOR ENTITIES AND REPORT GENERATION 263

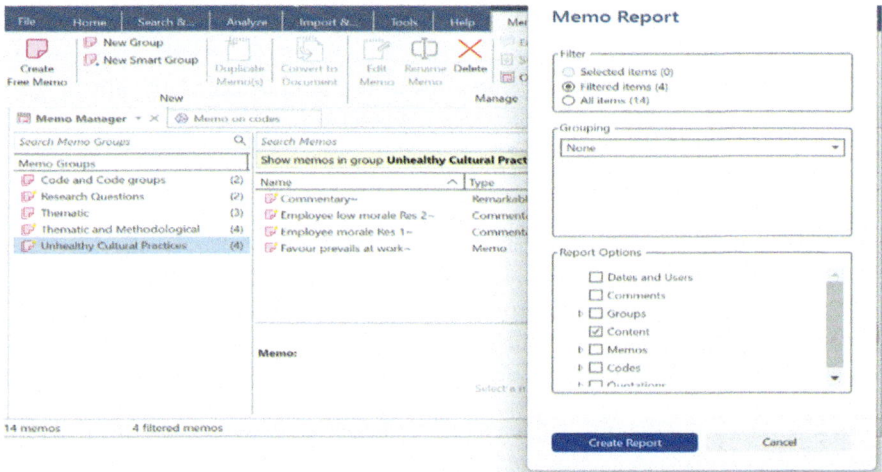

Fig. 9.17 A report for memo group

In Fig. 9.18, the 0 in the grounded column means that the memo is not supported by a quotation. Such memos are called free memos. In the same figure, there are four memos for the memo group "Unhealthy Cultural Practices." The memos are shown in the network along with their groundedness.

In the memo, "Employee low morale," the number 1 in the Grounded column means that the memo has 1 quotation. The Density value means the number of codes linked with the memo.

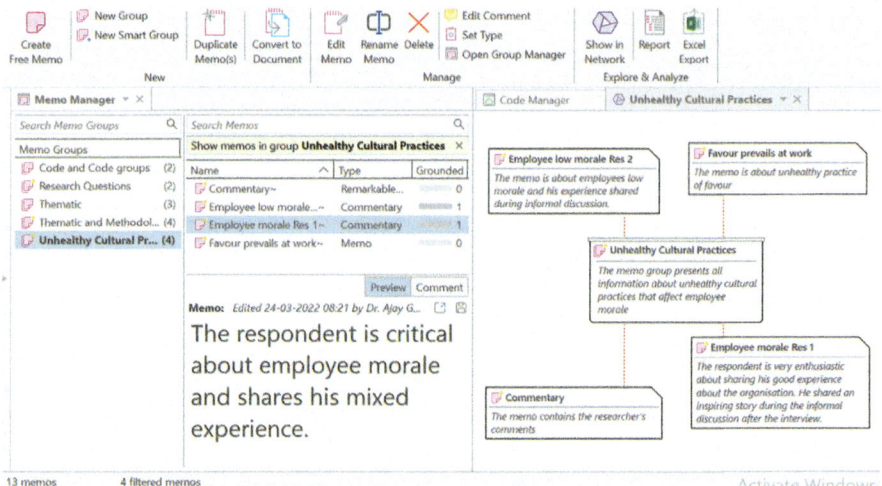

Fig. 9.18 Memo group and interpretation of network

A network was created for the memo group. For creating a network of memos, one must select Group and create a network using right click option. Next, associated entities can be imported using Add Neighbors. This applies to any entity: document, code, memo, and network.

Network Groups

Networks can be viewed with their contents and connections using the "Networks" icon (shown by the arrow) on the menu bar at top (Fig. 9.19). One can import the associated concepts, codes, memos, quotations, etc., into the network. Depending on the requirements of the study, the researcher (or user) can also create network groups. Alternatively, if the researcher wants to show research findings in a network view, multiple networks can be shown as part of one network group. Figure 9.18 shows how to create, open, rename, and comment on networks and network groups.

Creating a Network Group

There are two methods of creating network groups.

First method

1. Open Network Manager (Fig. 9.20)
2. Open a new group and give it a name. One can create as many networks as required.

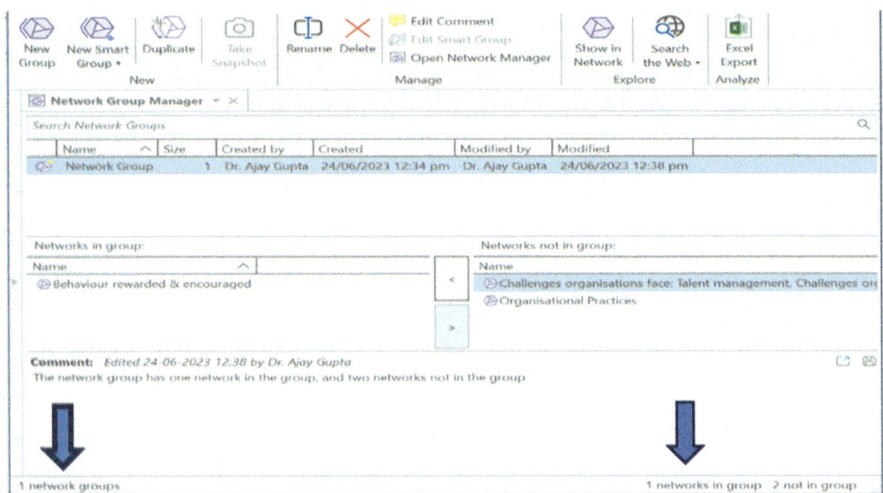

Fig. 9.19 Network group and comment

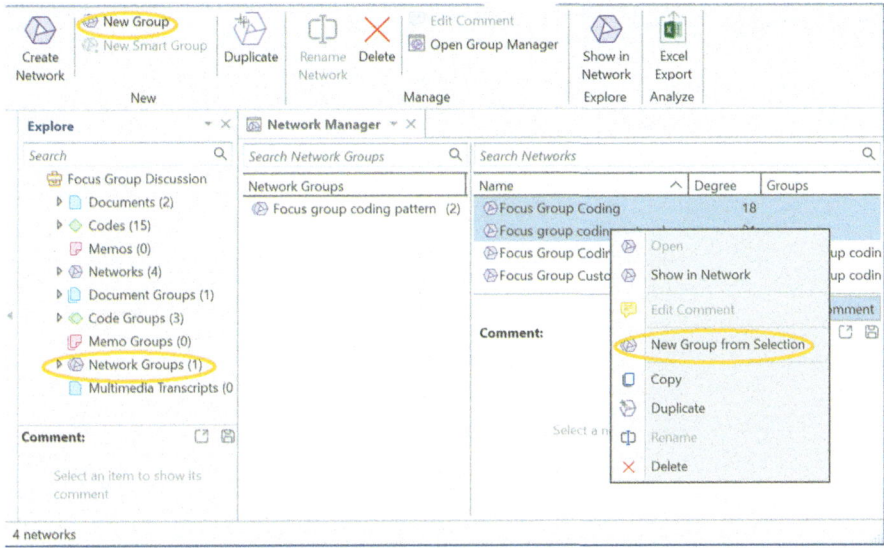

Fig. 9.20 Creating a network group

3. The network groups formed will be shown on the upper windowpane, while the networks are displayed at bottom right.
4. In Network Group Manager, click once to select a network group, browse the bottom pane on the right side, which is named 'Networks Not in Group'. From this list select a network or networks for adding to the new group (shown by arrow in Fig. 9.19)
5. Double click on the selected network.
6. The number of networks in the network group can be seen.

Second method

1. Select from the networks shown in the bottom right pane by holding down the Ctrl or Shift key.
2. Right click to create a new group. Alternatively, click on New Group option (top left corner).
3. Give a name to the network group.
4. The number of members of the network group can be seen.

Networks of Network Groups

A network of network group is useful for finding connections and relationships among various entities. Figure 9.20 shows a network group of the networks of Focus group custom coding.

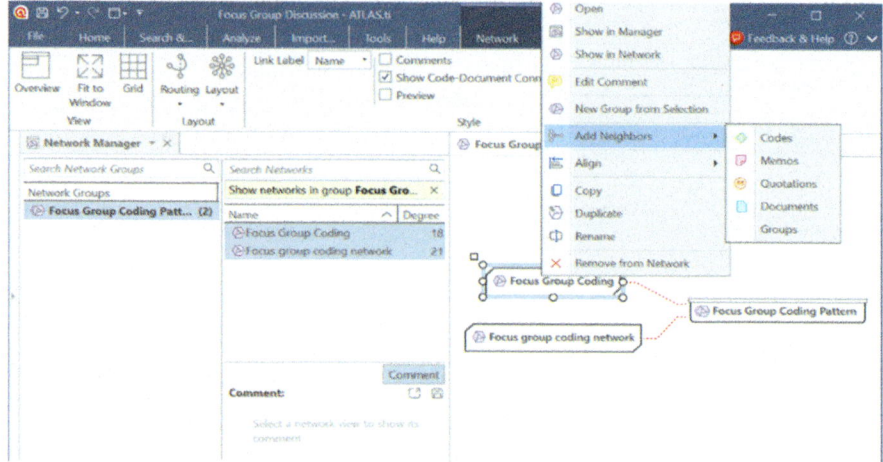

Fig. 9.21 Adding neighbors to a network

Showing a network group as a network helps the researcher to visualize the association of two (or more) networks in the network group. By using the Import option (Fig. 9.10)/Add Neighbors (ATLAS.ti 23), the researcher can link the information present in the network (i.e., codes, code groups, memos, memo groups, and so on). Figure 9.21 shows a network group consisting of networks. The user can link memos, quotations, relationships among quotations, descriptions of the networks and the network group. With a right click on a linked entity, one can import associated codes, memos, quotations, and code groups into the network group (Fig. 9.21).

Network of a Network Groups

1. Click on Networks.
2. Right click. A context-menu will display on the right.
3. Select 'Open Network/Show in Network'. The network is ready.
4. Go to 'View' and check comments. Descriptions of the networks and network group will be displayed.
5. Right-click on network and select Add Neighbors (quotations, code, or memo, etc.). A context menu with appear.
6. Select 'Add Neighbors'. Then select the entity for importing into the network (Fig. 9.21).
7. Use Layout and Routing to adjust network's design, shape and colours.
8. Delete items that are not required.

Reports for Networks

Reports for networks are like those for other entities. However, there is one important difference: only Excel reports are possible. Text reports are not possible. The reports can be generated for one or more networks by making the appropriate selection. Reports on networks are generated using Network Group Manager.

REPORT FOR SINGLE NETWORK

1. Open Network Manager by clicking on Network in Explorer or in the main menu.
2. Select the network (or networks) for which a report is required. Open Excel Export.
3. Select the option and generate report.
4. The Excel report will be shown along with the comment, degree, and network group.
5. By selecting all, one can get a report for all networks.

NETWORK GROUPS

One can generate a network group report in the Network Group Manager menu. The report will include only the networks linked with the network group. In the project in Fig. 9.22, the network group "Focus Group Coding Pattern" consists of two. The network has four degrees, meaning that it has four nodes.

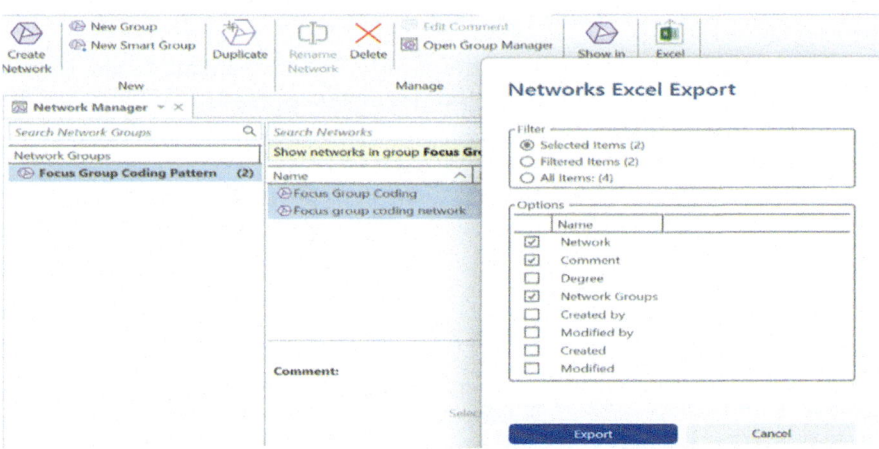

Fig. 9.22 Excel report for network group

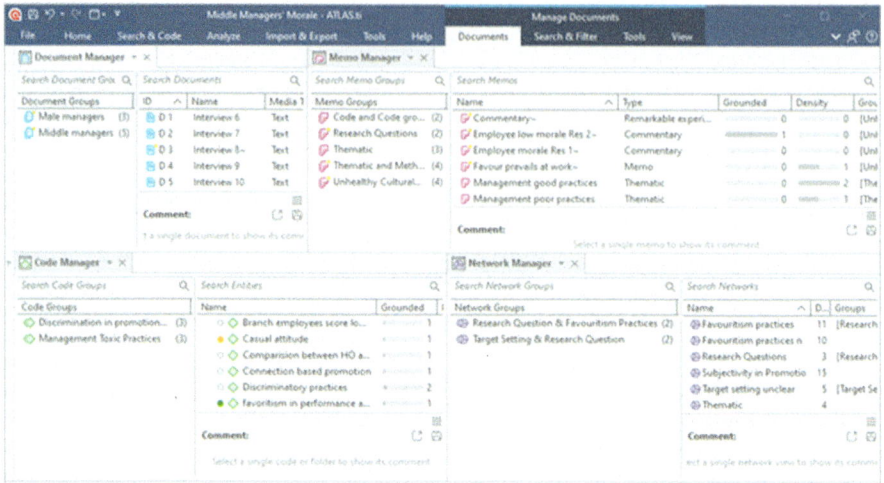

Fig. 9.23 Data classification window

1. Open Network Group Manager.
2. Highlight the Network Group and open Excel Export (Fig. 9.22).
3. Select the option and generate report.
4. The report will include only the member networks in that group, along with the comments for the networks.
5. In this report, the 'Selected Items' option was used (Fig. 9.22).
6. The report shows the networks included in the report.

Data Classification Window

Figure 9.23 shows the data classification window for Document, Code, Memo, and Network managers. There are two document groups, each with 3 and 5 documents. There are two code groups, each group having 3 codes; 5 memo groups with a number of memos in each; and two network groups.

The user can open all windows and generate reports. The window gives the researcher an overview of the document, memo, code and network groups in the project.

Recommended Readings

Bell, E. (2022). *Business research methods*. Oxford University Press.
Bryman, A. (2016). *Social research methods*. Oxford University Press.
Creswell, J. W., & Poth, C. N. (2016). *Qualitative inquiry and research design: Choosing among five approaches*. Sage Publications.
Denzin, N. K., & Lincoln, Y. S. (Eds.). (2011). *The Sage handbook of qualitative research*. Sage.

Glaser, B. G., & Strauss, A. L. (2017). *The discovery of grounded theory: Strategies for qualitative research*. Routledge.
Miles, M. B., & Huberman, A. M. (1994). *Qualitative data analysis: An expanded sourcebook*. Sage.
Saldaña, J. (2021). *The coding manual for qualitative researchers*. Sage.
Strauss, A., & Corbin, J. (1990). *Basics of qualitative research*. Sage Publications.
Strauss, A., & Corbin, J. (1994). Grounded theory methodology: An overview.
Yin, R. K. (2009). *Case study research: Design and methods* (Vol. 5). Sage.
Yin, R. K. (2011). *Applications of case study research*. Sage.

CHAPTER 10

Survey Data Analysis

Abstract This chapter discusses qualitative and quantitative data analysis in ATLAS.ti using survey tools. Preparation of data before its import into ATLAS.ti for analysis is necessary. The author explains the various types of survey data, and how it should be collected by researchers. Based on the researcher's preferences, codes and code groups, documents, and document groups are formed. The various data analysis tools and processes in ATLAS.ti are explained with the example of a research project on people's perceptions, which was conducted by the author. In this study, data was collected from respondents using open-ended questions in Google Forms. Code co-occurrence and code document tables facilitate the calculation of c-values and the generation of themes. The themes are discussed and interpreted at preliminary (code), special (code group), and advanced (scope) levels. Reports of data analysis can be generated in text, Excel, and network formats.

An Overview of Survey Data

ATLAS.ti is used for both qualitative and quantitative data analysis of survey data. Survey data has features that are different from other forms of data. Here, the data is collected using purpose-designed survey forms, usually in Google, Survey Monkey, or other digital formats. The responses are downloaded in Excel form and then imported to ATLAS.ti. Unlike creating quotations in text data that require the researcher to read the data thoroughly before selecting significant segments of information, ATLAS.ti automatically selects responses in the survey data as quotations and codes them with the descriptions written in the questions. The codes can be modified as needed by the researcher.

Thus, ATLAS.ti treats the survey questions as the code's name, the contents of the questions as the respective codes' descriptions, and the responses as quotations. Such survey data are suitable for sentiment, thematic, qualitative/quantitative content, and opinion-mining analyses. This is different from the conventional approach, where all types of data first are read by the researcher who then identifies and codes them, followed by a second stage of coding (axial coding, categorizing, selective coding or creating themes). Where the data size is big, more stages of coding may be required for generating themes.

However, researchers should note that only responses are automatically converted into quotations which further demands reading, identifying and selecting relevant information with reference to research question and should be coded. Automatically converted quotations are only responses. They do not show the relevant information that can be used for data analysis. Researchers should code the automatically created quotations for survey data analysis. ATLAS.ti facilitates the analysis of survey data with an auto function that can be used in certain kinds of data: auto import/extraction of information from surveys, focused group data, social media comments, Twitter data, and so on. It identifies responses as quotations and automatically codes them with a suitable code.

Close-ended survey data can also be analyzed in ATLAS.ti. However, certain kinds of statistical analysis like regression are not possible. However, ATLAS.ti facilitates the export of coded data to other software like SPSS or RStudio in which advanced statistical analyses can be performed.

Data in Excel format (.xls or.xlsx files) can be imported into ATLAS.ti for analysis. Survey data is first downloaded in an Excel form and then exported to ATLAS.ti using the import and export function. In addition, data attributes can also be imported. The data can be categorized in various ways, such as those based on the respondents' professional, social, demographic, personal, and other attributes. The attributes are converted to document groups. Both open-ended and close-ended response data can be used for analysis.

ATLAS.ti offers a range of options to help the user select the information that must be included in (or excluded from) the analysis. ATLAS.ti 22 & 23 will ask users to select the information that will be used for analysis (Fig. 10.1). The steps in the analysis of survey data are shown in Fig. 10.1.

Figure 10.1 shows the data for import into ATLAS.ti versions 22 and 23 in the Export and Import windows. First, responses should be downloaded in excel form and saved on computer as a response folder. Using Import & Export function of ATLAS.ti, appropriate responses should be imported for data analysis.

Important Import of data in ATLAS.ti versions 22 and 23 do not require prior formatting of data in Excel as in prior versions of ATLAS.ti.

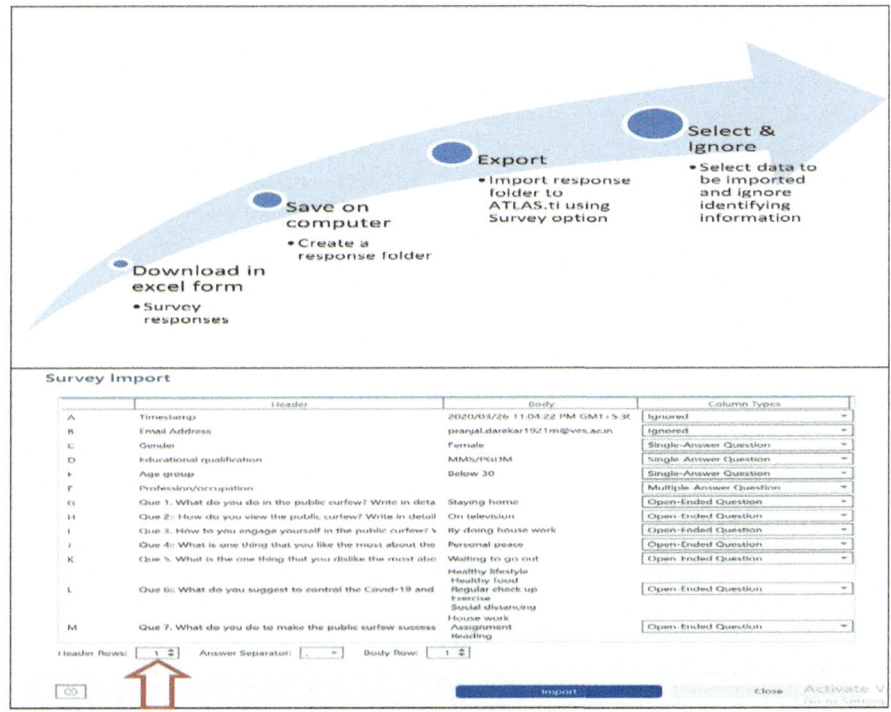

Fig. 10.1 Importing survey data into ATLAS.ti 22 & 23

As shown in Fig. 10.1, the user is only asked to select the information what must be imported. The user selects the column(s) and clicks the Import button to automatically transfer the data. Data collection doesn't need to be complete until the last response. ATLAS.ti allows the import of data (the responses) as and when they are received. The rows that were imported earlier are ignored in partial data imports and new rows are created for additional data. In other words, already existing data remains intact when additional data are imported.

Types of Data that Can Be Imported

Besides questions, survey forms also usually include fields for capturing demographic information about the respondents (the participants in the study). The nature of the study (qualitative or quantitative) determines the type of questions that are asked in the survey. In a quantitative survey, the respondent selects from a list of answer options. On the other hand, in qualitative studies, questions are open-ended, meaning that the respondents can reply in their own words and as much detail as they want to. For qualitative data analysis, ATLAS.ti automatically creates quotations, codes, document groups, and categorizes data based on the researcher's selections in the Excel spreadsheets on which the responses are captured. In Fig. 10.2, the researcher has decided not to import the Timestamp and Email addresses of respondents and, therefore, has selected the option 'Ignored' (Fig. 10.1) when importing the Excel spreadsheet into ATLAS.ti.

Table 10.1 shows the concepts that can be mapped by ATLAS.ti.

Example A question in the study of employee morale in Public Sector Banks (PSBs) will help the reader understand how ATLAS.ti selects responses, converts them into quotations and codes them automatically while importing responses in an Excel form:

Question: What is your opinion about employee motivation in your organization?

Timestamp	Email Address	Que 1: What is the bigg	Que 2: How do people	Que 3: What behaviour	Que 4: What behaviour	Qu
3-4-2023 15:14:23	khandolkarritu17@gma	Attrition. Employees ar	Training Programs, Me	Integrity, Collaborative	Dishonesty, Lies	Ca
3-4-2023 15:14:36	anshika.richa03@gmail	People Management	Leaning new technologi	Creativity, hardwork, d	Cheating, avoiding worl	He
3-4-2023 15:14:58	gulamfarooqu@yahoo.com		While there is lot if data	Ownership and account	Arrogant behaviour, nor	Ho
3-4-2023 15:16:45	sandipansankarr@gma	Employee Rentention: t	By Switching Organizat	Yes Manship and doing	Innovation, discussions	Co
3-4-2023 15:16:57	roopesh86@gmail.com	Talent attraction, retent	There is an entire progr	Performance centric behaviour, taking additiona		Pr
3-4-2023 15:17:00	nitish.kumar.mec09@itl	The biggest challenge i	Presently, I feel it is mc	Being innovative and gi	Questioning the decisio	It i
3-4-2023 15:17:11	mansiparinda@gmail.cc	Retention. Ensuring the	By looking for opportun	Innovation. Reduction ir	Dishonesty.	Ne
3-4-2023 15:17:26	saumyasonpaltiss@gm	Politics- employees or	Fabulous performance	Initiative and collaborati	Integrity issue	Gu
3-4-2023 15:17:44	nikshi1285@gmail.com	Talent management	Through IJPRole Chang	Self developmentPerfor	Back bitingRules breaki	Le
3-4-2023 15:18:05	deepalirawat22@gmail	Right talent Manageme	Regular upskilling and t	100 % attendance, Am	Unethical behaviour, lov	Le
3-4-2023 15:18:20	saurabh.yewalkar@gm	Hiring the right talentKe	By taking on challenging	Attitude of doing things	UnethicalNon subordina	pa
3-4-2023 15:18:46	paresh.brilliance@gmai	Culture - Biggest challe	Quantitative output with	Proactive thinking, prob	Unethical practices & a	Le
3-4-2023 15:18:46	shilpa.kerkar03@gmail.	Revenue Growth Crunc	Involved in New Project	Organisation should rev	Misuse of Value Integri	Le
3-4-2023 15:19:26	2207veena@gmail.com	I belive Attrition is one	By having measurable	Honesty,integrity,achiev	Unethical behaviour	Ha

Fig. 10.2 Importing survey data into ATLAS.ti versions 22 & 23

Table 10.1 Survey concepts mapped in ATLAS.ti and their meanings

Survey concept	*The concept in ATLAS.ti*	*What it means*
Open-ended questions Question Answer	 Code (and code comment) Content of a quotation	A question along with an answer becomes the quotation and is coded with the title of the question
Binary choice: 0 or 1	Document Group	Document groups with 0 and 1 are created separately
Single choice: more than 2 options	Document groups from question plus value	Document groups with 0,1,2, automatically created
Multiple choice	Document groups from question plus value	Document groups with many choices are automatically created

Answer: I think the employee motivation in my organization is quite high, and they work their duty with a sense of high responsibility and sincerity.

In ATLAS.ti, the question becomes the code, and the answer is the quotation associated with the code. Depending on the selection made by the user, the question may be shown in the code itself or the comment for the code. The software codes the questions, and the answers become the linked quotations (Fig. 10.3).

When the answer options are 0 or 1, as in a close-ended survey instrument, the respondent must choose from these values. In such a case ATLAS.ti creates two document groups with 0 and 1 separately. When there are more than two response options to choose from, document groups are created from questions and the value (Table 10.1). In a survey where the respondent can select answers from a list of options (multiple-choice), document groups are created from question plus value. In other words, "a separate document group is created for each response option selected. For example, if the answer options for a question in a survey are 0 or 1, separate document groups are created for the 0 and 1 selections".

Important By adding two colons (::) between the question number and the content, the question is split into a code label and comment. In this case, the portion of the question before the '::' becomes the code, and that after becomes the code comment. In Fig. 10.3, questions 2, 4, and 6 have the sign '::' in the Excel sheet. The '::' converts the contents in

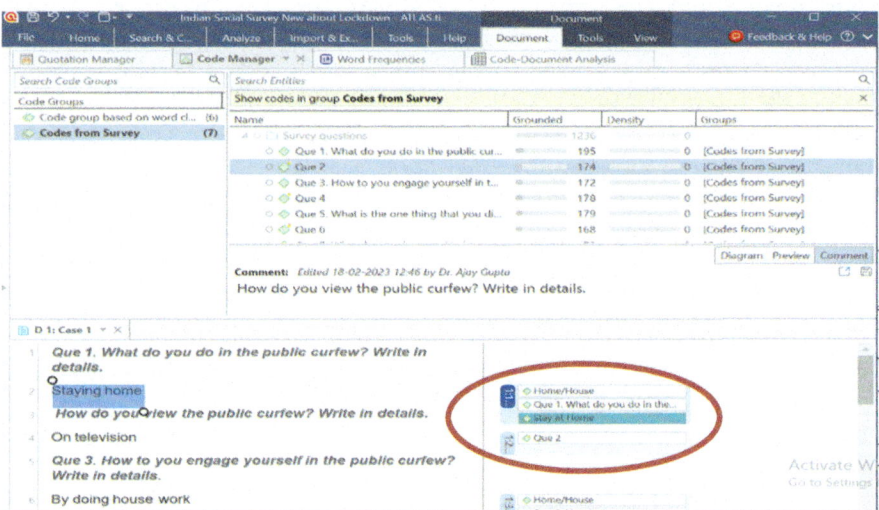

Fig. 10.3 Codes and comments in code manager in ATLAS.ti 22 & 23

> the code comment section. The codes for questions 1 and 2 are shown separately in the bottom part of the window in ATLAS.ti.

Using the '::' sign after the question separates the question and question content. The contents of the question appear in the comments section. Question 2 is a code and the content "How do you view the public curfew? Write in details", is the comment for the code, which is shown in the comment section.

Coding Options for Survey Data

Data can be coded using AI Coding, Text Search, Regular Expression Search, Sentiment Analysis, or Quotation Manager options (Fig. 10.4). Using the AI Coding option, data is automatically coded, categorized, and grouped under the 'AI Codes' code group. With AI Summaries, a summary of each document is created as a memo which can then be converted into documents for further coding and analysis. Documents converted from memos can be used for coding and data analysis if the researcher feels that deeper insights can be obtained from the documents.

AI Codes group code categories, each code category having its sub-codes. Researchers should review the sub-codes and merge them if necessary, based on their characteristics and sentiments. The process helps researchers identify sub-codes in a code category, which can be further used to create code-to-code and code-document tables. A code-to-code table shows the codes' strength and relationship concerning the survey questions. The code-document table shows the theme coverage across gender and other parameters in the data (code-to-code and code-document tables are discussed in Chap. 7). The researchers can report themes and interpret them concerning the survey questions. Screenshots, tables, Sankey diagrams, and networks can be provided to support the interpretations.

Researchers who are not confident with AI Coding can code their data manually. They can also use the Quotation Manager to determine usage of key words in quotations. A word frequency list can be created by selecting the quotations with a the right click. The words used in the responses are displayed along with their weightage which can then be used to code the quotations. Researchers should remember that a Word list may contain synonyms and

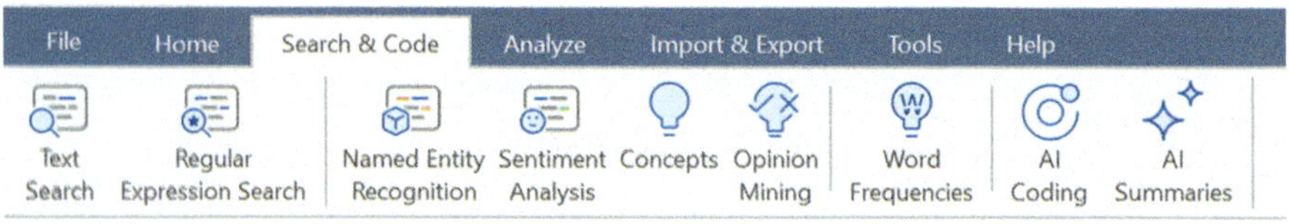

Fig. 10.4 Coding options In ATLAS.ti 23

related words, which should be brought together while coding. The Text Search or Regular Expression Search options can be used while coding similar and related words. Text Search uses 'AND' and 'OR', where 'AND' selects search term that must occur in the selected context, whereas 'OR' selects either information i.e. any of the search term should occur in the selected context. Thus, in the example shown in Fig. 10.5, the operator Leadership OR Employee selects 'Honesty' and 'Yesmanship' separately. Applying the operator Honesty OR Yesmanship selects either 'Honesty or 'Yesmanship' in sentences or paragraphs, but not both together. Regular Expression Search uses (|, a vertical line which means 'OR') to collect Leadership and Employee separately.

In manual coding, the researcher reads the data, identifies the segments of interest and codes them. Generally, manual coding is more time consuming and hence auto-coding should be the preferred coding option, especially where large data sets are involved. In the example being discussed here, both auto-coding and manual coding methods were used.

For auto-coding, the first step is to determine the frequencies of words in the data. The Word Cloud, Word list and Treemap functions help to organize the words in the ascending or descending order of the frequency of their occurrence in the data. In the example shown in Fig. 10.6, a Treemap was used. The number shown with the word is the frequency or number of times

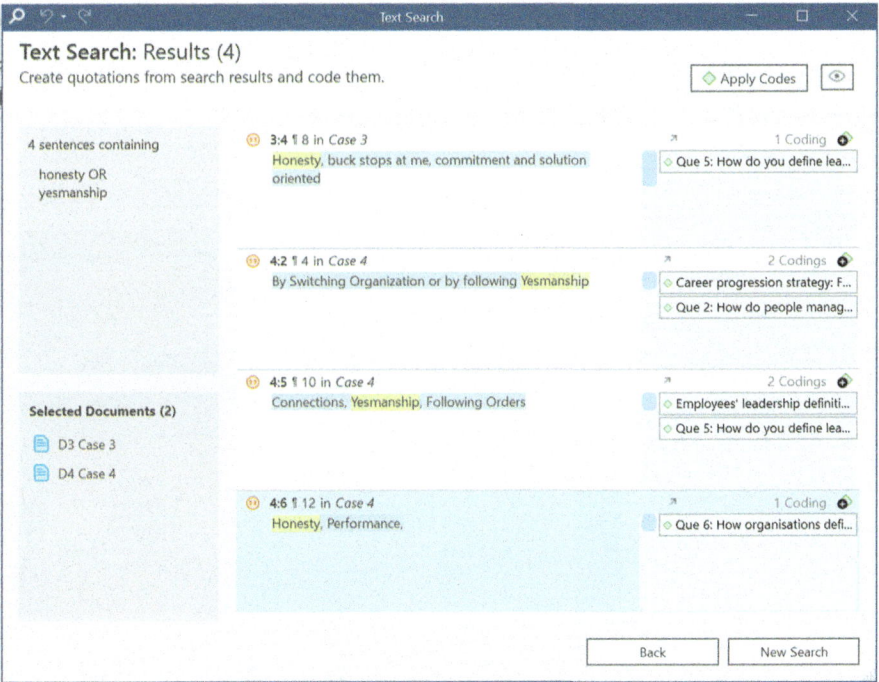

Fig. 10.5 Search text 'or' option

the word is used in the document being analyzed. A Treemap can be used for one or many documents. The Treemap in Fig. 10.6 was created for the checked documents on its left.

After auto-coding of data, the next step is analysis. The code document table and Query Tool is used in combination with the Scope option and the code co-occurrence table. The code-document table helps to understand distribution of codes across various response categories (see Chap. 7).

It is good practice to use the word cloud (or word list) before auto-coding for identifying the words that are used frequently. However, users should also remember that Word cloud, Word list or Treemap may not give accurate results.

The words representing the concepts are then used to code the segments. After auto-coding (see Chap. 15), the code document table helps to identify the emerging pattern. Then, Query Tools and operators can be used to create smart codes and smart groups to generate reports.

Important Users should identify 20–25 words in order of the frequency of occurrence. This is necessary for creating a sufficient number of codes for further analysis.

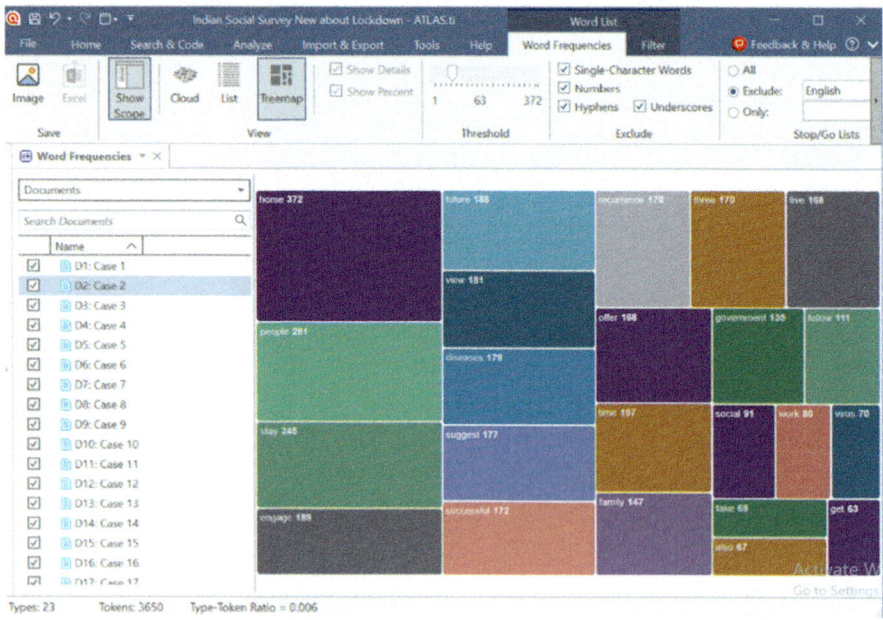

Fig. 10.6 Treemap in ATLAS.ti 23

Using Stop and Go Lists

Before coding survey data, the researcher must ensure that words mentioned in the survey questions are excluded from Word Clouds/Frequency Counts/List and Treemap. This precaution is necessary as a count of the words in survey questions can affect the search for meaningful information in the responses. In the sample project, the words used in survey questions:

1. First, open windows of the Word frequency count of documents and that of Quotation Manager alongside each other (Fig. 10.7). The Word frequency list or Treemap (as selected by the researcher) will display the words used in documents, including the ones used in the survey questions.
2. Delete the words from survey questions from the word frequency list. This can be done in two ways:
   a. One does a right click on the word and select Remove from option list. With a single click the word will be removed from the list. The user selects 'Select Add to Stop List' option and the word will be added to the Stop List (Fig. 10.7). The word or words selected for exclusion will no longer appear in the word frequencies, list, or Treemap.
   b. The second method for excluding a word is by using the Stop and Go List. Here, the user can remove the words that appear in the survey questions. New word in the 'Stop and Go Lists' can filter the word in word cloud, word lists, treemap. In other words, words added in

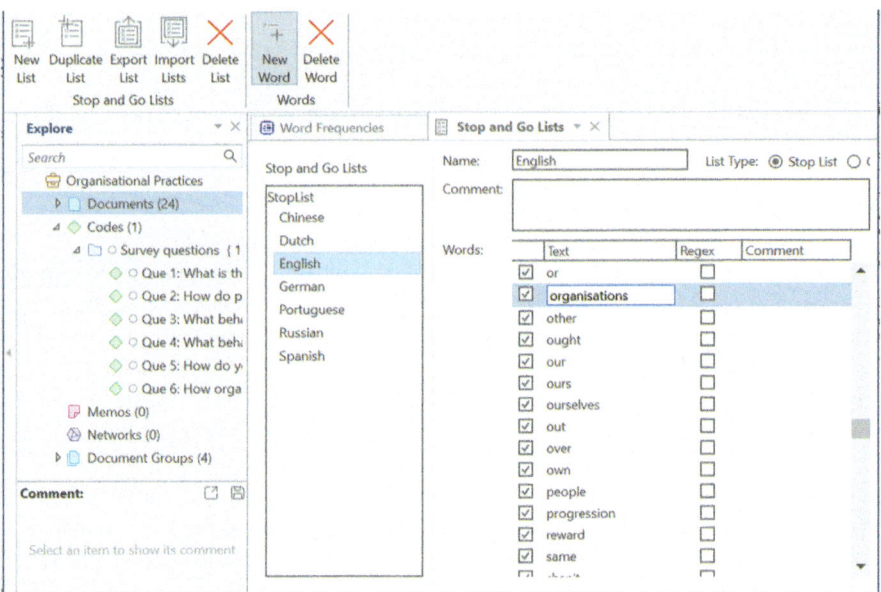

Fig. 10.7 Stop and go lists

the 'Stop and Go List' will not appear in the word list. The users can open word frequencies and select the Edit option. The Stop and Go lists will be displayed. By pressing the New Word option, a new line is created. Users can write the new word and check the box. The word cloud, list, and treemap will exclude the word selected by the user. In Fig. 10.7, the word 'organisations' was added to the Stop and Go Lists as an example.

The 'Stop and Go Lists' option is useful for information retrieval. Irrelevant words, symbols and alphanumeric words and expressions can be added to the list.

Steps in Data Analysis

After importing data, the first task is to verify that the data has been properly and completely imported, the survey questions are coded, and the appropriate document groups created. The next step is to code the responses (quotations) with reference to research questions. Appropriate coding decisions should be made based on research types. The next step is to code categorization. Codes should be reviewed to ensure that similarities and code attributes are covered in code categorization. Code to code relations can be established depending on the code sharing attributes. The data is ready for analysis and researchers can use code co-occurrence analysis to identify themes based on research questions. C value can be used to find code to code relationships (theme) and their interpretation. Further, code-document analysis can be used to find relative significance of themes between categories of respondents and parameters available in the data. Themes should be reported using code co-occurrence analysis and code-document analysis table. Query tools should be used to present themes in several combinations.

Figure 10.8 shows the steps followed in analysis of survey data.

Analysis of Survey Data

- Open Word Cloud and Quotation Manager window side by side.
- Select a word from the Word Cloud and write in the search box in the Quotation Manager window.
- Select the quotations shown in the Quotation Manager, and code them appropriately.
- Find similar words/synonyms words in the Word Cloud, search in Quotation Manager, and bucket them in the code that was already created.
- Search in the Quotation Manager for word using the first character, first with lower case and then capitals to capture all associated quotations.
- Review the coded quotations and ensure they are appropriately coded.
- Create code-to-code relationships using the analysis window to find the significant code-to-code relationships.

Fig. 10.8 Steps in survey data analysis

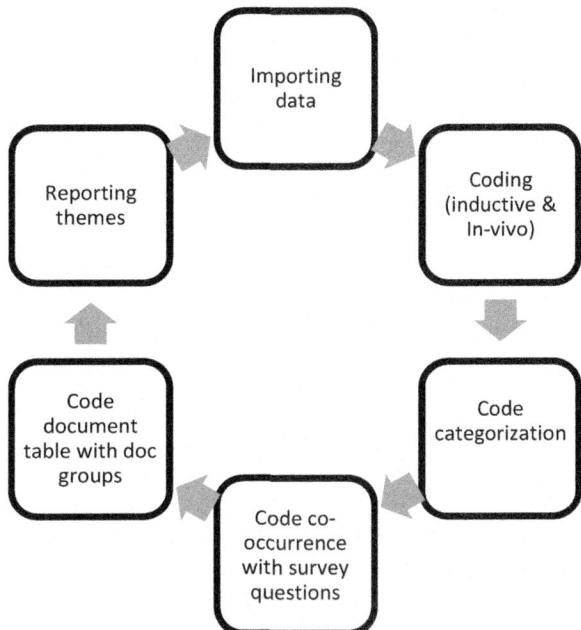

- The code-to-code relationships with significant values are the thematic codes.
- Create code-to-code relationships with the research questions and thematic codes.
- Create a code document table with thematic codes and document groups (i.e., male, female, and other parameters).

SAMPLE RESEARCH PROJECT

A qualitative study was conducted to understand people's perceptions about the curfews imposed during the Covid-19 pandemic. Descriptive responses (n = 195) to open-ended questions were collected using a Google form.

The Objective of the Study

A survey of people's experiences during the curfews imposed during the Covid-19 pandemic was conducted for the study. The survey, which required descriptive responses from the respondents, aimed at understanding people's perception during the pandemic. A google survey form was used to collect responses from the participants. Descriptive responses (n = 195) which were downloaded in the Excel format for analysis in ATLAS.ti. The research questions for the study are listed in Table 10.2.

Survey Questions

Table 10.2 The survey questions

SQ1: What did you do during the curfew? Please describe in detail
SQ2: How did you view the curfew? Please describe in detail
SQ3: What activities did you engage yourself during the curfew? Please describe in detail
SQ4: What is one thing that you liked the most about curfew? Please describe in detail
SQ5: What is the one thing that you disliked most about curfew? Please describe in detail
SQ6: What do you suggest for avoiding curfew? Please offer at least five suggestions
SQ7: What do you do to make the curfew successful? Mention at least three points

---

Data Collection Process

In the example, a total of 195 descriptive responses were collected by the survey team. The responses were saved in an Excel format before being imported into ATLAS.ti 23. Figure 10.3 shows the use of the double colons '::' in Questions 2, 4, & 6 in the Excel table. While importing survey data using the Import & Export function, a flag will open asking the researcher which information from the survey should be included or ignored for data analysis (ATLAS.ti 22 & 23). By clicking the 'Import' button ATLAS.ti will import all data.

Figure 10.9 shows the imported data as displayed on ATLAS.ti. The user will note that each question has become a code, while the corresponding answers are the quotations linked with the code. The content of each code is the code description. Depending on the prefix selected by the user, the code description will appear either in the code line or in the comment section. In this example, the survey questions 1, 3, 5, and 7, along with the contents, are shown in the code line. The survey question is prefixed with Que followed by the corresponding contents.

The imported data consists of responses from 95 male and 100 female respondents in Document Manager. The auto-coded questions (Table 10.2) are shown in Code Manager.

There are 1236 quotations for the seven survey questions. These quotations are shown with the prefix Que for each question (Fig. 10.9). In the Document Manager, three document groups can be seen: female, male, and imported survey data. Document groups help in disaggregating the research's outcomes. In this example, the document groups help the researcher to analyze data for males and females separately.

On importing the data, ATLAS.ti converted Que 4 to a code. The content of the question became the quotation. In the view tab of the Network window, the characters G 178 mean that there were 178 responses to this question.

Figure 10.10 shows the code, quotations, and the network for Question 4. In the network, the characters G 136 for the code 'Family' mean that 136 quotations were coded 'Family' for the survey question 4. The same quotation was also assigned the code 'Time'. Therefore, there is a code co-occurrence between Question 4 and Family & Time.

10 SURVEY DATA ANALYSIS 283

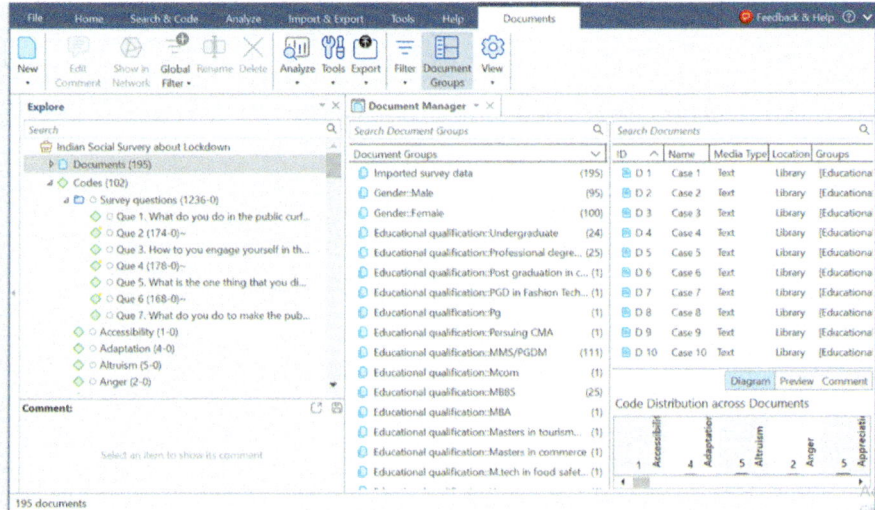

Fig. 10.9 Survey data in ATLAS.ti 23

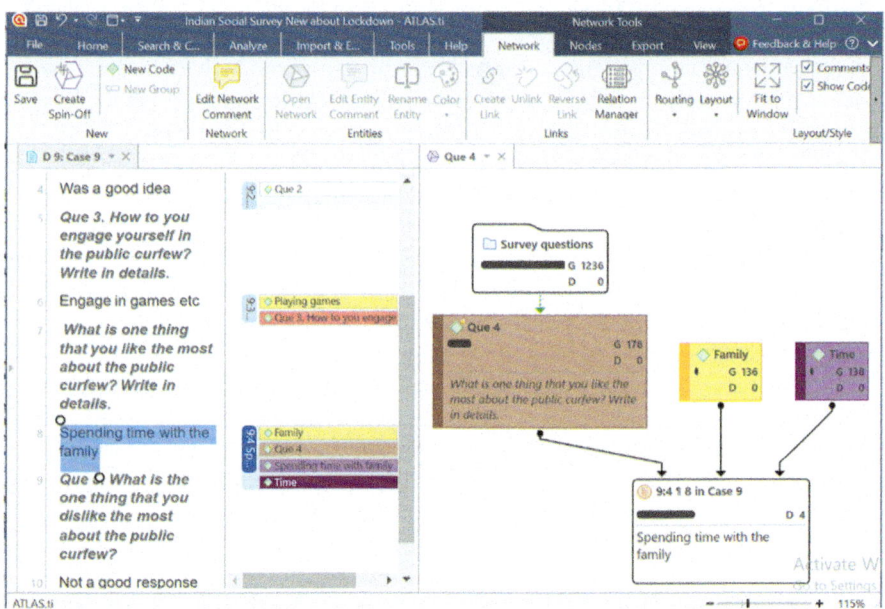

Fig. 10.10 Interpreting survey data

By selecting a survey question in Quotation Manager, the associated quotations are also displayed (on the right in Fig. 10.11). The user can select a quotation and, with a right click, can see the context menu. Click on Word Frequencies to visualize the words in the data—Word List, Word Cloud and

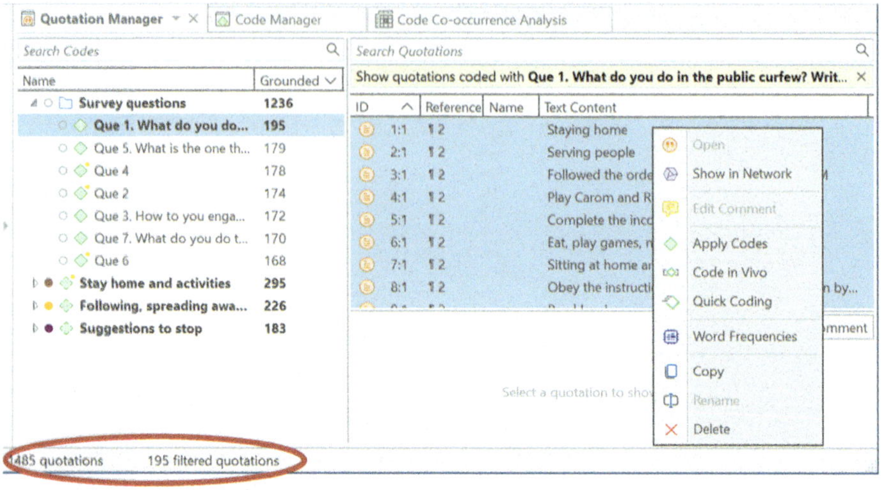

Fig. 10.11 Quotation manager and word frequencies

Treemap. Word frequencies help the user understand how often a word is used in the data (the responses). With a double click on a word, one can see the quotations and thus, understand the contexts in which the word is used. A text search or regular expression search option can be used to refine the search option. After coding, code categories can be formed. The user should note that using text search or regular expression search option use document as a context whereas in quotation manager, the search will only consider quotations.

In Fig. 10.11, the survey questions are shown in the code folder, and the sub-codes in code categories. The total number of quotations is shown at bottom left. From 195 responses, 1485 quotations were created. These are shown in the Quotation Manager window.

'Stay home and activities' is a code category. It includes the sub-codes 'stay at home', 'reading books', 'watch movies and television', 'work from home', 'play games', 'cooking activities' and 'spending quality time at home'. The sub-codes are prefixed with the name of the code category.

A code category indicates a significant theme emerging from responses. As the responses show, the participants in the survey were engaged in various activities while they stayed at their homes during the curfew. The three most common activities, based on the groundedness, were staying at home, reading books, and watching movies and television. Other code categories can be similarly interpreted.

In this example, the code categories were grouped as a theme. The network for the theme is shown in Fig. 10.12. Table 10.3 shows details about the code group, code categories, sub-codes, groundedness and density. The total number of quotations (grounded) for each code category are given, which

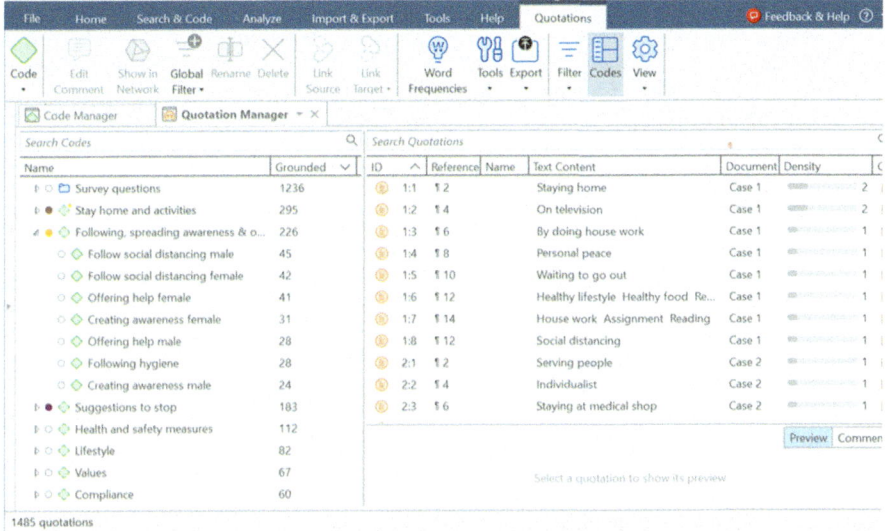

Fig. 10.12 Code group, code categories and sub-codes

are further divided into sub-codes. For example, 'Following, spreading awareness & offering help' is a code category with 226 quotations. The code category consists of seven sub-codes. The quotations for each sub-code is shown in Table 10.3. The themes emerging from code categories can be interpreted based on the number of quotations. In this example, 'Stay home and activities' is the most significant theme followed by 'Following, spreading awareness & offering help' and 'Suggestions to stop'.

Figure 10.13 shows the network of a theme. To show the perceptions of male and female respondents separately, a code document table was created. The quotations are shown on the right. For each significant code, the associated quotations were shown separately for males and females. Other parameters can be similarly coded.

Table 10.3 Themes, code categories, sub-codes and groundedness

Code categories	Sub-codes	Grounded	Density	Code groups
Following, spreading awareness and offering help		226	2	Themes
Following, spreading awareness and offering help	Creating awareness female	31	3	
Following, spreading awareness and offering help	Creating awareness male	24	2	
Following, spreading awareness and offering help	Follow social distancing female	42	2	
Following, spreading awareness and offering help	Follow social distancing male	45	2	
Following, spreading awareness and offering help	Following hygiene	28	0	
Following, spreading awareness and offering help	Offering help female	41	2	
Following, spreading awareness and offering help	Offering help male	28	2	
Stay home and activities		295	2	Themes
Stay home and activities	Cooking activities	21	0	
Stay home and activities	Play games	32	0	
Stay home and activities	Reading books female	35	0	
Stay home and activities	Reading books male	26	1	
Stay home and activities	Spending quality time at home	12	0	
Stay home and activities	Stay at home female	89	0	
Stay home and activities	Stay at home male	86	1	
Stay home and activities	Watch movie and TV male	25	0	
Stay home and activities	Watch movies and TV female	24	0	
Stay home and activities	WFH Female	23	0	
Stay home and activities	WFH Male	20	0	
Suggestions to stop		183	2	Themes
Suggestions to stop	Avoid moving out and contacting female	32	2	
Suggestions to stop	Avoid moving out and contacting male	36	2	

(continued)

Table 10.3 (continued)

Code categories	Sub-codes	Grounded	Density	Code groups
Suggestions to stop	Curfew-good initiative female	52	3	
Suggestions to stop	Curfew-good initiative male	42	2	
Suggestions to stop	People don't follow rules female	20	2	
Suggestions to stop	People don't follow rules male	5	2	

CODE DOCUMENT TABLES

Figures 10.14 and 10.15 show code document tables and Sankey diagrams for themes, as well as their relative significance for male and female respondents in the study. Similar tables and Sankey diagrams can be prepared for other document groups also.

One can create tables to find the patterns in various themes/codes across respondent categories (i.e., socio-economic background). In this study, the data collected was organized according to gender, occupation, and age. The data for this study was normalized, meaning that each document was considered to have an equal number of quotations so that code-relative frequencies are comparable. The themes show the relative weight of the information in male and female categories. Percentage shown against themes is the relative weight of themes between male and female respondents.

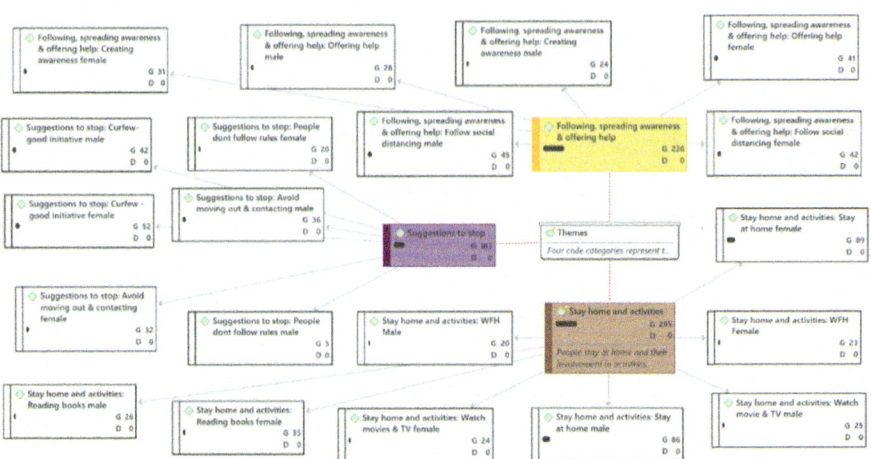

Fig. 10.13 A network of a theme

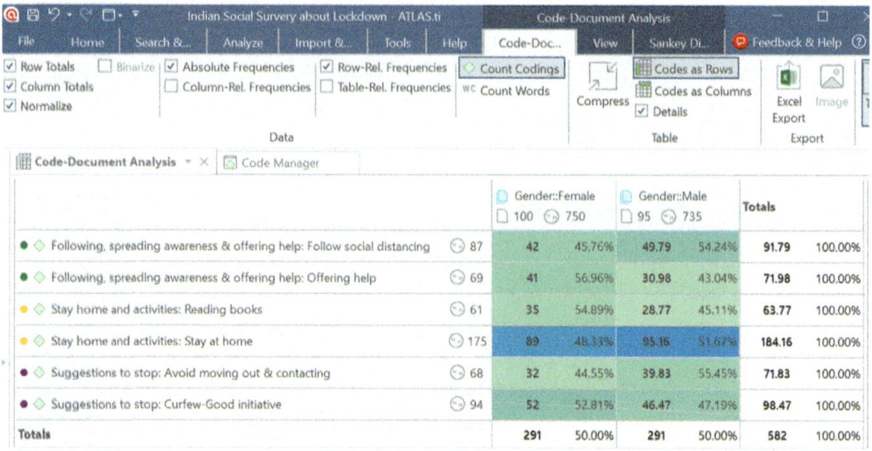

	Gender::Female (100, 750)		Gender::Male (95, 735)		Totals	
Following, spreading awareness & offering help: Follow social distancing (87)	42	45.76%	49.79	54.24%	91.79	100.00%
Following, spreading awareness & offering help: Offering help (69)	41	56.96%	30.98	43.04%	71.98	100.00%
Stay home and activities: Reading books (61)	35	54.89%	28.77	45.11%	63.77	100.00%
Stay home and activities: Stay at home (175)	89	48.33%	95.16	51.67%	184.16	100.00%
Suggestions to stop: Avoid moving out & contacting (68)	32	44.55%	39.83	55.45%	71.83	100.00%
Suggestions to stop: Curfew-Good initiative (94)	52	52.81%	46.47	47.19%	98.47	100.00%
Totals	291	50.00%	291	50.00%	582	100.00%

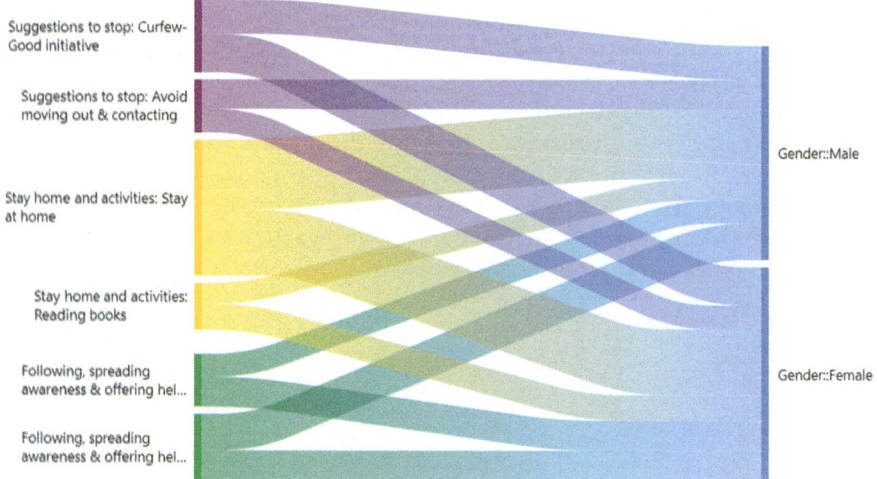

Fig. 10.14 Themes and their relative significance for male and female respondents

Fig. 10.15 Themes and their relative significance for male and female respondents in a Sankey diagram

Code Co-occurrence

Code co-occurrence was used to find code-to-code relationships and their significance. A Code co-occurrence table helps calculate the code-to-code significance based on C-value. As discussed in Chap. 7, the C value should lie between 0 and 1 (excluding 0 and 1). Any value tending towards 1 is significant. In Fig. 10.16, the C-value of 0.18 is relatively more significant than the C-values of 0.12 and 0.07. In the example the code category 'Stay at home

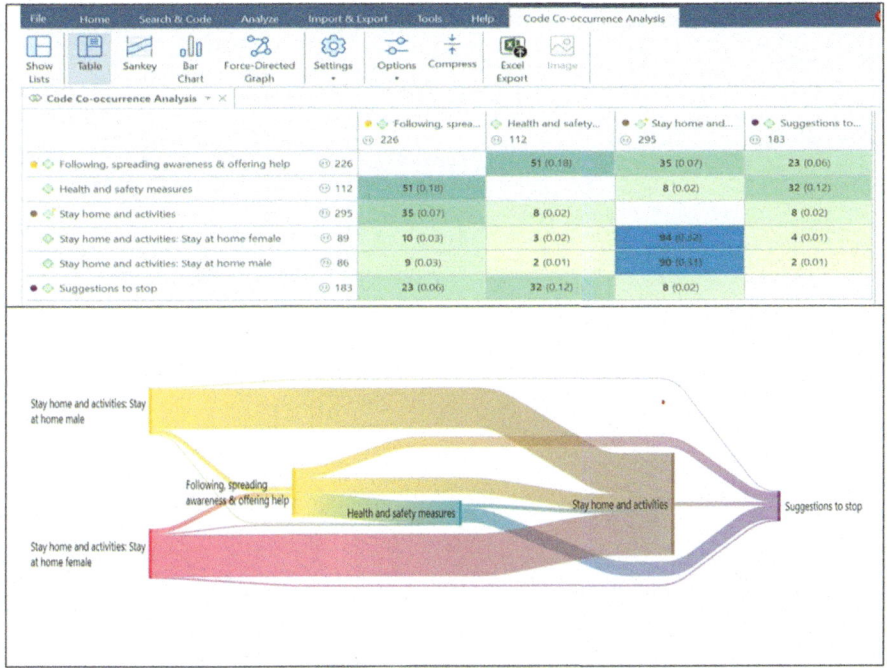

Fig. 10.16 Sankey diagram based on C values for themes

and activities' includes the sub-codes 'Stay at home female' and 'Stay at home male', whose c values are, respectively, 0.32 and 0.31.

The c-values for themes in Fig. 10.15 show that 'Stay at home female', 'Stay at home male', and 'Curfew good initiative female' are significant. The network in Fig. 10.16 shows the themes with reference to the survey questions. Survey questions appear on the right side and themes appear left side.

ANALYSIS OF SURVEY DATA USING AI CODING

The discussions in this section use data from a survey of organizational practices. Data was collected from managers and senior managers (n = 24) using Google Forms. There were six questions in the survey (Fig. 10.17). A total of 139 responses were received. The data was converted to an Excel format and imported into ATLAS.ti. Figure 10.17 shows the project window in ATLAS.ti 23.

A code co-occurrence table was created to identify the significant themes based on the C value (Fig. 10.18). The themes are shown for each question. A report can be generated for the significant themes, which can then be interpreted by the researcher. A network for themes can also be created (Fig. 10.19).

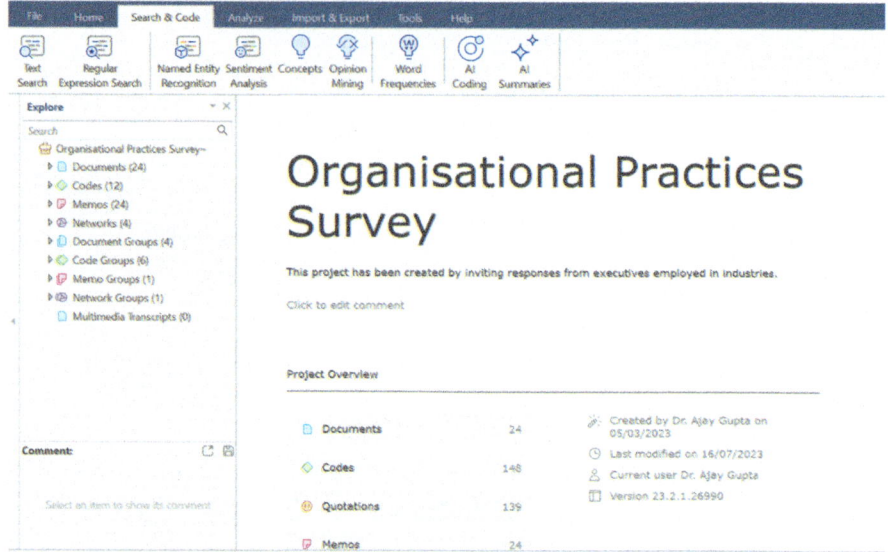

Fig. 10.17 Project window in ATLAS.ti 23

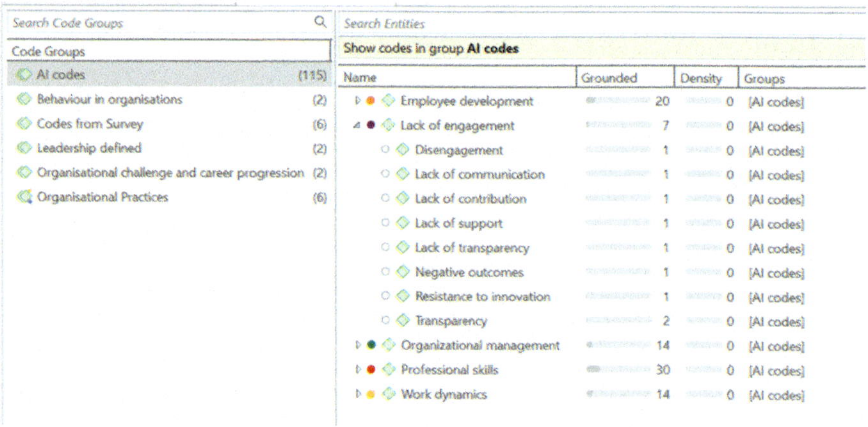

Fig. 10.18 Ai codes and code categories in ATLAS.ti 23

Figure 10.20 shows a network for organisational practices. Five themes have been shown as code categories, each category consisting of a few sub-codes. The sub-codes for two code categories are shown in the figure. A Sankey diagram for themes (Fig. 10.21) is useful for visualizing the relative significance of the themes with respect to the survey questions (Fig. 10.21), as well as the interconnection of these themes.

Figure 10.22 shows themes for two respondent categories and the survey questions for better understanding of the data. The code document table

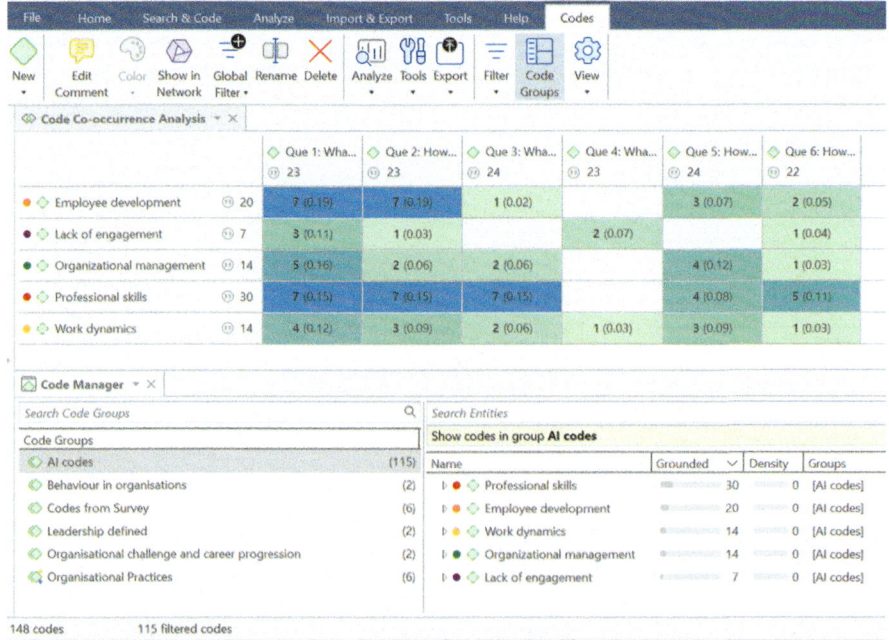

Fig. 10.19 Code co-occurrence and code manager

shows the patterns of the themes across respondent categories. This helps to understand relative significance of themes between managers and senior managers.

Figure 10.23 shows the Memo group, AI Summaries, and Memos, which were formed using the AI Summaries option. AI Summaries provides the summary of each document in memo which can be converted into documents for further analysis.

Figure 10.24 shows the option for converting memos into documents. After selecting Memos in Memo Manager, a right click opens a context window. Then, by selecting 'Convert to Document', all memos will be converted to documents and shown in the document list along with existing documents.

Generating a Text Report

Text reports for themes can be generated using the Query Tool and applying operators. Figures 10.25 and 10.26 show the steps that must be followed for generating thematic reports in the selected categories. In this example, the categories are male, female and both. The OR and Co-occurs operators were used to generate the report. The total number of quotations in the selection (1236) are shown at the bottom left of the window (Fig. 10.25).

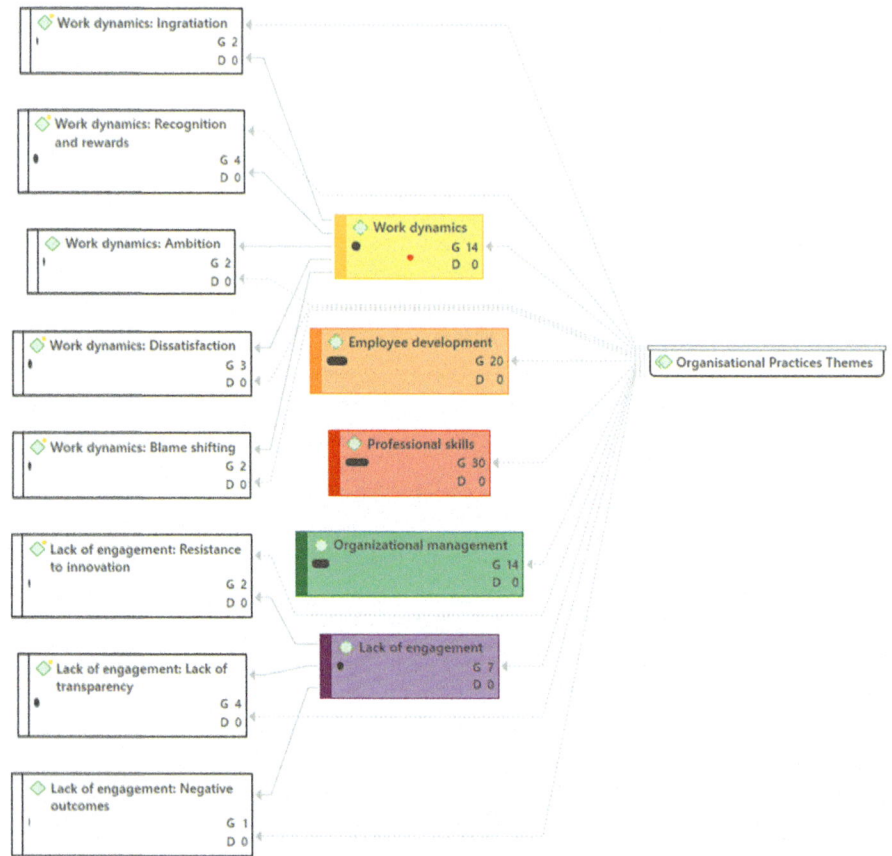

Fig. 10.20 A network of organisational practices

By clicking the Edit Scope option, the researcher can select the category for which the report is required. In Fig. 10.25, the categories male and female was selected for which the number of quotations (35) is seen at the bottom left of the window. Figure 10.25 shows the text report for the selection. The report shows that there are 35 quotations in the 'Male' and 'Female' category. Similar reports can be generated for any research question, themes, code, code groups under the selected document group.

If the researcher wants emerging insights for Research Question 2 in the Male category of respondents, the Edit Scope tool should be used to refine the search options:

1. Opening "Edit Scope", two options will be displayed: Add Document Groups and Add a Document.
2. To refine the search results to document level, use the 'Add a Document' option; otherwise, select the 'Add a Document Group' option.

10 SURVEY DATA ANALYSIS   293

Fig. 10.21 Sankey diagram of themes

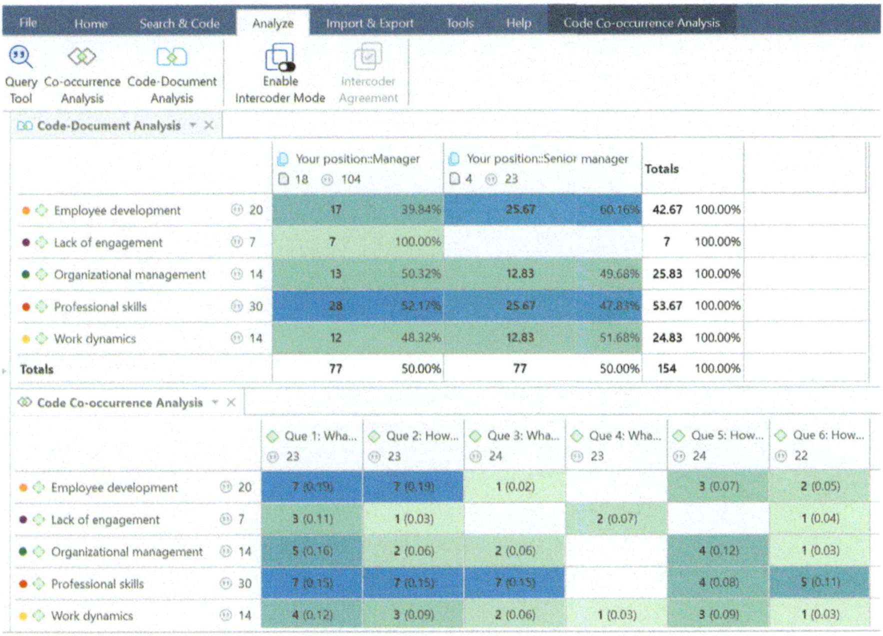

Fig. 10.22 Code co-occurrence and code-document analysis

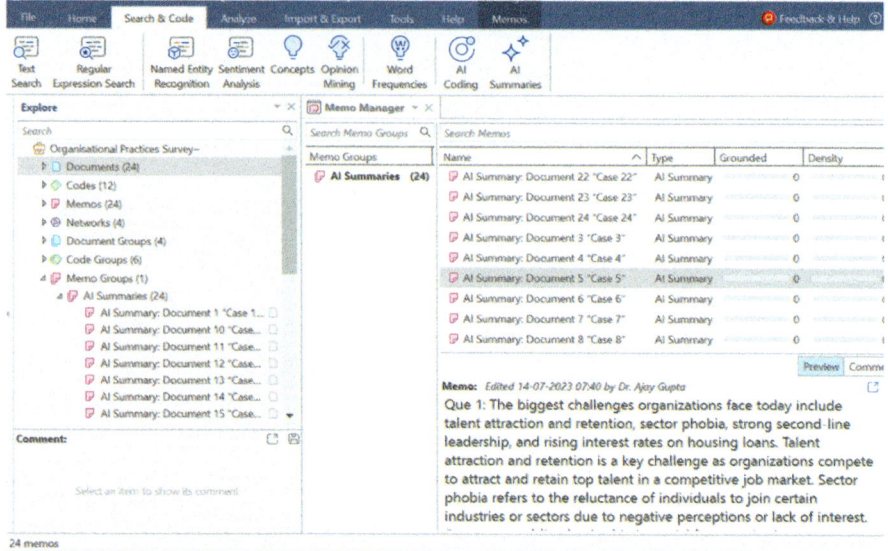

Fig. 10.23 Ai summaries and memos

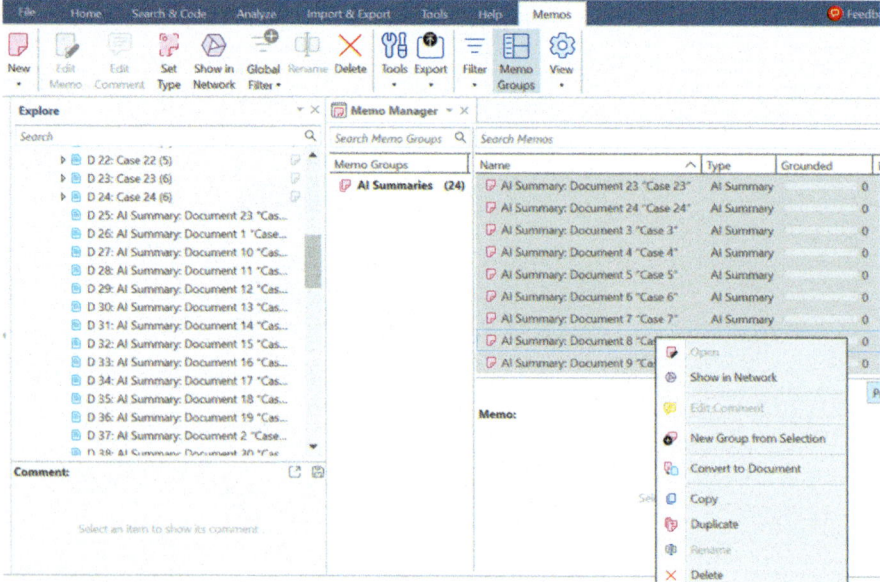

Fig. 10.24 Option to convert memos in documents

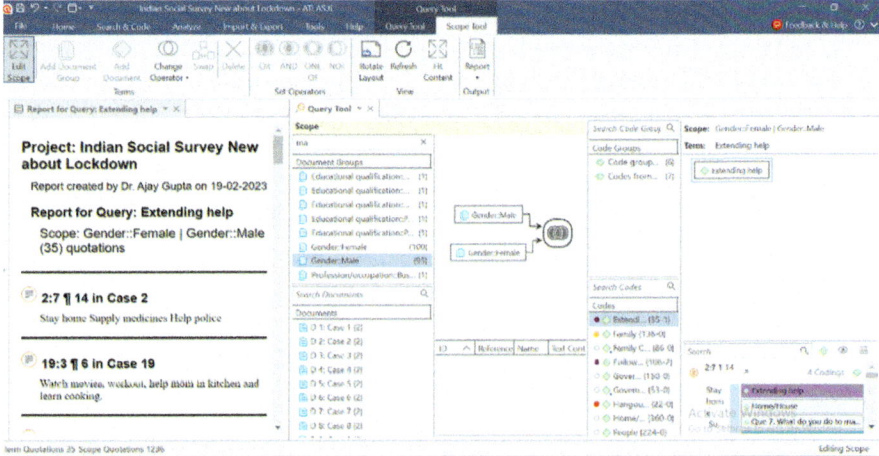

Fig. 10.25 Text report using the edit scope tool

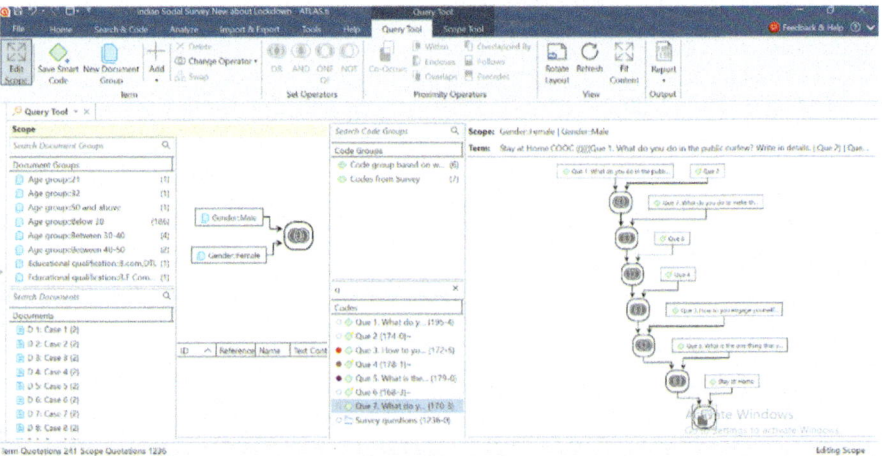

Fig. 10.26 Textual report using set and proximity operator

3. Depending on preference, click on either 'Document' or 'Document Groups'.
4. Add a document group or document.
5. The text reports can be generated in various formats: a list, a list with comments, full content, content with comments, and various others.

It is good practice to keep the text report in the exhibit for reference and write the interpretations in the findings section. This helps the researcher to present the findings of data analysis in a logical manner. The interpretations should be supported by the emerging concepts, c-coefficient values, quotations, and links with the research question. Networks can also be generated to provide visual support for the researcher's interpretations.

A network and bar chart are shown for the themes in Figs. 10.27 and 10.28 respectively.

Figure 10.28 shows the occurrence of themes on x-axis in number. Themes appear on y-axis with their relative significance. Themes can be presented using bar chart for data analysis.

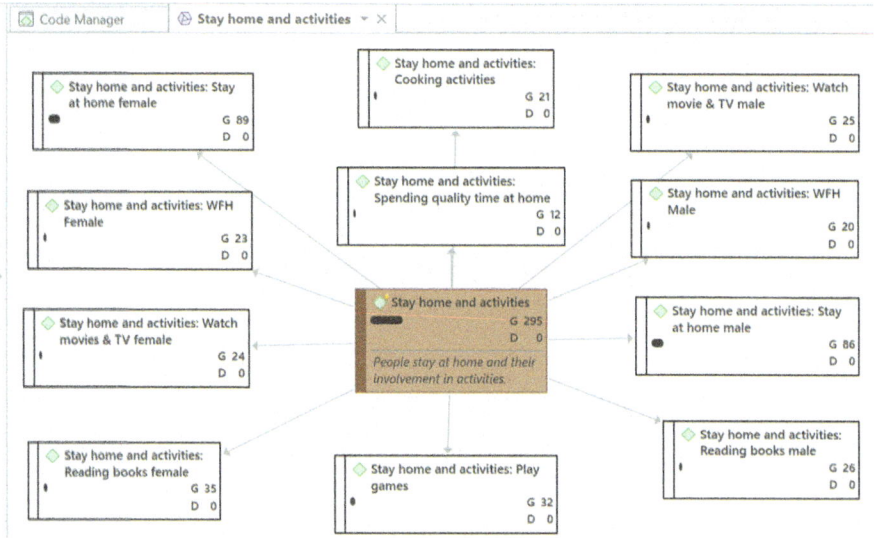

Fig. 10.27 A network of a theme and sub-themes

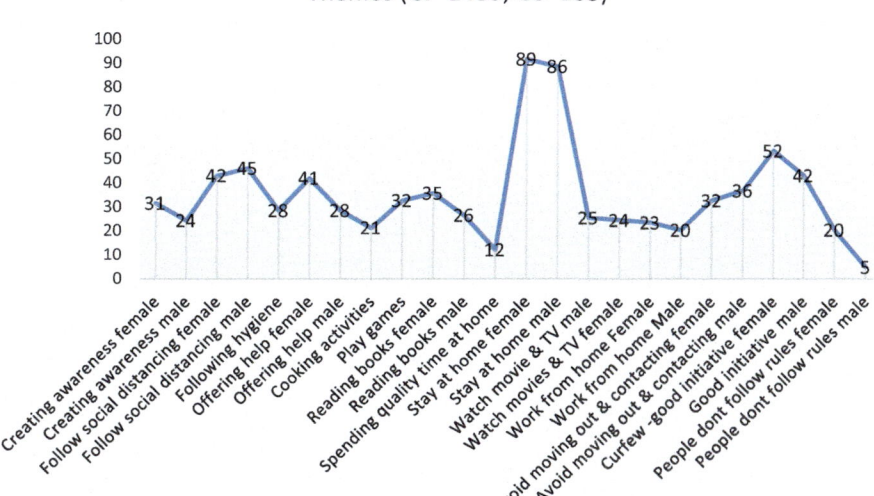

Fig. 10.28 Bart chart of theme

Recommended Readings

Abdullah, M., Madain, A., & Jararweh, Y. (2022, November). ChatGPT: Fundamentals, applications and social impacts. In *2022 Ninth International Conference on Social Networks Analysis, Management and Security (SNAMS)* (pp. 1–8). IEEE.

Galura, S. J., Horan, K. A., Parchment, J., Penoyer, D., Schlotzhauer, A., Dye, K., & Hill, E. (2022). Frame of reference training for content analysis with structured teams (FORT-CAST): A framework for content analysis of open-ended survey questions using multidisciplinary coders. *Research in Nursing & Health, 45*(4), 477–487.

Hertlein, K. M., & Ancheta, K. (2014). Advantages and disadvantages of technology in relationships: Findings from an open-ended survey. *The Qualitative Report, 19*(11), 1–11.

Jackson, K. M., & Trochim, W. M. (2002). Concept mapping as an alternative approach for the analysis of open-ended survey responses. *Organizational Research Methods, 5*(4), 307–336.

Knekta, E., Runyon, C., & Eddy, S. (2019). One size doesn't fit all: Using factor analysis to gather validity evidence when using surveys in your research. *CBE—Life Sciences Education, 18*(1), rm1.

LaDonna, K. A., Taylor, T., & Lingard, L. (2018). Why open-ended survey questions are unlikely to support rigorous qualitative insights. *Academic Medicine, 93*(3), 347–349.

Lupia, A. (2018). How to improve coding for open-ended survey data: Lessons from the ANES. In *The Palgrave handbook of survey research* (pp. 121–127).

Mann, C., & Stewart, F. (2003). Internet interviewing. In *Inside interviewing: New lenses, new concerns* (pp. 241–265).

Stewart, F., & Mann, C. (2000). Internet communication and qualitative research: A handbook for researching online. In *Internet communication and qualitative research* (pp. 1–272).

Thorne, S. (2000). Data analysis in qualitative research. *Evidence-Based Nursing, 3*(3), 68–70.

Williams, C. (2007). Research methods. *Journal of Business & Economics Research (JBER), 5*(3).

CHAPTER 11

Social Network Comments and Twitter Data Analysis

Abstract Increasingly, social media posts, and the information shared on them, are receiving the interest of qualitative researchers as they are sources of rich information about trends and behaviours. ATLAS.ti allows the import and analysis of data from eight platforms: Facebook, Twitter, Instagram, YouTube, TikTok, VK, Twitch, and Discord. Data is imported to ATLAS.ti using ExportComments.com. ATLAS.ti's advanced features help the researcher analyse this data. The Tweets by Twitter users contain rich information. Analysis of Twitter data begins with a search for concepts in Twitter posts (tweets), and then selecting that information for import into ATLAS.ti. The imported data is coded in ATLAS.ti followed by the formation of code groups, documents, and document groups. Coding may be done manually or by auto-coding. Text or Excel reports can then be generated, which are then interpreted with reference to the needs of the study. This chapter also includes a discussion of data analysis of YouTube comments, and the process followed for identifying themes and then interpreting them. This is followed by a discussion of sentiment analysis of YouTube comments and the presentation of outcomes in network forms.

PREPARATION OF TWITTER DATA

Twitter offers easy access to large volumes of data on a wide range of topics and issues. It has introduced new rules about data mining, etc. Analysis of Twitter data can provide deep insights into the subject, phenomenon, or persons of interest to the researcher. Twitter data is valuable for researchers because it is current and hence useful for studying global trends about a topic. The data is present in the tweets and can be analysed using ATLAS.ti.

© The Author(s), under exclusive license to Springer Nature
Switzerland AG 2024
A. Gupta, *Qualitative Methods and Data Analysis Using ATLAS.ti*,
Springer Texts in Social Sciences,
https://doi.org/10.1007/978-3-031-49650-9_11

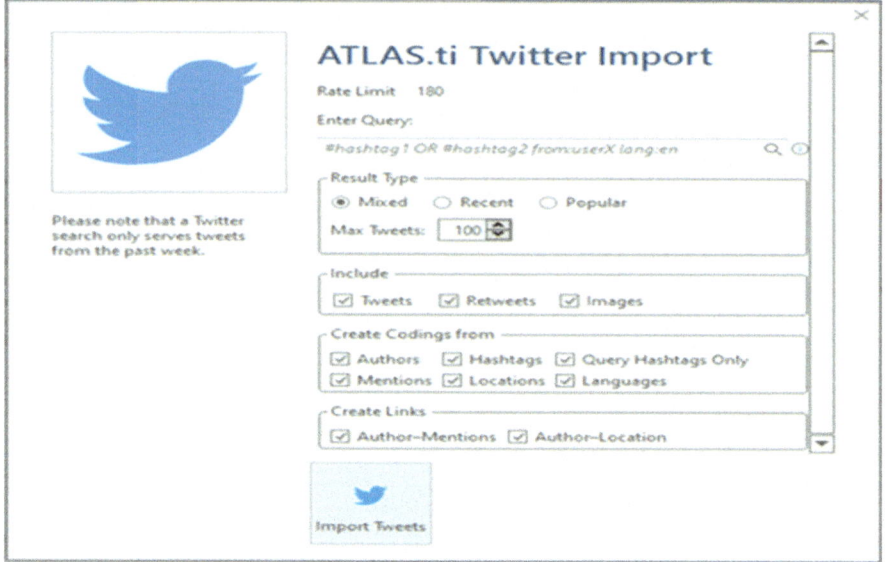

Fig. 11.1 Import tweets window in ATLAS.ti 22

ATLAS.ti can collect tweets of the past one week only. Therefore, researchers must timestamp the data that they are importing. While importing data, the researcher should apply keywords or concepts associated with the topic of interest. Proper use of the keywords will help the researcher to extract rich data.

Direct import from Twitter into ATLAS.ti 23.2.3.27778 has been disabled due to changes made to the Twitter API and licensing conditions. Twitter data is imported into ATLAS.ti version 23 using the 'Social Network Comments' option (Fig. 11.2). Previous versions of ATLAS.ti had 'Twitter import' option (Fig. 11.1).

Importing Twitter Data

1. Open the Import & Export function of ATLAS.ti to the main menu. A window will appear as shown in Fig. 11.1.
2. Write the search preferences using the prefix #, separated by vertical line | if the researcher has more than one search preferences (Fig. 11.1).
3. Open Twitter option (top left window). The resulting display is shown in Fig. 11.1.
4. Check the box for which the researcher wants to import information and click Import Tweets. ATLAS.ti will import all tweets of one week prior (Fig. 11.1).
5. For importing data that is older than a week, different dates and times should be used for the same key word. For example, if the researcher

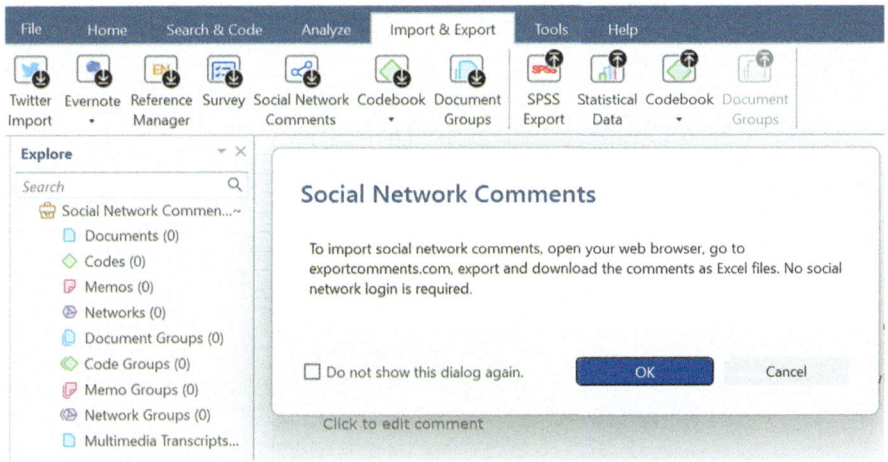

Fig. 11.2 Import and export window of ATLAS.ti 23

wants to import tweets on education (#education) for one month, they can import four times (in slabs of one week) for a month. For each import for the same project, ATLAS.ti creates a separate document. For example, data imported twice will show two documents, D1 and D2.

6. ATLAS.ti automatically codes the data based on the researcher's selection while importing the Tweets.

Importing Data into ATLAS.ti

Each tweet is coded with the selections ("Tweets", "Retweets", and 'Images") made by the researcher (Fig. 11.3). Each tweet becomes a code and the segment containing the keyword for search becomes a quotation. The code groups are based on the researcher's selection (see the two checked boxes in Fig. 11.1). On opening Code Manager, the code list, code group, document, the date, and time of the tweets can be seen. Using the Edit option, the researcher can insert the date and time of tweets. This function is necessary for identifying when the tweets were imported tweets for different dates.

The codes and code groups in Fig. 11.3 were based on the selections shown in Fig. 11.3. Tweets on Education using the # (hashtag) were imported. In all, five hundred tweets on education were selected for import using #Education|#education in the Enter Query option (Fig. 11.3).

Figure 11.4 shows the imported tweets that were converted into codes, quotations and code groups.

The highlighted tweet in Fig. 11.4 has four codes because the researcher had selected language, location, authors, and mentions. The tweets, retweets, and images become quotations and selection becomes codes. After importing

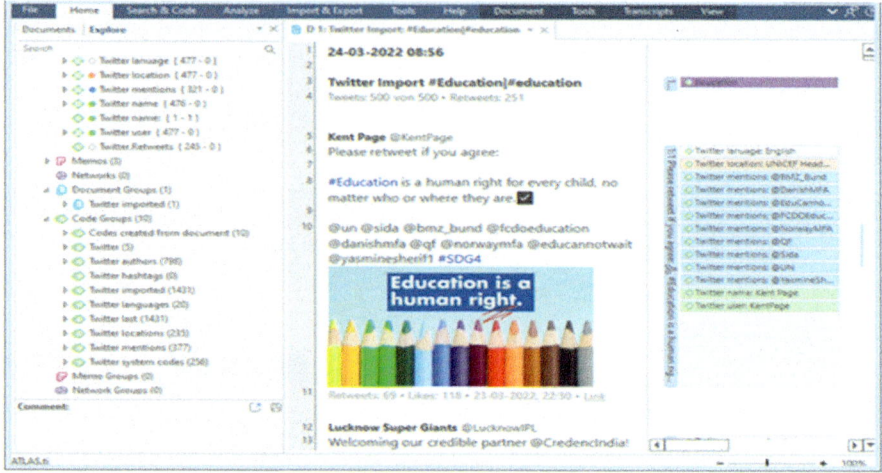

Fig. 11.3 Tweets, codes, and code groups

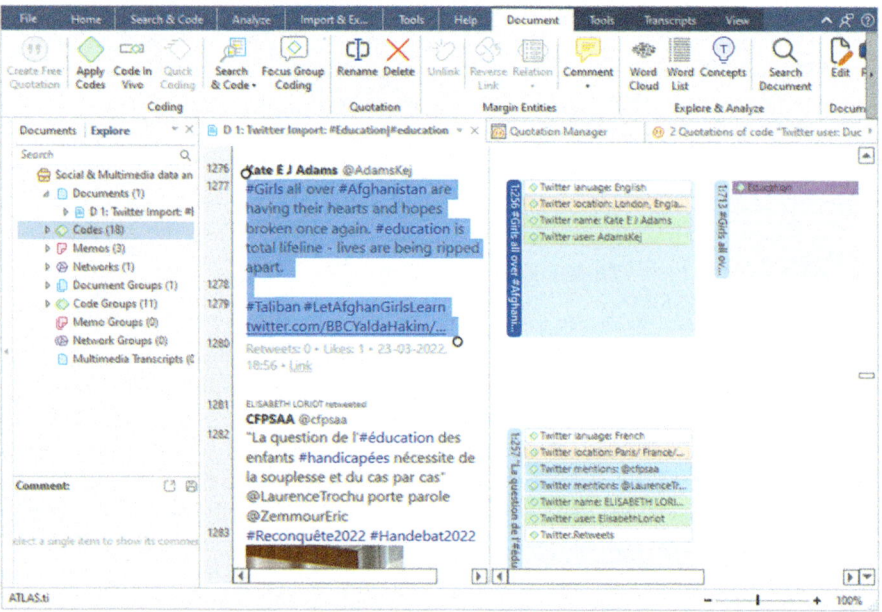

Fig. 11.4 Coding tweets in ATLAS.ti

the twitter data, the user can read, identify significant information from tweets (quotations) for analysis. It is important to know that ATLAS.ti crea3tes codes based on selection i.e., Author, Location, Language and other options and tweet become the quotation (Fig. 11.4). Quotation should be coded to extract meaningful information and relevant contents. Codes should be used to identify categories, group and themes. The themes emerging from imported twitter data can be interpreted with reference to research objectives.

> **Important** Tweets contain information about the phenomenon under investigation. Tweets become quotations based on search selection. Quotations are used for searching and identifying significant pieces of information related to search options. Researchers should code them using appropriate coding techniques. After coding, code categories, code groups, smart codes should be created based on research objectives. The next step is to analyze data using code co-occurrence analysis, code-document table, and query tool to generate reports and interpret accordingly.

Coding Tweets

Twitter data is selected for import based on research objectives. In search option, several words can be used with # as a prefix to import tweets. Tweets based on search selection get converted into quotations. Researchers should code tweets (quotations) based on research objectives and process follows as in text data analysis. After importing data into ATLAS.ti, the researcher codes the relevant segments. One can use Word Clouds, Word List, Concepts, Treemap in ATLAS.ti 22 and 23 for determining the usage of the key words.

The concepts option in ATLAS.ti 23 shows the most frequently used words, along with their contexts and the number of times they appear in the text (Fig. 11.5). For example, as shown in Fig. 11.6, on selecting the word 'leadership', all concepts associated with 'leadership' are shown, as well as the quotations linked with them. By clicking once on a word, the quotations are shown on the right. Here, on selecting #leadership, seven quotations were shown. The researcher then read these quotations and wrote their comments for each.

> **Important** While using the import and export function of ATLAS.ti to import Twitter data, the user can import data using 180 requests in a 15-min period.

The tweets can be saved using the 'Save as Memo' option in the Concepts menu. On saving, all the tweets will be converted into one document. By using 'Apply Proposed Codes', the quotations linked with a concept will be

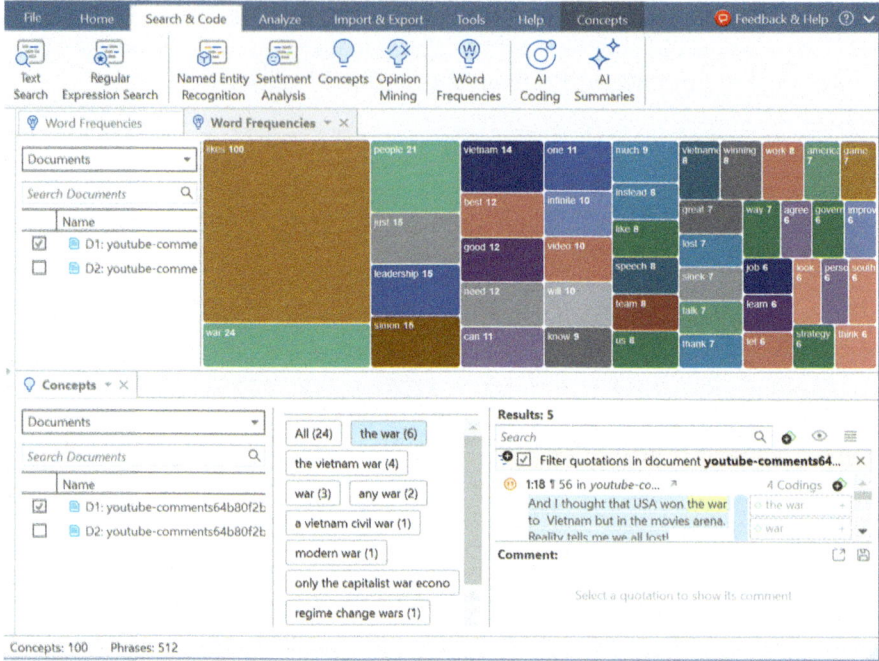

Fig. 11.5 Word tree, and concepts in ATLAS.ti 23

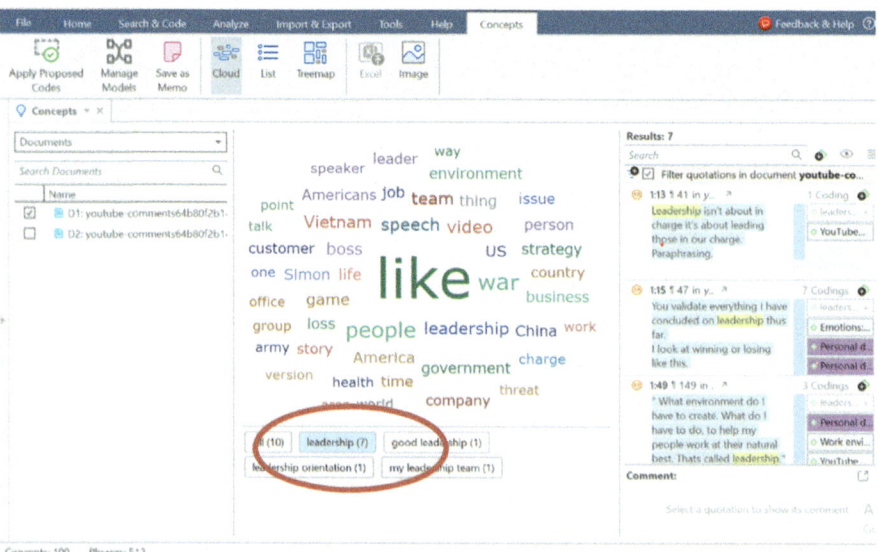

Fig. 11.6 The concepts window in ATLAS.ti 23

automatically coded with the word appearing below the Concepts cloud. This means that the concept will become the code.

Coding is followed by data analysis. Using code co-occurrence, the researcher can ascertain the relationships between codes. After selecting codes that have significant relationships with each other, the query tool is used to generate textual report after applying the appropriate operator (Chap. 9). The researcher then reads and analyses the report and makes interpretations.

ANALYSIS OF TWITTER DATA

Twitter data can be analyzed in various ways—by qualitative content, quantitative content, narrative, sentiment, and thematic analyses. Figure 11.7 is a schematic representation of the process for data collection and analysis of Twitter Data.

Data is first imported and then coded with the most used words and concepts. After data coding, a code-to-code co-occurrence table is generated which helps to find code-to-code relations. With the c-values, the researcher finds quotations coded with both codes. This is followed by data analyses, identification of themes, generating text reports, and interpretation of results, which are explained in Chap. 8.

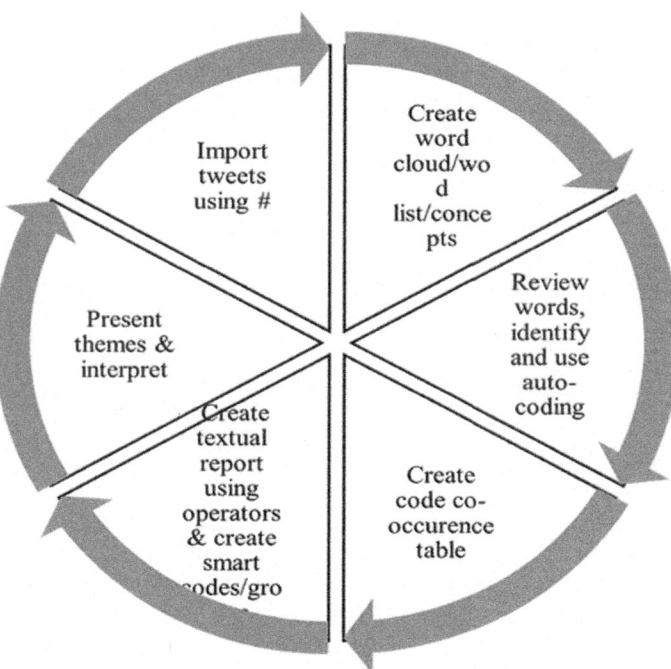

Fig. 11.7 Collection and analysis of twitter data

In this example of analysis of Twitter data, the association of the code 'education' with the codes 'learning' and 'schools' has significant c-values (0.07 and 0.06 respectively). The number outside the bracket in Fig. 11.8 shows the co-occurrence. Three quotations were coded with 'education' and 'learning'. The number 3 shows the number of quotations coded with 'education' and 'schools'. By clicking on the numbers, the quotations will be shown on the right.

A smart code was also created (Fig. 11.9) using which a report can be generated (see Chap. 9 for smart codes).

Figure 11.9 shows the code co-occurrence, with significance of code-to-code relationships indicated by the code coefficient. A Sankey representation of the co-occurrence of code 'Education' with codes shown in Fig. 11.10.

> **Important** Twitter data and Social Network Comments should be analyzed using the methods applied in text data analysis. The researchers should clearly define their study's objectives and the data analysis methods (i.e., sentiment, qualitative content, narrative, thematic or opinion mining, etc.). Themes should be reported and interpreted based on their significance.

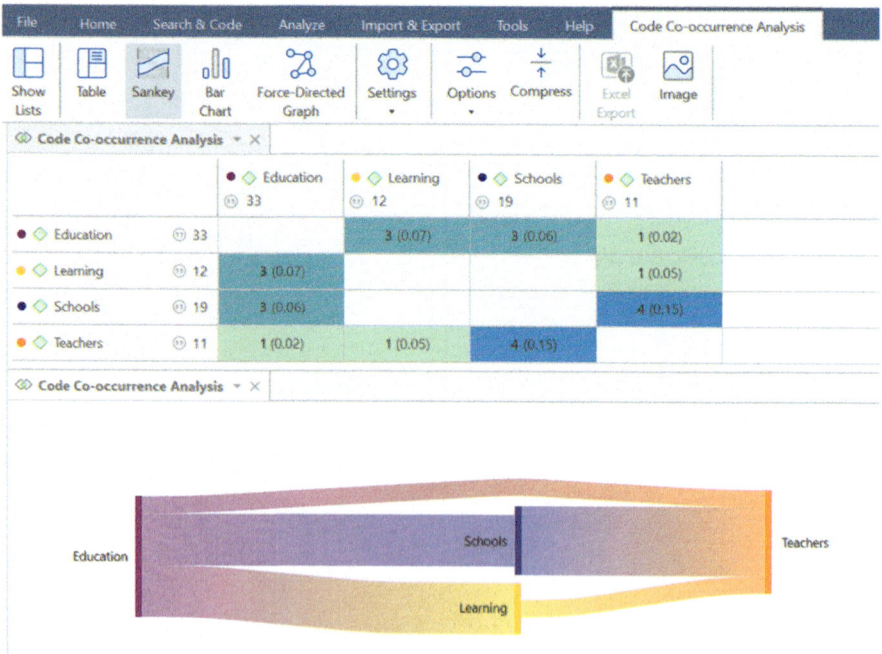

Fig. 11.8 Code co-occurrence table, bar chart and quotations

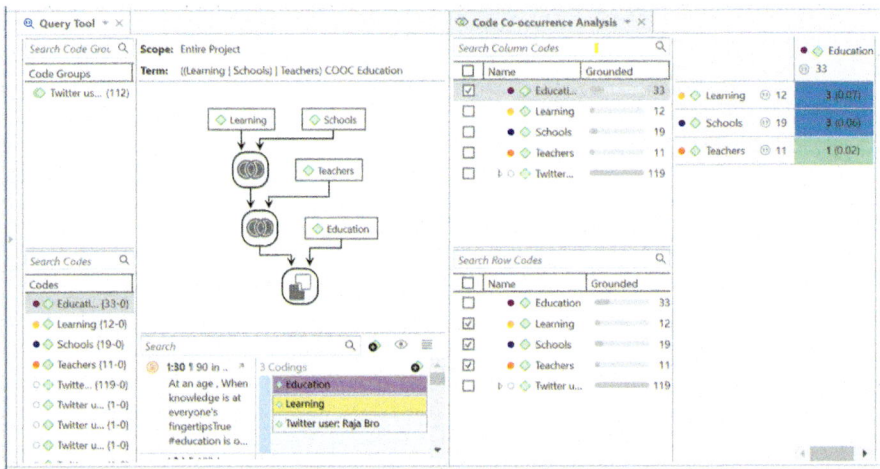

Fig. 11.9 Code co-occurrence table, and smart code

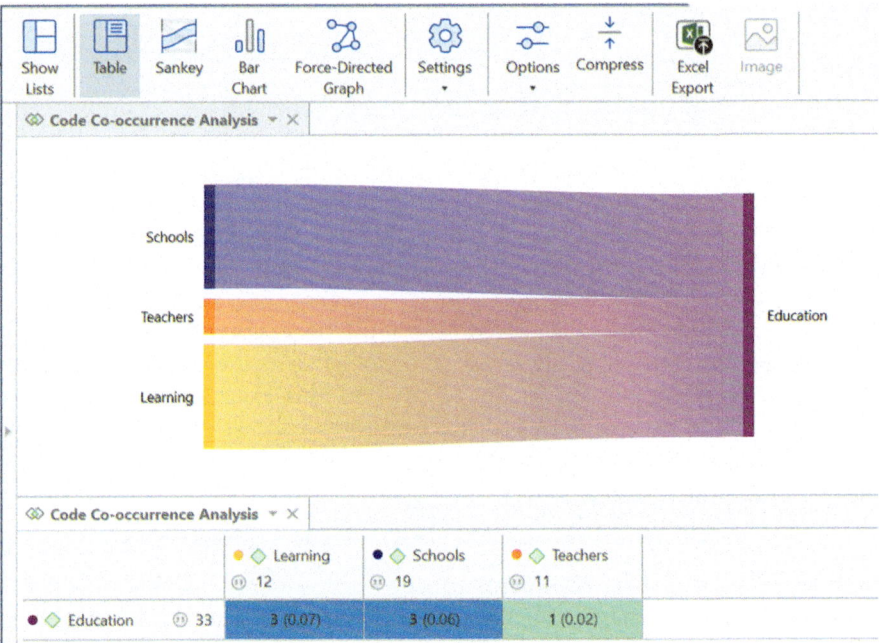

Fig. 11.10 Theme generation and Sankey diagram

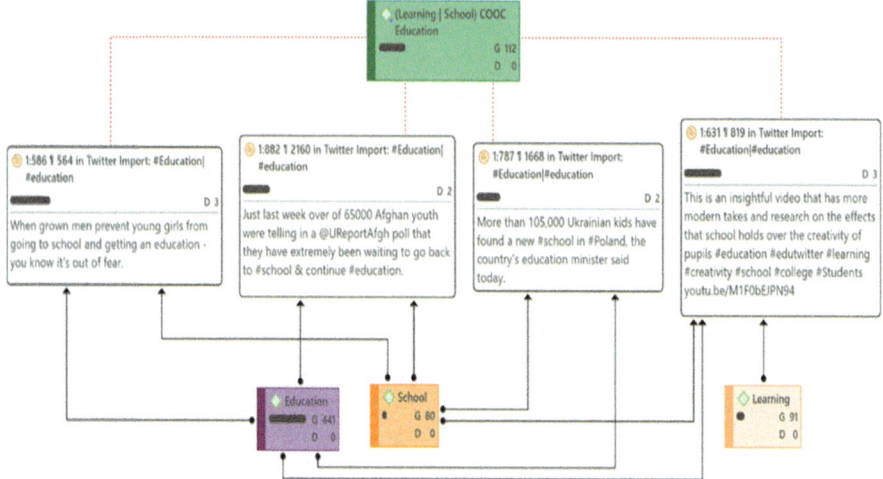

Fig. 11.11 A thematic network

Interpretation of Themes from Twitter Data

Text reports are generated using the Query Tool. In the example being discussed in the chapter, a theme may contain insights into education, learning and school. The steps for creating themes may involve the creation of sub-codes which are nested in codes. Reports should be interpreted with reference to clearly defined research objectives and the data analysis method used for data analysis. To understand how the code co-occurrence helps the researcher, we will discuss with an example.

The generation of themes should follow the process explained in Chap. 7. They can be represented as a network of codes, such as the one for 'education, school and learning' shown in Fig. 11.11. In this example, the theme has 112 quotations linked to three codes. The network shows the quotations with significant information about the theme.

ANALYSIS OF SOCIAL NETWORK COMMENTS

Analyses of social media comments can provide valuable insights into public perceptions and sentiments concerning a wide range of issues and topics, business, politics, media, celebrities, environment, sports, and so on. ATLAS.ti versions 22 and 23 have features that facilitate the analysis of comments on social networks.

1. Open web browser.
2. Go to exportcomments.com. One can export 100 comments for free per link. The comments are exported in excel form.

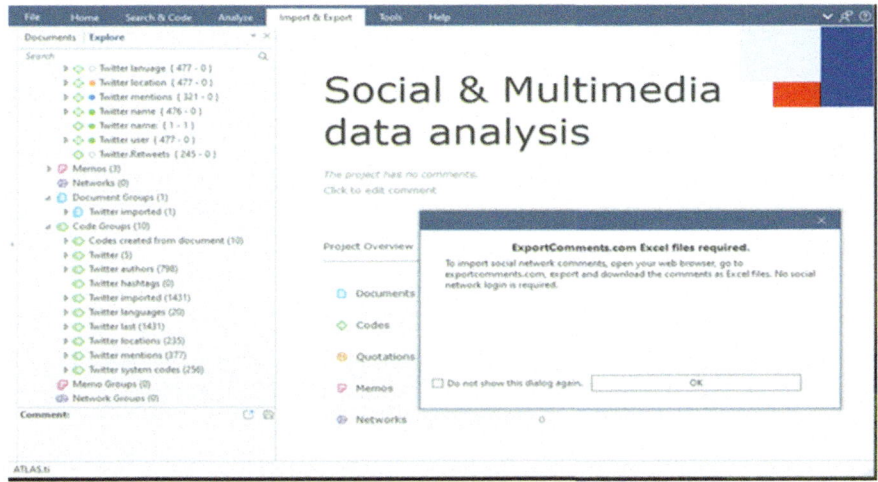

Fig. 11.12 Social network comments window

3. In the Import and Export function, select the social network comments option for importing the data to ATLAS.ti. Figure 11.12 is the screen display after clicking on the social network comments option. On clicking the OK button (Fig. 11.12), ATLAS.ti will open a window and ask to export the comments.

Figure 11.13 shows the eight social media platforms from which data can be imported to ATLAS.ti. They are Facebook, Twitter, Instagram, YouTube, TikTok, VK, Twitch and Discord. Unlike Twitter data, the other social media comments do not require # to export data. One only needs to copy the URL of the social media website for exporting comments in Excel form to ATLAS.ti. The comments are first saved on the computer and then imported to ATLAS.ti for analysis.

ATLAS.ti automatically codes the imported data in a social network and places them in the appropriate categories. In other words, contents are turned into quotations and users for the content becomes code (Fig. 11.14). The codes are then grouped after which they are analysed using various methods (content, narrative, sentiment and thematic analysis) depending on the objectives of the research.

ANALYSIS OF VIEWER COMMENTS ON YOUTUBE

In one study, the researcher exported comments on a YouTube video of Sheryl Sandberg speaking on 'Why we have too few women leaders | Sheryl Sandberg'. Out of 4793 comments, 100 (the limit for free export) comments were exported to ATLAS.ti (Fig. 11.15).

Fig. 11.13 Supported social networks

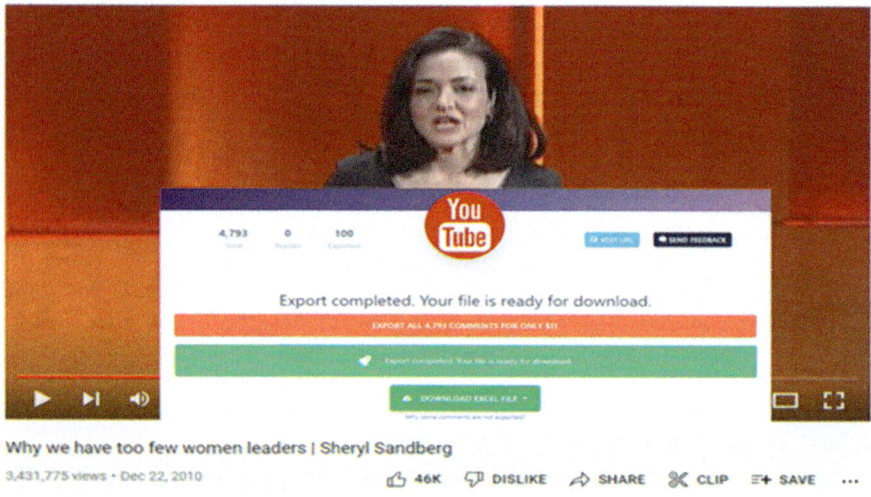

Fig. 11.14 Comments from Youtube

In Fig. 11.15, the code group 'YouTube user' has 100 quotations in the form of comments. Using qualitative content analysis, the researcher can identify contents, code, categorize and group them to identify themes.

Where necessary, the researcher can provide comments on code groups and codes. Similarly, researchers can read comments in quotation managers and write their comments (Fig. 11.15).

Figure 11.16 shows the code group 'YouTube users' that has 94 codes. ATLAS.ti has created codes from the users' names. The number in the Grounded column is the number of quotations associated with a code.

11 SOCIAL NETWORK COMMENTS AND TWITTER DATA ANALYSIS

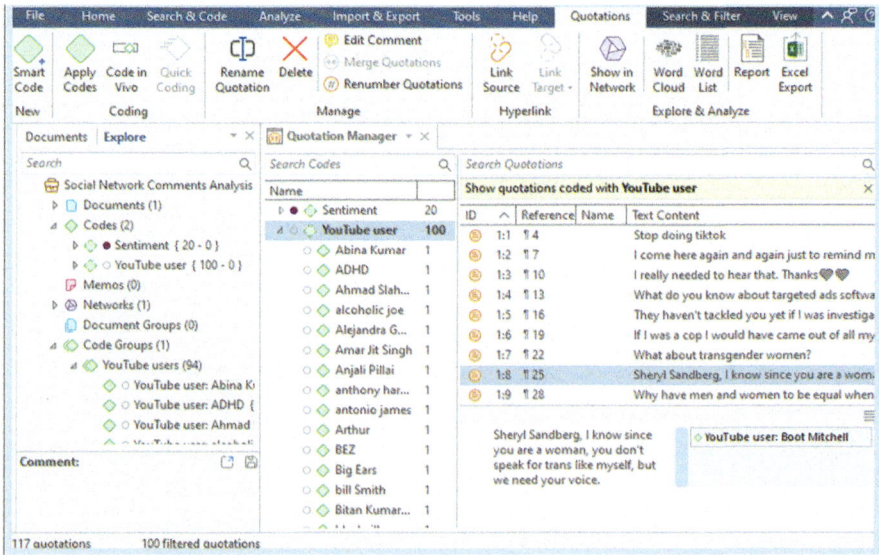

Fig. 11.15 Social comments in quotation manager

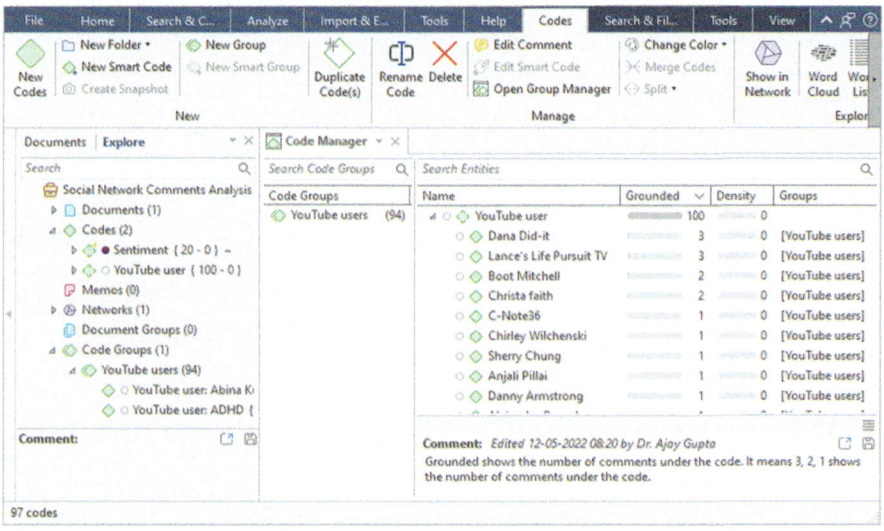

Fig. 11.16 Code groups and comments

Figure 11.17 is the display on the screen display the comments are imported into ATLAS.ti 22. It is good practice to insert the date and time of import in the document. Figure 11.18 shows the code group along with the code list. There are 94 codes—each username is treated as separate code—and 94 quotations.

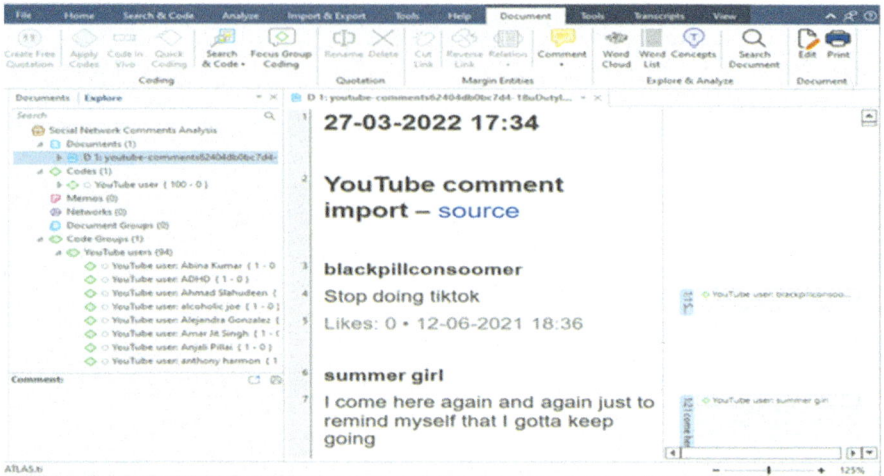

Fig. 11.17 Youtube comments, codes and code groups

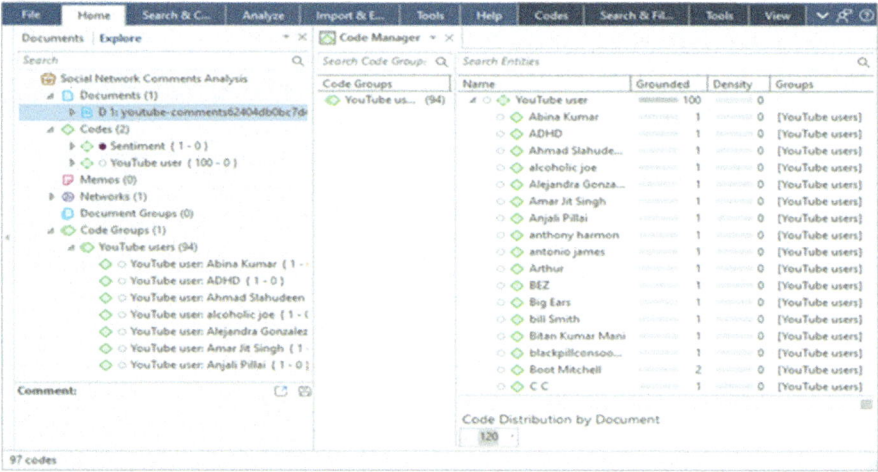

Fig. 11.18 Code groups and codes

After the import of comments into ATLAS.ti, the Word Cloud, Word List or Concepts options are used to find frequency of use of the words or concepts (Fig. 11.19). Whenever the auto-coding option is used, it is good practice to review the codes and comments before analysis. It must be remembered that manual coding is always better than auto-coding because the researcher has more control over the process.

The steps followed for analysing the comments are like the ones in Twitter data analysis. In the present example, sentiment analysis was done

Fig. 11.19 Concepts and quotations

for which only the positive sentiments were coded. Twenty quotations indicating positive sentiments were coded (Fig. 11.20) and shown in a network (Fig. 11.21). Other methods of analysis like sentiment, narrative, qualitative content, network, and thematic etc. can also be applied.

In this example, the researcher has coded only the positive sentiments included twenty quotations which were coded (Fig. 11.20). A network of the selected quotations is shown in Fig. 11.21. Figure 11.22 shows the text report

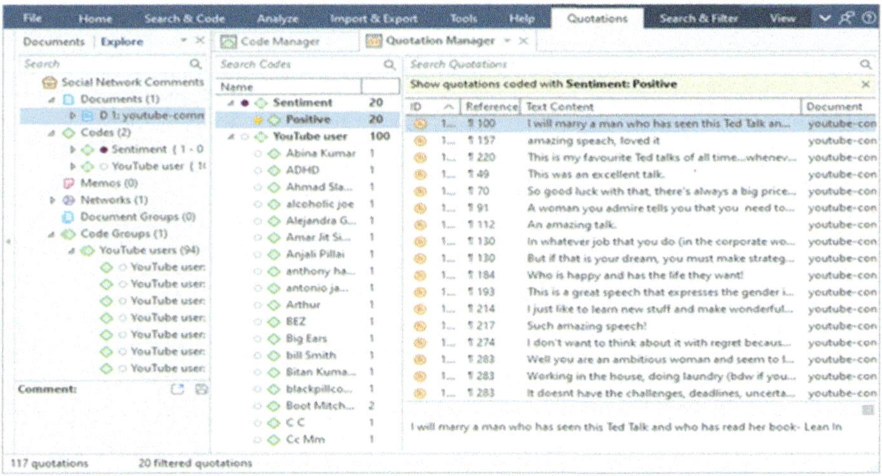

Fig. 11.20 Quotation manager for positive sentiment

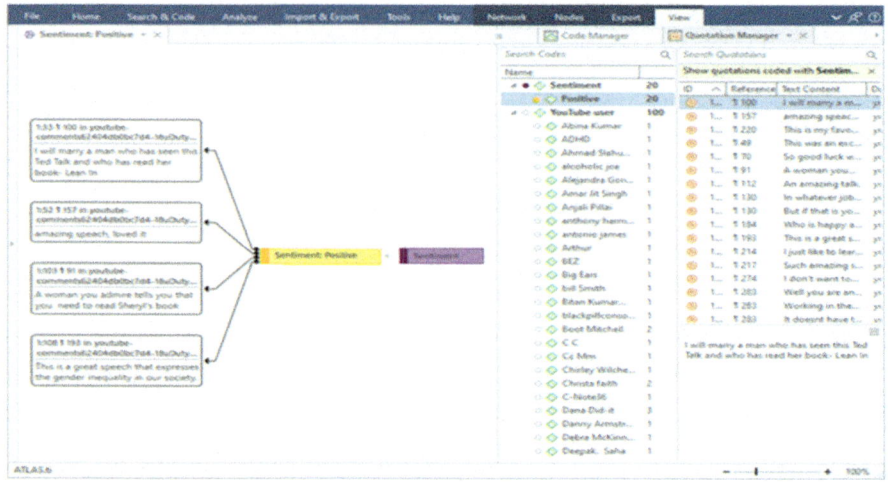

Fig. 11.21 Network for positive sentiment

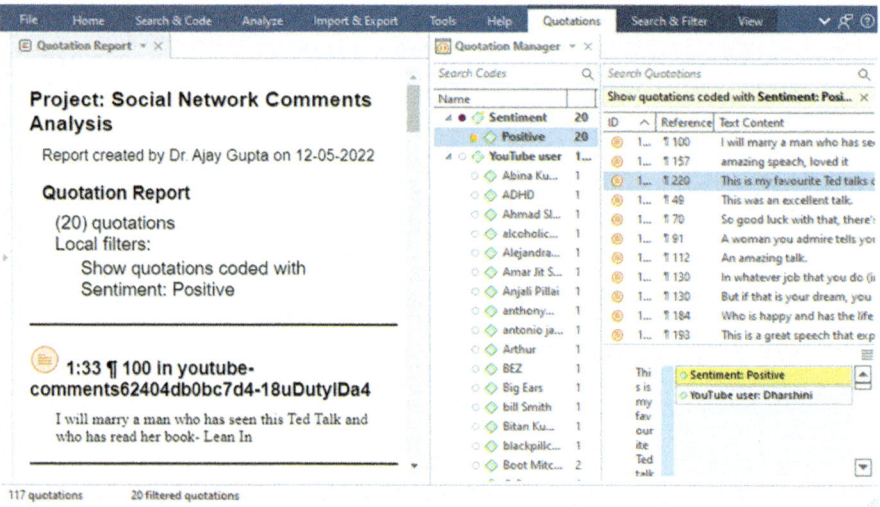

Fig. 11.22 Text report for positive sentiment

for positive sentiments. The number of quotations is shown on the right side in the display, and the report on the left.

Analysis of YouTube Comments Using AI Coding

Comments in the YouTube video, "5 fundaments of leadership-leadership development|Simon Sinek, were accessed through its URL and downloaded in Excel using 'exportcomments.com'. Then, with the help of Using Social

Network Comments in the Import & Import function, the comments were imported into ATLAS.ti.

Figure 11.23 shows how ATLAS.ti treats social network comments. Usernames—the names under which the comments are posted—become the codes, while the comments become the quotations for the code. The users whose comments were downloaded for analysis have been grouped as "YouTube users", the name of the code group.

Social network comments can be coded using the coding options explained in Chap. 10 (see section on Coding options for Survey Data), where AI coding is discussed. The Grounded value for each code shows the number of comments made by the user. For comparison, the YouTube comments were imported twice and saved in two documents, D1 and D2. Data in D1 was coded, while no coding was performed in D2. Through the use of Global filter, we can demonstrate the functionality of AI coding, which will help researchers understand good practices. The screenshots Fig. 11.23 (for D12 and 11–24 (for D1) compare the two documents.

It is necessary to understand the limitations of AI coding. AI codes comments within specific word limits with the usernames only (Fig. 11.23). Other comments are coded along with the username, as shown in Fig. 11.24. The comments, 'Amazing video', and 'I am so excited right now. GOD is up to something', were coded with the usernames only. Both comments show positive sentiments and can be coded accordingly. However, these were missed

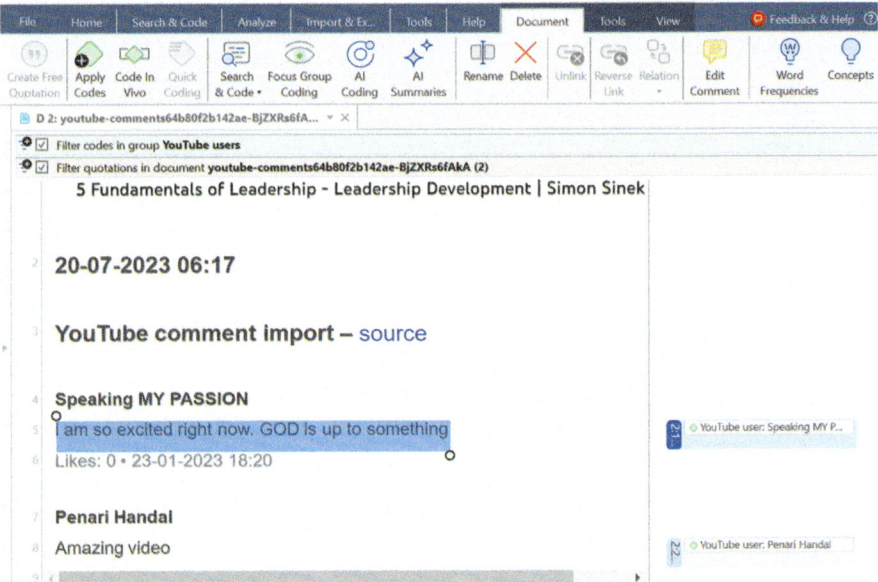

Fig. 11.23 Youtube comments in ATLAS.ti 23

by AI Coding. Therefore, it is recommended that researchers pay particular attention to comments that are coded under the user's name.

Figure 11.25 shows the AI code group with code categories and codes grouped under 'AI codes. Each code category includes several connected codes. These should be reviewed, linked with the appropriate relationships, and merged or split as necessary.

In the code category 'Personal Development, there are several codes linked to 35 quotations. After review, many codes can be merged based on their relationships. The codes in the category, 'personal development' is shown in Fig. 11.26. Leadership essentials, leadership defined, criticism, relative perspective and self-esteem are sub-themes under the theme 'personal development'. Leadership concept is the code group that has code categories. Figure 11.27 shows the report generation option under the Export menu.

A network for the theme 'Leadership concepts' is shown in Fig. 11.28. The sub-themes under the code 'personal development' are shown in purple.

Sentiment analysis was carried out to segregate the positive and negative comments (shown in the code category 'Sentiment' in Fig. 11.29). A network of sentiments is also shown in Fig. 11.30.

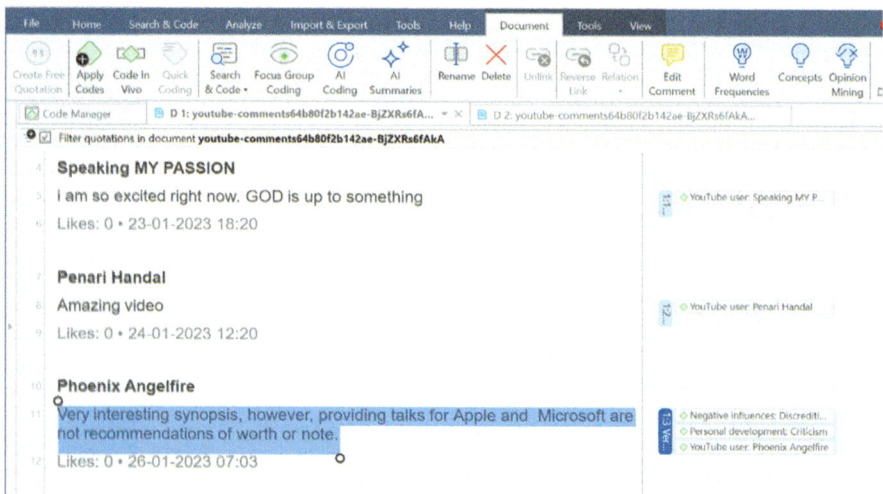

Fig. 11.24 YouTube comments with AI coding

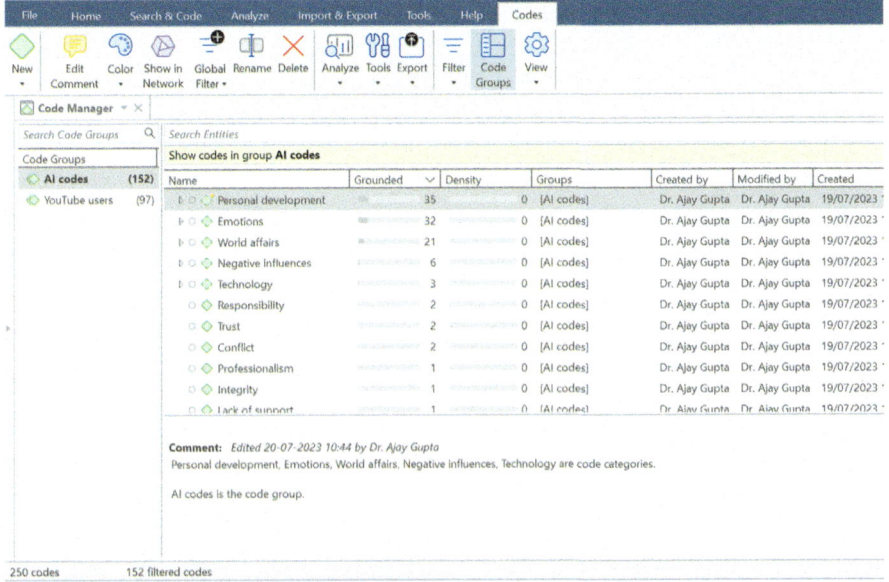

Fig. 11.25 Ai codes, code categories and group

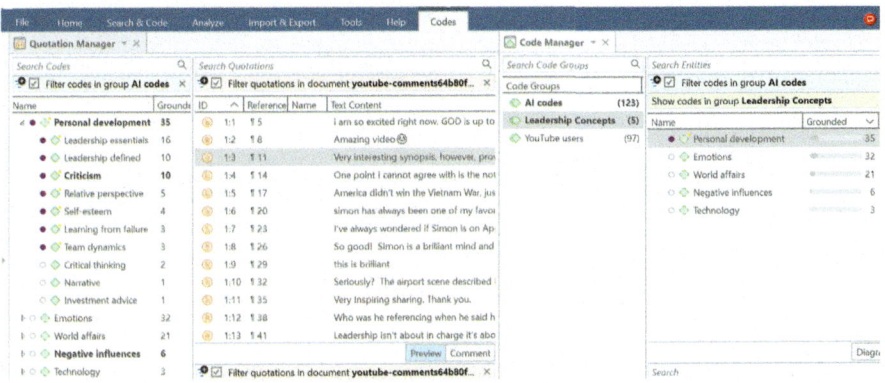

Fig. 11.26 Code category and codes

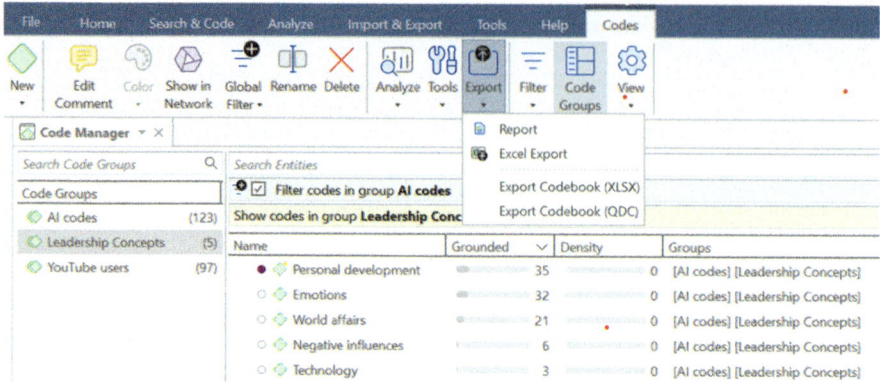

Fig. 11.27 Themes and report option

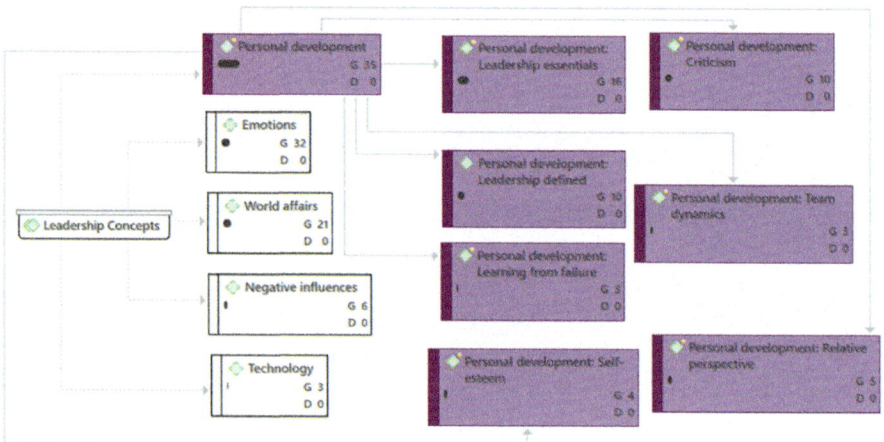

Fig. 11.28 A theme network

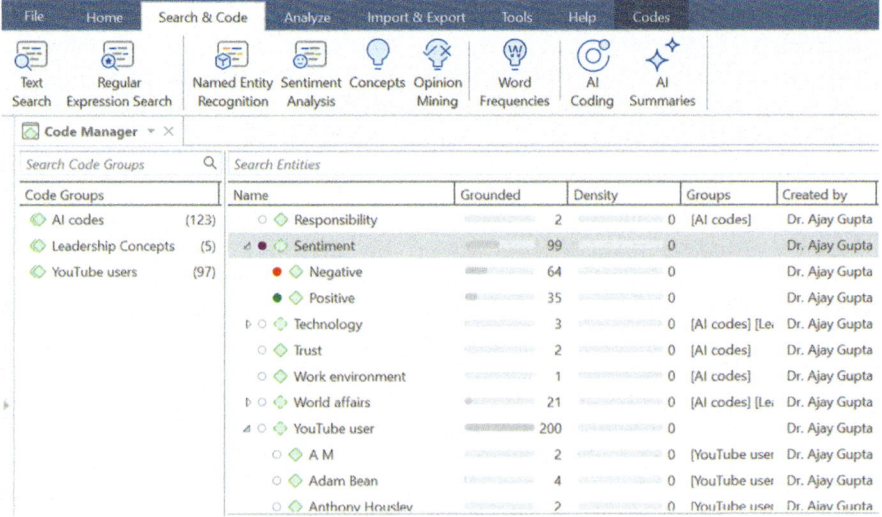

Fig. 11.29 Sentiment analysis

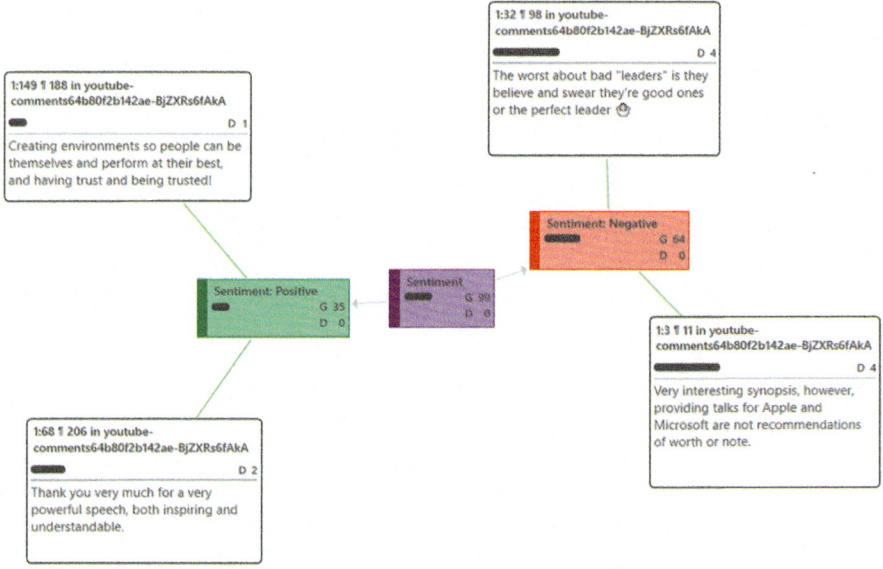

Fig. 11.30 A sentiment network

Recommended Readings

Choe, Y., Lee, J., & Lee, G. (2022). Exploring values via the innovative application of social media with parks amid COVID-19: A qualitative content analysis of text and images using ATLAS.ti. *Sustainability*, *14*(20), 13026.

Friese, S. (2019). *Qualitative data analysis with ATLAS.ti*. Sage.

Mohamad, A. M., Yan, F. Y. Y., Aziz, N. A., & Norhisham, S. (2022, May). Inductive-deductive reasoning in qualitative analysis using ATLAS.ti: Trending cybersecurity twitter data analytics. In *2022 3rd International Conference for Emerging Technology (INCET)* (pp. 1–5). IEEE.

Paulus, T. M., & Lester, J. N. (2016). ATLAS.ti for conversation and discourse analysis studies. *International Journal of Social Research Methodology, 19*(4), 405–428.

Smit, B. (2021). Introduction to ATLAS.ti for mixed analysis. In *The Routledge Reviewer's Guide to Mixed Methods Analysis* (pp. 331–342). Routledge.

Soratto, J., Pires, D. E. P. D., & Friese, S. (2020). Thematic content analysis using ATLAS.ti software: Potentialities for researchs in health. *Revista brasileira de enfermagem, 73*.

Woods, M., Paulus, T., Atkins, D. P., & Macklin, R. (2016). Advancing qualitative research using qualitative data analysis software (QDAS)? Reviewing potential versus practice in published studies using ATLAS.ti and NVivo, 1994–2013. *Social Science Computer Review, 34*(5), 597–617.

CHAPTER 12

Analysis of Focus Group Data

Abstract This chapter explains the analysis of data from focus group discussions and piggyback focus groups. Types of data, frameworks for data collection and analysis have been discussed. ATLAS.ti offers several user-friendly ways of analysis to help the researcher arrive at rich insights into the study topic and present their findings. Focus group coding, auto-coding, thematic networks, the use of symbol in focus group data, Word Cloud, and Concepts are explained with the help of flow charts and screenshots and, wherever required, in simple steps.

Focus Group Discussion

The concept of focused interviews was introduced in the 1940s by Merton and his colleagues (Merton et al., 1956). The word focus indicates the direction of the group's discussion of a particular topic or issue. Focus groups have a few distinct features (Litosseliti, 2003). The participants in a focus group discussion are required to share their views, perspectives, and opinions on the subject of the discussion only. A focus group discussion provides the social space for individuals to interact with each other. In the process, data and insights are generated that would not otherwise be accessible to the researcher (Duggleby, 2005; Morgan, 2010).

Unlike interviews, in a focus group discussion, the researcher takes a peripheral, rather than a centre-stage role (Bloor et al., 2001; Hohenthal et al., 2015). Various studies have used focus group discussion to clarify and extend findings, such as motivations for different resource use regimes (Harrison et al., 2015; Manwa & Manwa, 2014), qualify or challenge data collected through other techniques such as ranking results through interviews (Harrison et al., 2015; Zander et al., 2013); and to provide feedback to research participants (Morgan et al., 1998). However, the use of focus group discussion technique is not recommended when there is a risk of raising participants' expectations that cannot be fulfilled or where "strategic" group biases are anticipated (Harrison et al., 2015).

In such settings, group dynamics play a crucial role in generating information, with the moderator playing a deterministic role. Since focus group discussion depends on participants' dynamics, it should be avoided where participants are uneasy with—or antagonistic towards—each other, or where social stigmatisation due to the disclosure may arise (Harrison et al., 2015).

Focus group discussions are valuable for generating ideas. Merton et al. (1956) argue that a focus group discussion widens the range of responses. Further, they also help to discover information that would otherwise be difficult to obtain from other methods, such as interviews and observations. Focused Group Discussions also encourage participants to overcome barriers to communicating.

Focus group discussions are used in several disciplines of study. Although the use of focus group discussion as a research technique has been dominant in the other disciplines such as sociology and psychology, its use has recently grown in the conservation social science research (Bennett et al., 2017; Paloniemi et al., 2012).

Conflicts among participants can—and do—arise in a Focus Group Discussion. The researcher must expect differences in opinions and perspectives, or a situation in which the discussions touch upon personal issues. Properly managed, differences and disagreements in a focus group discussion can yield rich information. Therefore, researchers should expect and be prepared for differences of opinion among the participants. They should prepare for the discussions and create the conditions for meaningful interactions and communication among the participants and between the participants and moderators (Miranda, 1994). Further, participants should be responsive to environmental factors. The researcher should make the environment conducive for communication and discussion on the subject. It is necessary to set guidelines for what to avoid in the discussion that might create inter-participant conflict.

Krueger and Casey (2000, p. 10) identified five characteristics of focus groups. In a focus group discussion, the role of the researcher is reduced (Madriz, 2000), and the interactions sometimes become unruly (Wilkinson & Kitzinger, 2000). This interaction between participants can enable them to challenge each other, effectively revealing more private, 'backstage' behaviours and allowing the discussion to move deeper into the topic area (Hyde et al., 2005; Robinson, 2009). On the other hand, the interaction might also lead to silence on a topic (Smithson, 2000) as the group may collude with, or intimidate other members, of the group to not say anything of importance.

It is important to note that focus group discussions do not take place in natural settings. The choice of location for a focused group discussion is determined by various factors, such as the convenience of access, the absence of distractions, and several others. The researcher must make every effort to make the participants feel at ease. For example, they can offer tea, soft drinks, and snacks. They can also engage in informal conversations with the participants before the commencement of the discussion. Openness and interaction depend on the nature and sensitivity of the issues being discussed.

Researchers and moderators should be aware of biases that are likely to affect group dynamics and influence discussions. Various types of dynamics are likely to be at play in a focus group discussion, examples of which are the dominance effect, halo effect, and groupthink (Mukherjee et al., 2015). The researcher should be on the alert and address these issues.

In a focus group discussion, members of the participating group interact and share their experiences and insights. Group processes and the moderators play a significant role in making the discussion productive. There is no thumb rule to start the discussion; however, the moderator may explain a few guidelines for maintaining decorum and motivation. The discussion begins with the researcher inviting each participant to share their opinions on the topic.

The participants are not likely to know each other before the focus group discussion. Scholars suggest that a focus group discussion should have a minimum of 2–3 and a maximum of 10 participants. However, the size of a focus group is decided by the researcher after considering the objectives and complexity of the study, the diversity of perspectives and opinions among the participants, and various other factors and constraints. If the aim of the study is to understand behavioral issues, it is better to invite fewer people. On the other hand, if it is a pilot study to understand emerging issues, inviting more people to participate may be more appropriate.

Motivation and experience of participants is vital. Those who are experienced and motivated are more likely to share information than those who are not. Managing a larger group can be challenging at times as fewer questions or opinions can be discussed.

Participants in a focus group discussion must be explained the purpose of the study, as well as the importance of their role in it. In addition, the researcher should also share as much detail as possible, the processes that being followed and the possible benefits of the study. Participants' experiences and knowledge must be appreciated. There may be expectation of reward or some form of compensation for participation, which can pose ethical issues. Therefore, there must be as much transparency as possible in the communication process. Though there are no guidelines, the way ethical issues are addressed are highly dependent on the settings of the study. For example, reimbursing travel expenses and arranging for refreshments are regarded as good practices. Researchers must be aware of the possibility that payments and incentives can influence participant behaviour and can compromise the objectivity of the study.

Table 12.1 shows the steps followed for collecting data in a focus group discussion and its subsequent analysis (Nyumba et al., 2018).

Table 12.1 Conducting a focus group discussion

Research design	Objectives-purpose, key questions, ethical protocol
	Identify and recruit participants-ensure homogeneous composition (gender, education, language, etc.), decide number of participants, recruit a facilitator, decide on number of focus groups
	Identify suitable locations-select venue away from distractions, arrange materials (recorder, name tags, consent forms etc.)
Data collection	Pre-session preparation-familiarize with script, group dynamics, seating preferences, record duration
	Facilitation during meeting-introduction (self, consent, group rules, confidentiality), start discussion (record, observe, probe, pause, observe non-verbal cues), track questions and follow up on themes of discussion, conclude (acknowledge participants)
Data analysis	Options include-ranking, coding, content, discourse, and conversational analysis
Results and reporting	Decide on target audience-academics, policy makers and practitioners, participants of the study

The researcher should define the purpose of the focus group approach and identify the participants for the discussion. They should invite participants at the specified location after getting informed consents from them. The researcher should decide the number of focused group discussion he intends to conduct. The researcher and facilitator set the guidelines for discussion and explain the subject and ethics involved in the research.

After a focus group discussion has ended, the data is transcribed and uploaded to ATLAS.ti. The data is then coded with the appropriate methods (Table 12.1 and Figs. 12.1, 12.2, 12.3, 12.4 and 12.5). Analysis of focus group data mostly adopts the thematic, content, conversational and discourse types of approaches. The results are interpreted with reference to research objectives and shared with the participants in the study.

NATURE AND TYPES OF FOCUS GROUP DATA

Focus group data is largely in transcription form. Focus group discussion data is collected through audio or video tape. Many types of data emerge from the focus group discussion. Since data is recorded during the discussion, moderators take notes, and researchers also capture information based on observations. Audio or videotaped data can be either fully transcribed or only selected segments should be transcribed. Generally, it takes several hours to transcribe a focus group discussion and may result in several text pages. Researchers may choose to create an abridged transcript for the selected clipping. For this, researchers listen to the audio/video tape and selectively transcribe the portion of recorded information and attach the transcript with the selected clippings.

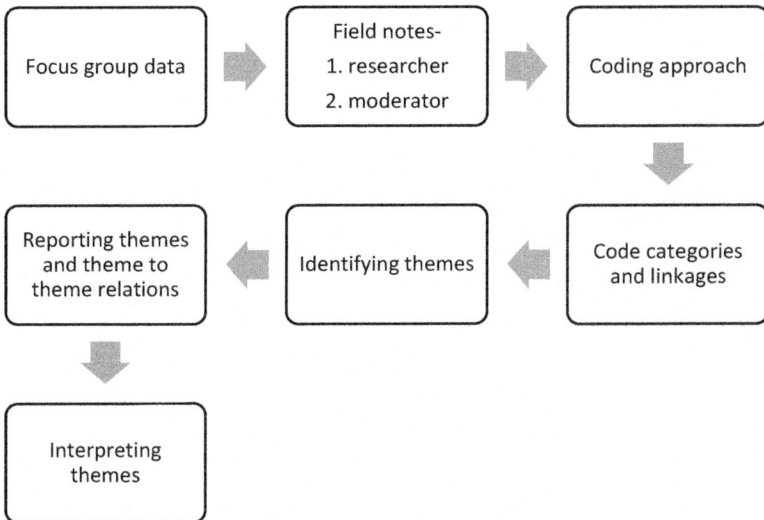

Fig. 12.1 Focus group data analysis framework

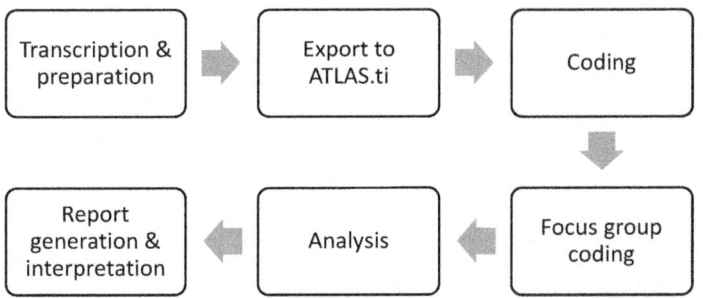

Fig. 12.2 Flowchart for focus group discussion analysis

Field base data involves the moderator's notes and notes based on the researcher's memory. Both play a vital role in capturing information, including non-verbal communication, group dynamics, facial expressions, excitement, shyness, avoidance, laughs, group conflict and emerging issues, etc.

Data can be organized and analyzed on three levels: individual, group, and group interaction (Duggleby, 2005). However, researchers don't have a consensus on the unit of analysis for the focus group data analysis. Some believe that the individual or the group should be the focus of analysis instead of the unit of analysis (Kidd & Marshall, 2000); however, Morgan (1997) proposes using the group as the unit of analysis.

Researchers should analyze emergent themes based on consensus and dissenters. This step will help us understand the phenomenon more and more broadly.

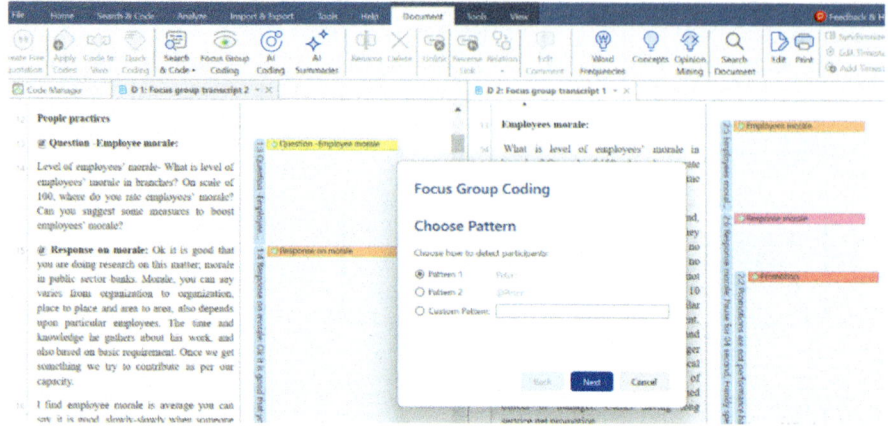

Fig. 12.3 Focus group coding pattern

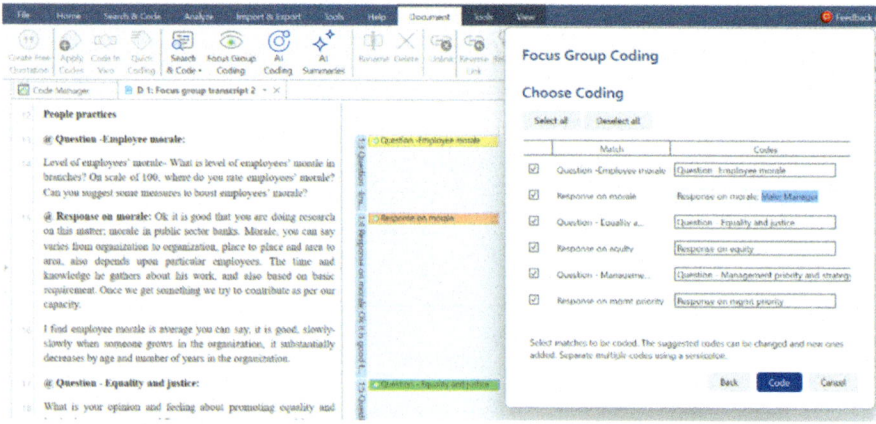

Fig. 12.4 Adding codes in focus group discussion coding

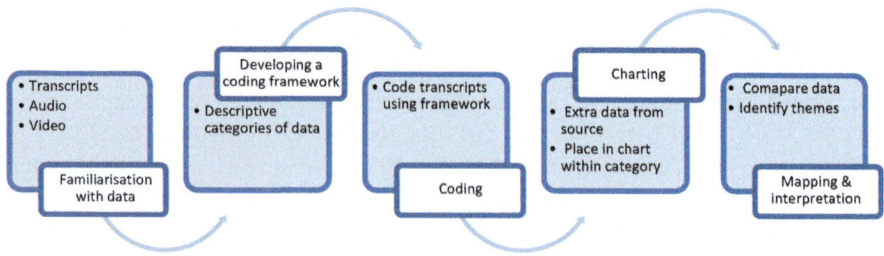

Fig. 12.5 Framework for data analysis (Ritchie & Spencer, 1994)

Types of Focus Group Discussion

There are five distinct forms of offline focus group discussions and two types of online focus group discussions (Nyumba et al., 2018). Each serves a unique function. Single focus group, two-way focus group, dual moderator focus group, duelling moderator focus group, and respondent moderator focus group are the types of offline focus group discussions. Mini-focus groups and online focus groups are online focus group discussions.

In a single focus group, participants and facilitators engage in interactive conversations about a subject in a single location (Morgan, 1996). In a two-way focus group, one group actively discusses the issue while the other group watches the first group (Morgan et al., 1998) while the first group is oblivious of the second group's presence. The second group watches and records exchanges that frequently result in divergent findings. Dual moderators play a distinct role inside the same focus group, as the name suggests (Krueger & Casey, 2000). In a focus group with duelling moderators, there are also two moderators, but they adopt opposing positions on a topic (Krueger & Casey, 2000). The purpose is to increase the depth of information disclosure (Kamberelis & Dimitriadis, 2005) by including opposing viewpoints in the conversation. In the respondent moderator focus group, the researcher recruits some participants to temporarily assume the position of moderators (Kamberelis & Dimitriadis, 2005), with the intention of influencing participants' perspectives and boosting the likelihood of honest responses.

Online focus groups are preferable in situations where participants are difficult to reach. These individuals are required due to their extensive knowledge (Hague, 2002). Mini focus groups consist of between two and five people. Online focus groups are conducted via chat rooms, conference calls, or other online methods. Some problems that the researcher could encounter include connectivity issues and the inability to capture nonverbal data (Dubrovsky et al., 1991).

Piggyback Focus Groups

Focus group discussions are usually arranged at a location that is accessible to, or convenient for the participants. However, for various reasons, this may not always be possible. In such cases, the researcher may adopt a piggyback approach. The literal meaning of piggyback is to ride on someone's shoulders or back; in the context of focus group discussions, it means taking advantage of a certain situation to complete a task. Thus, researchers can conduct Focus Group Discussions by 'piggybacking' on situations where people gather, such as, for example, on their morning walks, at shopping malls, weddings, and religious events, where the participants are likely to come together for another purpose. The researcher must also take care that the primary purpose of the participants' gathering is not disturbed. This strategy may work well when planned carefully.

Frameworks for Focus Group Data Analysis

ATLAS.ti helps analyze focus group data in several ways. It facilitates data analysis within and between focus groups. The themes can be presented and interpreted using significance values appropriate to qualitative research. Researchers have offered several frameworks to analyze focus group data. The chapter has shown focus group data analysis using real projects.

Focus group discussion helps the researcher arrive at rich insights into the study topic and present their findings. Focus group discussions are used to bring up issues that are not possible using interviews and other methods (Duggleby, 2005; Morgan, 2010). In focus group discussion, the researcher takes a facilitator or moderator role (Bloor et al., 2001), whereas the role of a researcher in an interview is that of an investigator. Group dynamics play a great role in generating information, and the role of moderator becomes the crucial part. Further, biases, personal differences and several issues are natural to emerge that researchers should address before and during the discussion.

Focus group interviews are suitable for respondents to cooperate when the time to conduct a one-on-one interview is limited, especially when respondents are hesitant to provide information (Krueger & Casey, 2009). However, focus group discussion is not preferred where group biases are anticipated (Harrison et al., 2015). Further, participants are hostile towards each other for several reasons, such as social stigmatization and chances of disclosure is possible (Harrison et al., 2015).

There are several frameworks to analyze focus group data. According to Duggleby (2005), data can be organized and analyzed on three levels: individual, group, and group interaction. However, researchers don't have a consensus on the unit of analysis for the focus group data analysis. Some believe that the individual or the group should be the focus of analysis instead of the unit of analysis (Kidd & Marshall, 2000); however, Morgan (1997) proposes using the group as the unit of analysis.

According to Ritchie & Spencer (1994), focus group data analysis comprises five processes: familiarization of data, identifying a thematic framework, indexing, charting, mapping, and interpretation. It is similar to the thematic analysis of Braun and Clarke (2006). It demands creating codes code categories, comparing, identifying themes, and interpreting them with reference to research objectives.

Krueger (1994) offered seven criteria as a framework for analyzing coded data: words, context, internal consistency, frequency and extent of comments, specificity of comments, the intensity of comments and key ideas. Krueger and Casey (2000) offered five criteria: frequency, motion, specificity of responses, extensiveness, and big pictures. Nyumba et al. (2018) offered data analysis that includes ranking, coding, content, discourse, and conversational analysis.

Morgan (1988) offered a "three-element coding framework," which refers to the two steps involved in the content analysis that yields quantitative results and the one step involving the ethnographic analysis that yields qualitative results. The first step involves coding and creating code categories (Charmaz, 2006). This step involves relationships between codes, code categories and their occurrences (grounded). The second step involves focused coding for recurring ideas and wider themes connecting the codes (Charmaz, 2006; Krueger, 1994; Ritchie & Spencer, 1994). This step involves the comparison across focus groups, group dynamics, and participants' statements (Carey & Smith, 1994; Morgan, 1995).

According to Leech and Onwuegbuzie (2007, 2008), qualitative analysis techniques that can be used to analyze focus group data include grounded theory analysis (Charmaz, 2006; Glaser, 1978, 1992; Glaser & Strauss, 1967; Strauss, 1987), content analysis (Morgan, 1988) and discourse analysis (Potter & Wetherell, 1987).

The result can be presented in a narrative or pointwise format. The report should capture participant information such as gender, age, education, key quotations, and emphasis and should be sent to participants for member checking and validating results (Birt et al., 2016; Lincoln & Guba, 1985). The process is not free from caveats.

Focus group data can be analyzed using ATLAS.ti 23 based on suggested frameworks by researchers. Data can be coded using inductive, deductive, and in-vivo codes based on research objectives. It follows the creation of code categories and linking them based on appropriate sharing relationships. Using thematic analysis automatically captures the attribute of content analysis. Therefore, thematic analysis with minor modifications appears more appropriate for focus group data analysis.

Reporting should be performed based on the occurrence of quotations and code-to-code relations (C value). This will lead to creating reports based on themes (theme-to-theme relations) and should be interpreted. Data analysis within and across focus groups can be reported and interpreted in ATLAS.ti. The framework for focus group data analysis is shown in Fig. 12.1.

The field notes during the interview should be used to probe questions after the focus group. The responses should be used for data analysis. Clues such as shyness, hesitation, excitement, avoidance, and worry should be noted in the field notes during the focus group discussion. The researcher and moderators both should take such notes. Later, this should be used to obtain a response, which will be used for data analysis. Screenshots, a Sankey diagram, and networks for the real project can be used to interpret findings.

Flowchart of Focus Group Discussion Analysis

Focus group data usually follows the process shown in the flowchart (Fig. 12.2). Data is prepared after selecting the transcription and coding patterns after which the transcripts are added to ATLAS.ti for data analysis. The approach to coding must be guided by the research objectives.

While focus group coding, the researcher should deselect data not required, add codes using the semicolon in the code box (Fig. 12.3). In the example, two new codes, 'Male' and 'Manager', were added using a semi-colon. By pressing Code, appearing bottom right, both codes and the code response on morale will appear.

After coding, the researcher can create code categories, code groups, or even smart codes as is done with text data. Data analysis is done with co-occurrence analysis in the Code Document Analysis option. If needed, Query Tools can be used to create reports. Interpretations should be supported by key quotations, screenshots, tables, networks etc.

Computer-Aided Analysis of Focus Group Data

Analysis of qualitative data begins with coding. The coding process and type is determined by the research type and objectives. It is the researcher who decides the most suitable coding approach: Quantitative content analysis, narrative analysis, or document analysis use quantitative measures like frequencies, counts etc. while coding.

According to Agar and MacDonald (1995), quantification of focus group discussion data is not the correct approach. However, Morgan (1995) and Shamdasani (1990) suggested cutting and pasting important data segments in content analysis which can be either qualitative or quantitative. Stewart and Shamdasani (1990), and Knodel (1993) suggested the use of computer-assisted software in analyzing focus group discussion data using inductive, deductive, or abductive coding methods.

Evidence on the influence of the role of the researcher on the relationship with the study's participants reveals a fundamental difference between interviews and Focus Group Discussions (Smithson, 2000). Interviews involve a one-to-one, qualitative and in-depth discussion, where the researcher performs the role of an "investigator." This means that the researcher asks questions, controls the dynamics of the discussion, or engages in dialogue with one individual at a time. In contrast, in a focus group discussion, researchers are "facilitators" or "moderators" of interactions between participants and not between the researcher and participants. Unlike interviews, the researcher must maintain a presence in a focus group discussion (Bloor et al., 2001; Kitzinger, 1994).

Coding Options for Focus Group Data

After data is imported into ATLAS.ti, Focus Group Coding is applied after choosing a pattern that best fits the transcript (Table 12.2).

Table 12.2 Focus group coding pattern

Focus group coding pattern	Style of writing respondent & response	ATLAS.ti coding
Pattern 1	Ajay:	Ajay will be the code (content before (:) will be code), and the response will become a quotation
Pattern 2	@Ajay	Ajay will be the code (content after (@)will be code), and the response will become a quotation
Custom pattern	Content appearing in the transcript gets coded	For example, morale in the custom pattern will code contents having the word morale with code-morale

> **Important** It should be remembered that while selecting a focus group coding pattern, aligning it with the transcript preparation pattern is important. For example, a transcript using @ choose focus group coding Pattern 2, which uses @ to code the data. Data will not be coded if the incorrect pattern is selected.

While coding focus group data, the researcher should deselect content that should not be coded. In the codes section, other codes can be added using a semicolon (;). Using @ as a prefix and: as a suffix in the question, response or speaker in a transcript can use both focus group coding patterns. Figure 12.3 shows the usage of semi colon, and two new codes 'Male' and 'Manager' has been added.

In Fig. 12.3, Document 1 has used the '@' as well as ':', meaning that coding patterns are used, whereas Document 2 has used only ':' (Coding pattern 1 in Table 12.2). Irrespective of the coding pattern selected, codes can be added using a semicolon (Fig. 12.4).

Coding Focus Group Discussion Data

Coding process started with the following steps (Fig. 12.4):

1. Inserting a semi-colon (;), adding a space after the default code.
2. Writing the new code (Male).
3. Inserting a semi-colon and a space after 'Male', and then typing the second code (Manager).

Table 12.3 Focus group coding patterns

Patterns	Symbol and code description	Interpretation
1	:	Used as a suffix for the speaker All text followed by a ':' will be quotation and codes preceded by ':' Additional information, such as gender, age, etc. can be added inserting a semi-colon (;) after code (Fig. 12.5). Each identifier then becomes a code
2	@ and:	The symbol '@' is used as a prefix and ':' as suffix for the speaker Text after ':' will be quotation. Demographic coding can be applied (Fig. 12.5)
Custom	Regular expression	Used for coding two or more segments of information or speakers on researcher's preference using regular expressions

4. Click on Create.
5. The new codes are shown with the default code (see Fig. 12.4).

ATLAS.ti provides three coding options for coding focus group data (Tables 12.2 and 12.3). The researcher can code data using certain identifiers for the participants. The documents supported by ATLAS.ti are doc, docx, rtf or txt.

The focus group discussions must be transcribed before coding. The next step is to upload the transcripts in ATLAS.ti using the Add Document option. The data is now ready for coding.

Two-Level Analysis of Focus Group Data

As with other methods of qualitative data analysis, the basis for addressing the research question is the actual words and actions of the participants. Data should be transcribed in analytic language. The researcher retains responsibility for the analysis's interpretation process. The study participants may not always respond to the researcher's queries. The study of data gathered from focus group discussions follows several phases:

1. Transcription of the audio or video data collected.
2. Coding (deductive, inductive, hybrid, in-vivo, affective, and value-based methods). Predetermined codes are deductive codes (Gale et al., 2013), codes emerging from the data are inductive codes (Charmaz, 2006), and codes referring to the group dynamics are also inductive codes (Barbour, 2014), which help to comprehend the group's behaviour regarding the investigated issues. This is also known as affective coding (Saldana,

2021), which investigates the subjective opinions, emotions, and attitudes of individuals. The values coding of a participant reflects his or her values, attitudes, and beliefs, thereby representing his or her worldview (Saldana, 2021).
3. In the two-step approach (Silverman, 2006; Wong, 2008), data analysis should initially be conducted at the group level, where the group serves as the unit of analysis. It contains phrases such as "many participants agreed with a certain point, many participants disagreed with a certain point, etc." Researchers should avoid using quantitative findings that do not contribute to the advancement of knowledge. The individual level is followed by the second data analysis of what individuals have said in an integrative and theoretic manner. The procedure entails identifying patterns, consistencies, themes, commonalities, and distinctions within and between diverse data sources. Lastly, comparisons between groups involved in the topic are made (Bromley et al., 2003).
4. Using consensus, coherence, triangulation, and reflexivity to establish validity and reliability. Cross-validation of focus group results with other techniques employed in the same or comparable study.

The most fundamental level of analysis consists of a descriptive account of the data: an explanation of what was stated, with no assumptions made. The second level of analysis is interpretative, which involves the comprehension of themes (or perspectives), the creation of connections between themes, the demonstration of how those themes emerged, and the development of a theory based on the data. It is encouraged to use a model to illustrate the relationship and reciprocal influences of each category and theme.

The outcomes should be presented in the context of the group's discussion, as opposed to a single focus group session. In interpreting findings, both the intensity of comments and the specificity of responses to probes must be considered. Although it has been suggested that reporting results of focus groups in numerical terms is inappropriate, it has been argued that some qualitative data can be analyzed quantitatively. If a theme appears repeatedly in the data, it is acceptable to quantify its frequency. Simple statistical frequencies can be used to describe the important characteristics of the themes.

Focus Group Data Analysis

There are two types of data produced by focus group discussion: spoken and unspoken. Spoken are verbal, while non-spoken is nonverbal data. Researchers should analyse both sets of data using qualitative methods, such as grounded theory analysis (Charmaz, 2006; Glaser, 1978, 1992; Glaser & Strauss, 1967, Strauss, 1987), content analysis (Morgan, 1988), and discourse analysis (Charmaz, 2006; Glaser, 1978, 1992; Strauss, 1987).

Initial coding (Charmaz, 2006) entails the development of several code categories codes, while focused coding identifies recurring ideas and larger themes connecting the codes (Charmaz, 2006; Ritchie & Spencer, 1994).

Five processes comprise framework analysis (Ritchie & Spencer, 1994): familiarization, identifying a thematic framework, indexing, charting, mapping, and interpretation (Fig. 12.5). It employs a thematic approach and allows themes to emerge from research questions and participant narratives.

Krueger (1994) offers seven established criteria that provide the following topics as a framework for analysing coded data: words; context; internal consistency; frequency and extent of comments; specificity of comments; intensity of comments; and key ideas.

Krueger Focus Group Analysis Tips

1. Words-Researchers should consider the actual words and their meaning used by participants. It includes phrases, local words, jargon etc.
2. Context-Researchers should find stimulus of comment in the light of that context. It includes tone, intensity and oral comments.
3. Internal consistency-Researchers should consider how participants offer consistent opinions or shift opinions and clues based on discussion.
4. Frequency or extensiveness- How often a comment is made decides the insight. Researchers should find how participants emphasize contents and their frequencies etc.
5. Intensity-Researchers should consider the depth of feeling in which comments are expressed including voice tone, speed, certain words, etc.
6. Specificity-Researchers should pay greater attention to experience based responses than hypothetical responses.
7. Finding big ideas-Researchers should show larger concepts from evidence that includes key important findings and big ideas.

Analysis of FGD Data in ATLAS.ti

The questions in a focus group discussion are generally unstructured. The interactions among the participants are guided by the moderator. The discussions and interactions surround the topic of interest, and are either audio or video recorded, depending upon the nature of consent obtained from the respondents. Moderation of discussions is important to fostering a natural and free flow of conversation about the research topic.

For computer-aided analysis, data from focus group discussions must first be transcribed into a digital form. The data will comprise opinions on the topic of discussion. It may not be prudent to group these opinions and perspectives based on certain characteristics; rather, the researcher should analyse individual opinions. Thus, the researcher can create a document group for each focus group discussion and then compare data and information between focus groups. Within a focus group, the researcher can compare the responses and opinions of individual participants to understand the emerging information.

Data from each focus group should be treated differently. By using the code co-occurrence table, one can compare participant responses within and across focus groups. The researcher can also compare data from individual participants or focus groups based on certain attributes, such as gender, age, status, experience, and so on. Comparing significant statements can be done in two ways:

- For Interview Data by using the "Code Document Analysis".
- For Focus Group Data, one can compare statements of different respondent groups using the "Co-occurrence Analysis".

Important While coding focus group data, researcher should also code the demographic profiles of the participants. The default option in ATLAS.ti is to code data at participant level. The respondent's name can be recoded later with a suitable identifier. For example, while assigning a code for a male respondent aged 35 years, who is experienced and a senior manager, each discrete bit of information can have only one code. Thus,

1. 'Male, 35 years, experienced person and senior manager' is one code.
2. 'Male'; '35 years'; 'senior manager'; 'an experienced person' are four codes (each code is separated by a ';').

In Fig. 12.6, the second method of focus group coding was used. In this example, the software had selected the code "Response on morale". However, it was felt that two additional codes were required: 'Male' and 'Manager'. These two codes were created, and process is described in more details in the next section.

Steps

1. Open the document. Use one of the two options for coding focus group data:
    I. Directly by opening the option on the top task bar or
    II. With a right click on the document and then selecting focus group coding option.

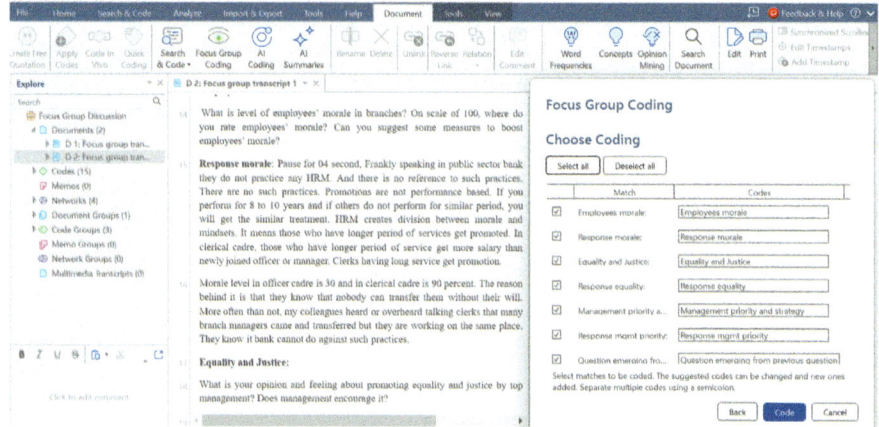

Fig. 12.6 Coding of focus group data

2. A flag will open (Fig. 12.3), showing three coding patterns:
   I. Pattern 1 will code speakers using the ':' symbol as a suffix.
   II. Pattern 2 will code speakers using the '@' symbol as prefix and ':' as suffix.
   III. Pattern 3 offers custom pattern functions on regular expression pattern.

   Depending on the pattern selected, the responses of speakers A and B will be coded as shown below:

3. Pattern 1. Speaker A: will code Speaker A as a code and contents associated with it as a quotation.
4. Pattern 2. @Speaker A: will code Speaker B as a code and contents associated with it as a quotation
5. Pattern 3. Speaker A| Speaker B will code all contents bearing Speaker A and Speaker B and contents associated with codes as quotations.

Examples of codes using patterns 1, 2 and custom coding are shown, respectively, in Figs. 12.3, 12.7, 12.8, and Table 12.3.

Any word, information, speaker, question, or response with a ':' will become a code in pattern 1. With '@' as a prefix and ':' as a suffix, it becomes a code in pattern 2. In Fig. 12.9, which shows custom coding, the researcher has coded 'Late sitting' and 'Transfer and Promotion' using custom coding pattern.

The researcher then created a code group for each type of coding. Figure 12.10 shows the code groups, and the network for the groups (Fig. 12.11).

After coding, the researcher can choose a method of data analysis that is appropriate for the objectives of the study. The most common methods are

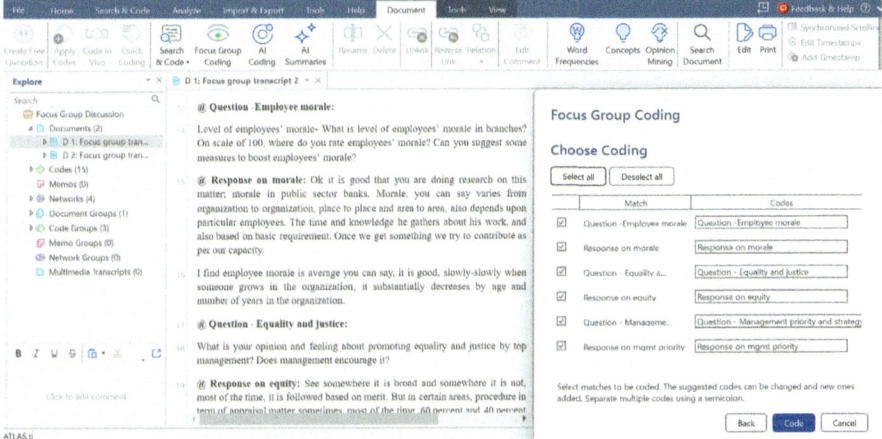

Fig. 12.7 Focus group coding with Pattern 1

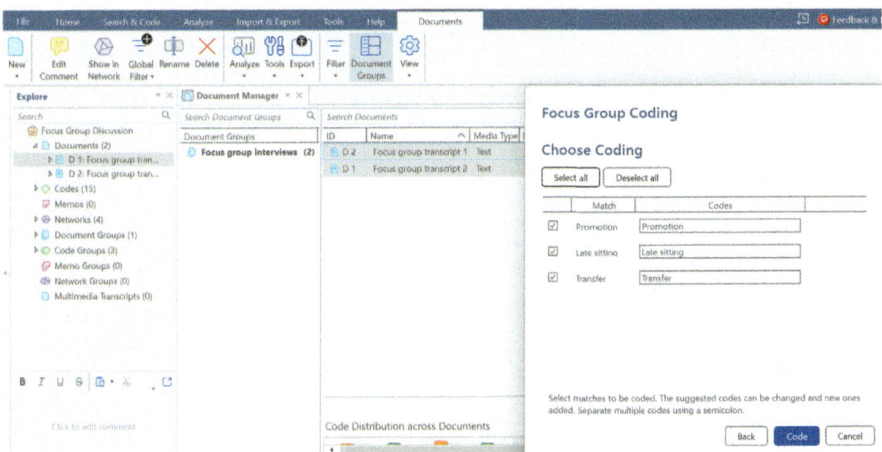

Fig. 12.8 Focus group coding with Pattern 2

thematic, content and discourse analysis. Code co-occurrence and code document tables are used to compare, analyse, and explore patterns within and across focus groups.

Analysis of Focus of Group Discussion Data in ATLAS.ti

We will discuss data analysis using ATLAS.ti with the help of an example. In a focus group discussion, there were two respondents (in practice, there are 6–8 in a focus group) The researcher guided the discussions with three questions, with each question asked after discussion of the previous one.

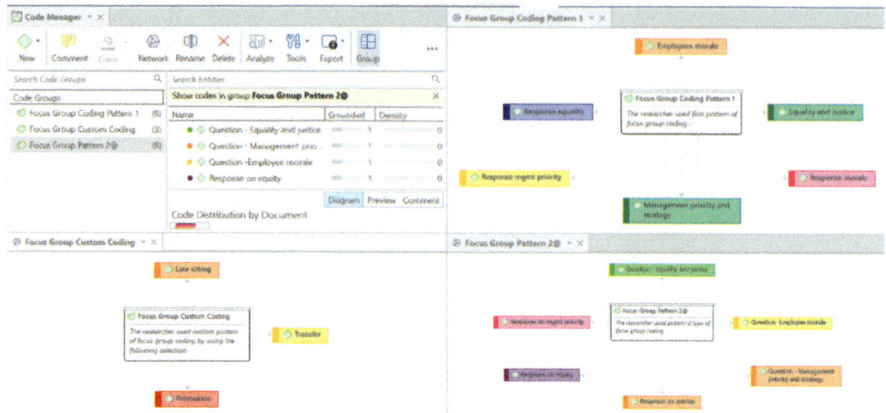

Fig. 12.9 Focus group coding-custom pattern

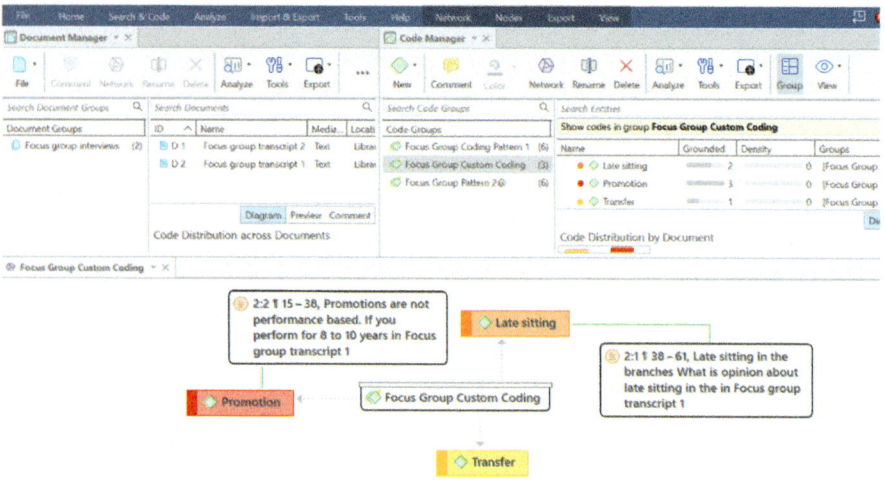

Fig. 12.10 Focus group coding network for custom coding

The respondents in the study were middle-level managers, Praveen and Sahgal (the names are fictitious), both males. The questions around which the discussions took place were:

1. What is your opinion about middle managers' morale in public sectors banks?
2. How fair is the process of promotion and transfer of middle managers? How does the human resource department play a role in it?
3. How does the target-setting process take place, and what do middle managers think about it?

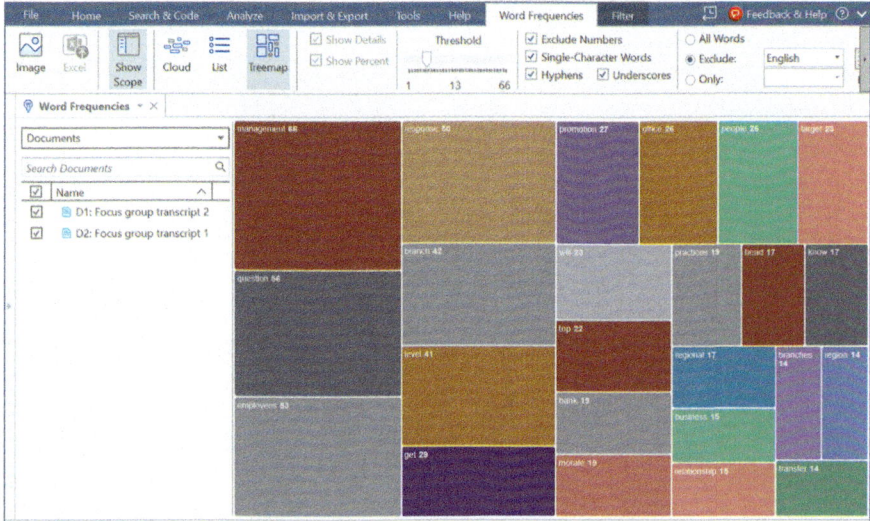

Fig. 12.11 Network of focus group codes

The responses of the participants were transcribed. The researcher also made notes of non-verbal communications, such as changes in the expressions and tone of the participants, and other tactile information. For analysis, however, only the data in transcripts were used for analysis. After a review of the transcripts by the researcher, the document was imported into ATLAS.ti. Emerging themes can be analyzed on individual and group level.

In the example discussed here, the Word Cloud function was used to determine the frequency of occurrence of certain words (Fig. 12.12). The word cloud helped to gain quick insights into the data. However, word clouds have a marginal role in qualitative data analysis. It only offers the usage of words in documents. However, the word cloud function can be useful for gaining an insight into the concepts present in the responses. The auto-coding option helps to automatically code the selected segments which can be reviewed and refined later. This step has advantages over manual coding especially when data size is large. Researchers can focus more on refining the codes and quotations.

Auto-coding was done after identifying the concepts which were indicated by the high frequency of use of certain words and expressions (after excluding common words like many, get, just, etc.). After scrutiny of the segments that reflected a concept, the researcher assigned the selected piece of information with new codes.

After coding, a code-occurrence table was generated (Fig. 12.13). The table shows the links between the research question, demographic profile, and emerging themes.

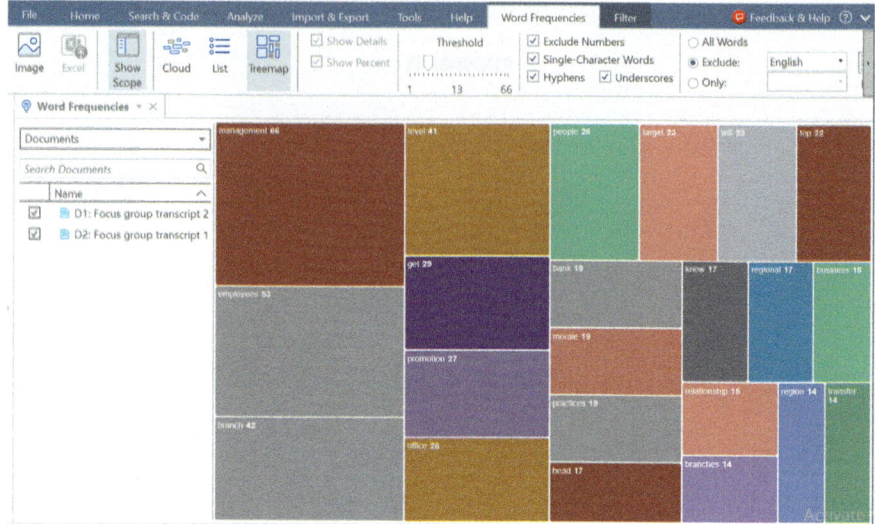

Fig. 12.12 Word tree view

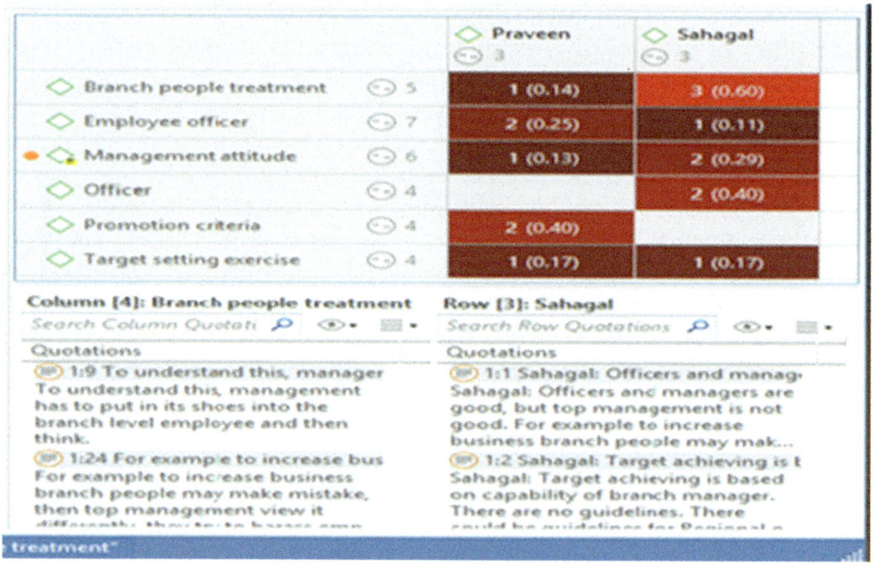

Fig. 12.13 Focus group code co-occurrence

The C-coefficient (discussed in Chapter…) shows the strength of the emerging concepts. It is also known as the C value and calculated using the formula,

$$\text{C-coefficient}(\text{C Value}) = N_{12}/(N_1 + N_2) - N_{12}$$

where N_{12} is the quotations that co-occur between two codes.
N_1 is the number of quotations in one code (groundedness).
N_2 is the number of quotations in the second code (groundedness).
(Note that only cluster quotations are used for calculating the C value).
A sample calculation is shown below.
C-coefficient between Sahagal and Branch people treatment is 0.60.
$N_1 = 3$ (There are three quotations for Sahagal).
$N_2 = 4$ (There are four quotations for treatment of branch employees).

Three quotations shared by the two codes. Using the formula, the C Value of **C-coefficient** in this example is $3/(3 + 5) - 3 = \mathbf{0.60}$.

The relative significance of opinion is more than the other respondent for the theme.

Recommended Readings

Agar, M., & MacDonald, J. (1995). Focus groups and ethnography. *Human Organization, 54*(1), 78–86.

Barbour, R. S. (2014). Analysing focus groups. *The SAGE Handbook of Qualitative Data Analysis*, 313–326.

Bennett, N. J., Roth, R., Klain, S. C., Chan, K., Christie, P., Clark, D. A., ... Wyborn, C. (2017). Conservation social science: Understanding and integrating human dimensions to improve conservation. *Biological Conservation, 205*, 93–108.

Birt, L., Suzanne, S., Debbie, C., Christine, C., & Fiona, W. (2016). Member checking: A tool to enhance trustworthiness or merely a nod to validation? *Qualitative Health Research, 26*, 1802–1811.

Bloor, M., Frankland, J., Thomas, M., & Robson, K. (2001). Trends and uses of focus groups. *Focus Groups in Social Research, 1*, 1–18.

Bromley, H., et al. (2003). *Glossary of qualitative research terms*. King's College, University of London.

Carey, M. A., & Smith, M. W. (1994). Capturing the group effect in focus groups: A special concern in analysis. *Qualitative Health Research, 4*, 123–127.

Charmaz, K. (2006). *Constructing grounded theory: A practical guide through qualitative analysis*. Sage Publications Inc.

Dubrovsky, V., Kiesler, S., & Sethna, B. (1991). The equalisation phenomena: Status effects in computer-mediated and face-to-face decision-making groups. *Journal of Human-Computer Interaction, 13*, 133–152.

Duggleby, W. (2005). What about focus group interaction data? *Qualitative Health Research, 15*(6), 832–840. https://doi.org/10.1177/1049732304273916

Gale, N. K., Heath, G., Cameron, E., Rashid, S., & Redwood, S. (2013). Using the framework method for the analysis of qualitative data in multi-disciplinary health research. *BMC Medical Research Methodology, 13*(1), 1–8.

Glaser, B. G. (1978). *Theoretical sensitivity*. Mill Valley, CA, USA: Sociology Press.

Glaser, B. G. (1992). *Discovery of grounded theory*. Chicago, IL, USA: Aldine.

Glaser, B. G., & Strauss, A. L. (1967). *The discovery of grounded theory: Strategies for qualitative research*. New Brunswick, NJ, USA: Aldine Transaction, A Division of Transaction Publishers.

Hague, P. (2002). *Market research* (3rd ed.). London: Kogan Page Ltd.

Harrison, M., Baker, J., Twinamatsiko, M., & Milner-Gulland, E. J. (2015). Profiling unauthorized natural resource users for better targeting of conservation interventions. *Conservation Biology: THe Journal of the Society for Conservation Biology, 29*(6), 1636–1646. https://doi.org/10.1111/cobi.12575

Hohenthal, J., Owidi, E., Minoia, P., & Pellikka, P. (2015). Local assessment of changes in water-related ecosystem services and their management: DPASER conceptual model and its application in Taita Hills, Kenya. *International Journal of Biodiversity Science, Ecosystems Services & Management, 11*(3), 225–238. https://doi.org/10.1080/21513732.2014.985256

Hyde, A., Howlett, E., Brady, D., & Drennan, J. (2005). The focus group method: Insights from focus group interviews on sexual health with adolescents. *Social Science & Medicine (1982), 61*(12), 2588–2599. https://doi.org/10.1016/j.socscimed.2005.04.040

Kamberelis, G., & Dimitriadis, G. (2005). Focus groups: Strategic articulations of pedagogy, politics, and inquiry. In N. K. Denzin & Y. S. Lincoln (Eds.), *The Sage handbook of qualitative research* (pp. 887–907). Sage Publications Inc.

Kidd, P., & Marshall, M. (2000). Getting the focus and the group: enhancing analytical rigor in focus group research. *Qualitative Health Research, 10*, 293–308.

Kitzinger, J. (1994). The methodology of focus groups: The importance of interaction between research participants. *Sociology of Health & Illness, 16*(1), 103–121.

Knodel, J. (1993). The design and analysis of focus group studies: A practical approach. *Successful Focus Groups: Advancing the State of the Art, 1*, 35–50.

Krueger, R. A. (1994). *Focus groups: A practical guide for applied research*. Thousand Oaks, CA: Sage Publications Inc.

Krueger, R. A., & Casey, M. A. (2000). *Focus groups, 2000. A practical guide for applied research*.

Krueger, R. A., & Casey, M. A. (2009). *Focus groups: A practical guide for applied research*. Thousand Oaks, CA: Sage.

Leech, N. L., & Onwuegbuzie, A. J. (2007). An array of qualitative data analysis tools: A call for data analysis triangulation. School Psychology Quarterly, 22, 557–584.

Leech, N. L., & Onwuegbuzie, A. J. (2008). Qualitative data analysis: A compendium of techniques and a framework for selection for school psychology research and beyond. *School Psychology Quarterly, 23*, 587–604.

Lincoln, Y. S., & Guba, E. G. (1985). *Naturalistic inquiry*. Sage.

Litoselliti, L. (2003). Using focus groups in research. *Using Focus Groups in Research*, 1–104.

Madriz, E. (2000). Focus groups in feminist research. *Handbook of Qualitative Research, 2*, 835–850.

Manwa, H., & Manwa, F. (2014). Poverty alleviation through pro-poor tourism: The role of Botswana forest reserves. *Sustainability, 6*(9), 5697–5713. https://doi.org/10.3390/su6095697

Merton, R. K., Fiske, M., & Kendall, P. L. (1956). *The focused interview*. The Free Press.

Miranda, S. M. (1994). Avoidance of Groupthink: Meeting management using group support systems. *Small Group Research, 25*, 1, 105–136.

Morgan, D. L. (1995). Why things (sometimes) go wrong in focus groups. *Qualitative Health Research, 5*, 516–523.

Morgan, D. L. (1996). Focus groups. *Annual Review of Sociology, 22*(1), 129–152.

Morgan, D. L. (1997). *Focus groups as qualitative research* (2nd ed.). Qualitative Research Methods Series 16. Thousand Oaks, CA: Sage.
Morgan, D.L. (1988). *Focus groups as qualitative research.* Newbury Park: Sage Publications.
Morgan, D. L., Krueger, R. A., & King, J. A. (1998). *The focus group kit* (Vols. 1–6). Sage Publications Inc.
Morgan, D. L. (2010). Reconsidering the role of interaction in analyzing and reporting focus groups. *Qualitative Health Research, 20*(5), 718–722. https://doi.org/10.1177/1049732310364627
Mukherjee, R., Wray, E., Curfs, L., & Hollins, S. (2015). Knowledge and opinions of professional groups concerning FASD in the UK. *Adoption & Fostering, 39*(3), 212–224.
Nyumba, O. T., Wilson, K., Derrick, C. J., & Mukherjee, N. (2018). The use of focus group discussion methodology: Insights from two decades of application in conservation. *Methods in Ecology and Evolution, 9*(1), 20–32.
Paloniemi, R., Apostolopoulou, E., Primmer, E., Grodzinska-Jurcak, M., Henle, K., Ring, I., Kettunen, M., Tzanopoulos, J., Potts, S., van den Hove, S., Marty, P., McConville, A., & Simila J. (2012). Biodiversity conservation across scales: Lessons from a science–policy dialogue. *Nature Conservation, 2,* 7–19. https://doi.org/10.3897/natureconservation.2.3144
Potter, J., & Wetherell, M. (1987). *Discourse and social psychology: Beyond attitudes and behaviour.* Sage Publications Inc.
Ritchie, J., & Spencer, L. (1994). Qualitative data analysis for applied policy research. In A. Bryman, & R. Burgess (Eds.), *Analysing qualitative data* (pp. 173–194). London, UK: Routledge.
Robinson, J. (2009). Laughter and forgetting: Using focus groups to discuss smoking and motherhood in low-income areas in the UK. *International Journal of Qualitative Studies in Education: QSE, 22*(3), 263–278. https://doi.org/10.1080/09518390902835421
Saldaña, J. (2021). *The coding manual for qualitative researchers.* Sage.
Silverman, D. (2006). *Interpreting qualitative data* (3rd ed.). SAGE Publications.
Smithson, J. (2000). Using and analysing focus groups: Limitations and possibilities. *International Journal of Social Research Methodology, 3*(2), 103–119. https://doi.org/10.1080/136455700405172
Stewart, D. W., & Shamdasani, P. N. (1990). *Focus groups: Theory and practice.* Newbury Park, CA: Sage.
Strauss, A. L. (1987). *Qualitative analysis for social scientists.* Cambridge, UK: Cambridge University Press.
Wilkinson, S., & Kitzinger, C. (2000). Thinking differently about thinking positive: A discursive approach to cancer patients' talk. *Social Science & Medicine (1982), 50*(6), 797–811. Ανακτήθηκε από. https://www.ncbi.nlm.nih.gov/pubmed/10695978
Wong, L. P. (2008). Focus group discussion: A tool for health and medical research. *Singapore Medical Journal, 49*(3), 256–260.
Zander, K., Stolz, H., & Hamm, U. (2013). Promising ethical arguments for product differentiation in the organic food sector. A mixed methods research approach. *Appetite, 62,* 133–142. https://doi.org/10.1016/j.appet.2012.11.015

CHAPTER 13

Multimedia and Geo Data Analysis

Abstract ATLAS.ti can be used for analysing multimedia data and geodata. Multimedia data consists of images, audio and video, as well as text. Images capture the expressions of the participants of the study. Geodata analysis helps in the 'discovery' of geographical spaces. Both data types play a significant role in the study of a phenomenon for which text data alone may not be sufficient. In the section on multimedia data analysis, the author discusses various transcription techniques (manual as well as software-assisted), transcription processes, and the challenges that must be addressed while working with video data. An example of a research project that involved analysis of video data is also described, beginning with the import of multimedia data into ATLAS.ti, coding and quotations, development of themes, and their interpretation. Similarly, geodata analysis is discussed with the help of a research project.

Multimedia Data

Multimedia data comprises graphics, images, animation, video, and audio objects, which contain information, expressions and feelings of the participants, and their experiences of the phenomenon under investigation. The findings of multimedia data analysis are supported by images, audio, and video clippings along with texts.

Multimedia data can be analyzed in ATLAS.ti along the same principles and processes followed for text data. The main challenge in multimedia data analysis, however, is in the management and organization of the data. As multimedia data contains more information—and more voluminous—when compared with text data, it is important to be familiar with data management strategies.

Besides the challenges faced in the analysis of multimedia data analysis, questions can also be raised over the methods of transcription, the segments of multimedia data that should be transcribed, and the transcription method (manual or machine-assisted). Several software options are available for automating the transcription process; however, the researcher must be familiar with the advantages and limitations of each before making the appropriate selection. For the reader's interest, references to more information on automated transcription are provided at the end of this chapter.

Preparing Transcripts

Working with multimedia data (audio, graphic, animation and video) and geodata is to a large extent like working with text documents. With audio and video data, it is crucial to have verbatim transcripts to maintain data quality. Researchers can create audio/video quotations while they are listening to the files. ATLAS.ti facilitates working with recording directly. One can also work with multimedia data to code them without first transcribing them. Transcription techniques and challenges have been discussed in detail in Chap. 2.

It is good practice to check the transcribed for synchronization with associated multimedia before analysis. Transcripts and multimedia data should be stored as individual files in one folder for easy access so that, when they are opened and the multimedia file is played, the corresponding text in the transcript is automatically highlighted.

Five Approaches to Multimedia Data Analysis

There are five approaches to analysing multimedia data:

1. Working with audio or video data only. Transcripts are not required or used.
2. Working with audio or video data with transcripts of selected clips only.
3. Working with audio or video data with manually prepared transcripts.
4. Working with audio or video data with software-assisted transcripts.
5. Working with audio or video data with automated transcripts.

Any or a combination of the above can be used. The first and second approaches are preferred when the size of data size large, manual transcription is difficult, and the researcher wants to avoid using transcription software because of the risk of losing important information. In such a situation, one can create multimedia quotations while listening to the audio or video file or viewing the video files. Working with transcripts from selected clips is usually the preferred option. While transcribing manually, it is good practice to first listen to the audio and/or watch the video as often as necessary. Whenever transcription software is used, the researcher must verify the accuracy and completeness of transcription.

Multimedia data can be directly added to ATLAS.ti using the Add Linked video/audio option in Add Documents. While coding, one can also write comments, name the quotations, and analyse the data following the processes for text data. Code categories, code groups, smart codes, and smart groups can be formed according to need and, with the appropriate selection (Chap. 4), reports generated in various formats (text, network, and Excel).

While working with transcripts, whether prepared manually or with the help of software, it is important to note that the multimedia data should be exported to ATLAS.ti before importing the transcripts. The transcript is imported into ATLAS.ti using the Import Transcript option in the Transcripts menu. When the audio or video document is opened, the transcripts option is automatically activated. Timestamped transcripts are automatically synchronized with the audio/video data and when the multimedia file is played, the text data between two timestamps is highlighted, showing the alignment of multimedia and text between two timestamps.

Figure 13.1 shows the text report for the data. The quotations are supported by images. Figure 13.2 shows the Quotation Manager and the output text report. Video data was added, and a new text document was opened. Video was transcripted in the text document. Transcript was coded and associated video clipping was linked using hyperlinks.

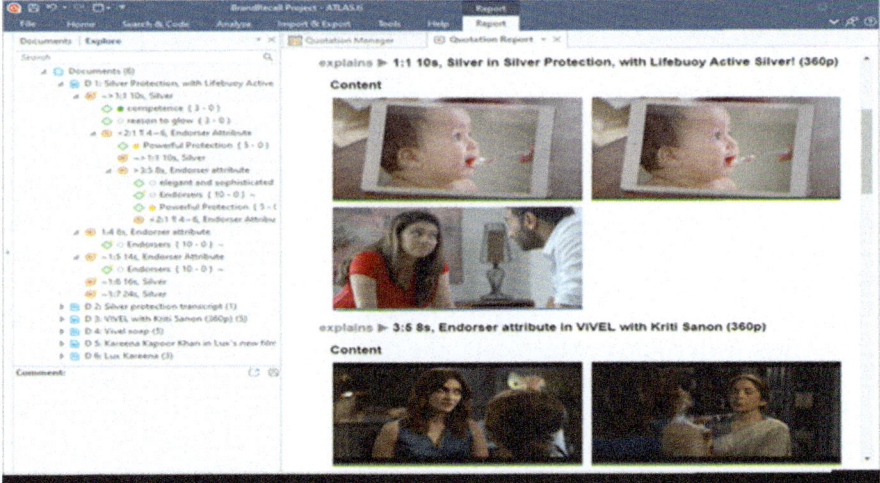

Fig. 13.1 Textual report of video data

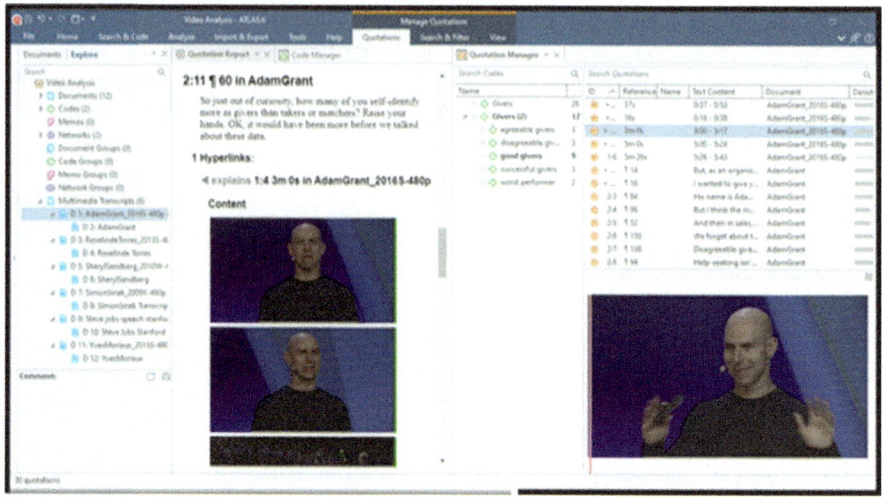

Fig. 13.2 Text report and quotation manager

The quotation report shows the quotation along with the picture. While creating a quotation from the text, the video for the text should be linked to the text quotation. When a report is generated, the associated video clipping will be shown in the report.

Transcription Software

One can use ATLAS.ti to transcribe audio or video data. Transcripts done offline, either manually or using software, can also be uploaded to ATLAS.ti. Regardless of the method, the researcher should verify the transcripts thoroughly with audio or video data to ensure authenticity, accuracy, and completeness.

ATLAS.ti recommends the use of transcription tools. Several software are available for automating the transcription process. However, most transcription softwares available at present are useful for transcribing in English only. Hence, they are of little help with other languages. ATLAS.ti 22 supports several transcriptions software: ELAN, Go Transcribe, Hyper Transcribe, oTranscribe, Scribie, Speechmatics, Spoken online, TranscribeMe, Transcriber Pro, Transcriva, Transcripto, Voicedocs, WebEx. The transcripts are saved in VTT or SRT files. After importing the transcripts into ATLAS.ti and synchronizing them with audio or video data, the text must be read by the researcher. The Edit option can be used to make the required changes and modifications.

Table 13.1 presents a list of transcription software their supported file formats.

Table 13.1 Transcription software compatible with ATLAS.ti

Transcription software (RTF or Word format)	Automatic transcription supported services (VTT and SRT format)
Easytranscript	MS Teams
f4 & f5 transcript	Zoom
Transcribe	YouTube
Inqsribe	Happyscribe
Transana	Trint
ExpressSribe, a.o	Descript
	Sonix
	Rev.com
	Panopto
	sonix
	Transcribe by Wreally
	Temi
	Simon Says
	Vimeo
	Amberscript
	Otter.ai
	Vocalmatic
	eStream

> **Important** Researchers should exercise care while using YouTube data to ensure that they do not harm enterprises, people, and other stakeholders. They must also disguise information with which entities can be identified to avoid adverse consequences for people and organisations.

Importing Multimedia Data

ATLAS.ti supports several multimedia file formats. Video files supported by ATLAS.ti are *0.3g2, 0.3gp, 0.3gp2, 0.3gpp, .asf, .avi, .m4v, .mov, .mp4, .wmv for window and .avi, .m4v, .mov, mp4 for Mac*. The recommended format is .mp4 files with AAC audio and H.264 video. Audio file formats supported by ATLAS.ti include *aac, m4a, mp3, mp4, wav*. They can be played on both Windows and the Mac OS. Timestamping formats used are #00:00:00-0# or [00:00:00]. The researcher can import a timestamped transcript for an already imported audio or video file. Before opening the project on ATLAS.ti, the folder containing the files for import should be ready. It is good practice to name multimedia data and transcript files for their easy location.

1. Open the Home tab and add documents (transcripts).
2. Add linked Video/Audio.

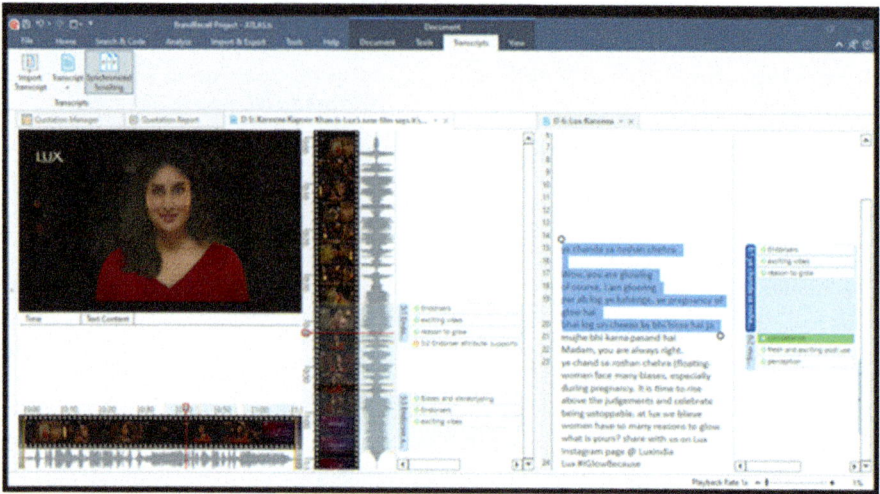

Fig. 13.3 Coding video data and transcript

3. Locate the data file and add it.
4. After adding the video file, the Tools (in ATLAS.ti 9) or Transcripts (ATLAS.ti 9 & 22) options will appear at the top of the screen (Fig. 13.3).
5. The imported transcript will be shown alongside the video data. With a double click on Transcript or Multimedia file in multimedia transcripts, the documents will be opened side-by-side.
6. To play the multimedia document and view the transcript at the same time, activate Synchronized Scrolling in the Tool tab and play the multimedia file.
7. One can also move the play head to a different position and play the document from there.
8. If the transcript of the multimedia file is not available, the user can create a blank word file and use the Tools or Transcripts option to import the transcript.
9. Next, activate the Synchronized Scrolling option, play the multimedia first, pause and write transcription in the document appearing alongside.
10. Use the Edit option before starting transcription. Keep saving the document as transcription progresses. The playback speed is adjustable (Fig. 13.4).
11. The process for coding, highlighting quotations, merging, and linking codes is the same as the one for coding text data (see Chap. 4).

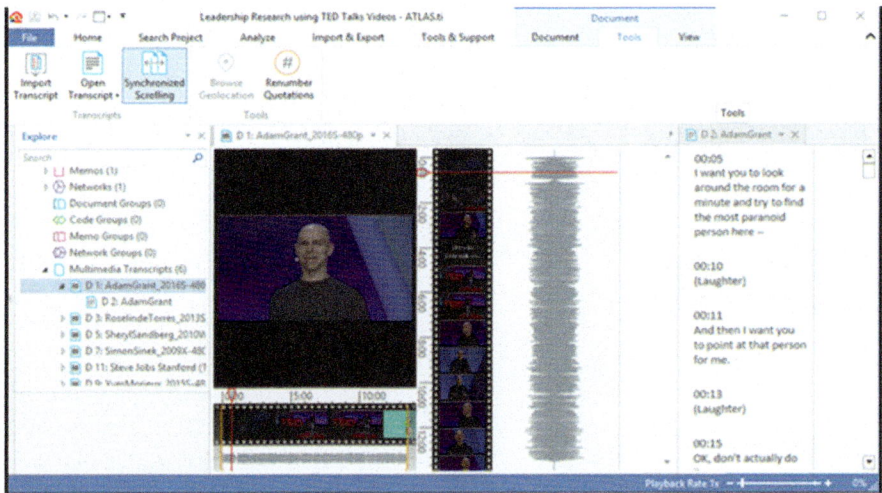

Fig. 13.4 Video document view in ATLAS.ti

Creating Quotations and Codes from Multimedia Data

The documents (video/audio or transcript) in 'Multimedia Transcripts' option will open simultaneously alongside each other. While playing back the multimedia file, wherever the researcher finds an important segment of information, the start of the segment is marked as the start of the quotation by using the signs '<' or ',' (comma). The end of the segment is marked with a '>' or '.' (the period sign). After identifying the segment, the quotation is created using the "Create Quotation" function (Fig. 13.2).

One can zoom to the time bar by dragging the image or audio frame. Using the appropriate buttons in the documents, the researcher can play the document forward or back, pause, fast forward or backward, etc., depending on need. The playback rate can be increased or reduced as required, a feature that is useful when working with manual transcriptions. Playback volume is adjusted using the volume key. The Zoom option is utilized to appropriate display of audio and video file (Fig. 13.5).

1. Play the multimedia.
2. Mark the segment of interest and create a quotation.
3. Code the quotations (explained in Chap. 4). One can either create free quotations first and code them later or create quotations and code them simultaneously.
4. It is good practice to name the quotation for easy identification. Right-click on quotation and name it, using the rename option.

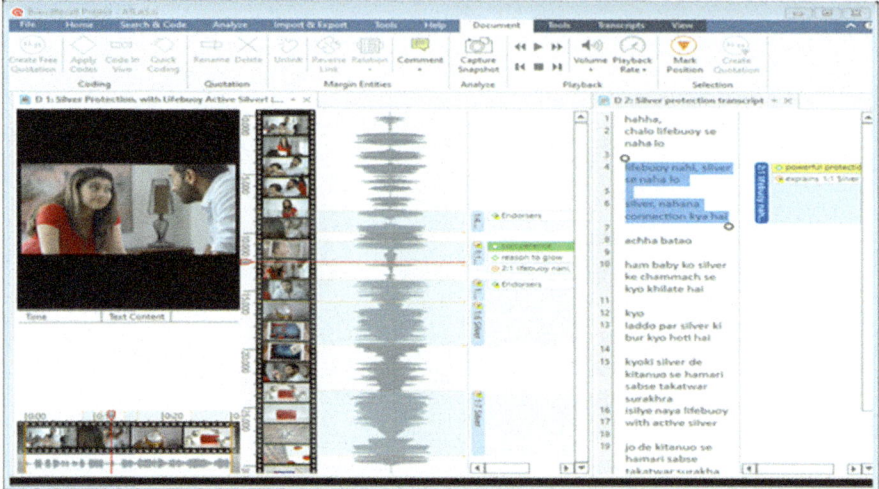

Fig. 13.5 Using manual transcripts in multimedia documents

When synchronized, multimedia data and the transcript will move together (Fig. 13.6).

1. To create a multimedia quotation, move the cursor pointer to the top of the audio wave and mark the start of the segment with a left click. Drag the cursor to the spot to mark the end of the segment. On releasing the cursor, the quotation is created.

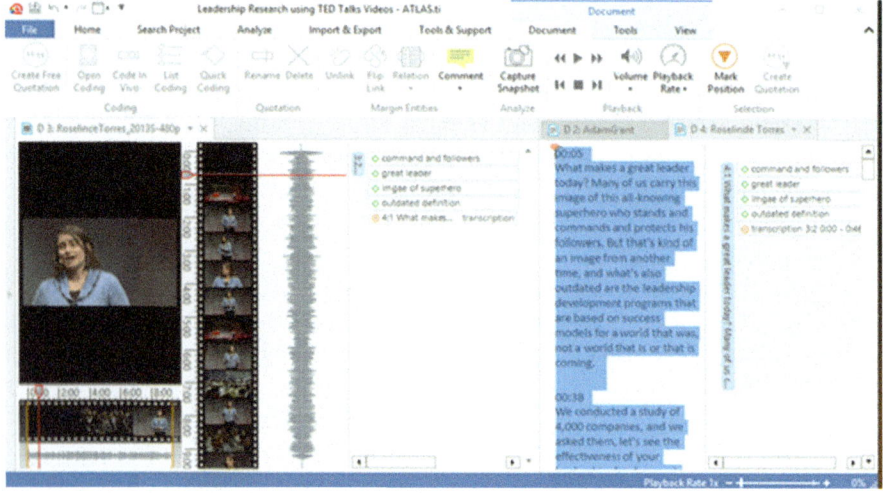

Fig. 13.6 Multimedia and automated transcripts

2. Users can also create quotations from multimedia data by

   (i) using the Mark and Create a Quotation button or
   (ii) the short-cut keys (< for start and > for end positions), or
   (iii) using "," and "." to mark the start end positions respectively.

3. The quotations are then renamed and coded in the process followed for coding textual data.
4. Multimedia quotations will be shown in Quotation Manager with their start and end time.
5. Multimedia quotations can be previewed in a network using the Add Neighbors and Quotations options.

Working with Multimedia Data and Ready Transcripts

Working with ready transcripts of multimedia data is relatively easier than working with multimedia data without transcripts because the files are already synchronized. Quotations can be created from both multimedia data and the transcripts, connected by hyperlinks and then coded. This can be done in two ways: linking the text quotations to the quotations in the multimedia data and then coding them; or by linking the multimedia quotations to the text quotations and then coding them.

Non-availability transcripts of multimedia data can create challenges for researchers who must first prepare the transcripts and then timestamp them for synchronization. This is done in two ways—by using transcription software that is compatible with ATLAS.ti or transcribing manually.

> **Important** One can analyze video frames in more depth by taking snapshots. Move the play head to the desired position in the video. Snapshots can be taken by right-clicking on the video and using the Capture Video Frame option or by clicking on the camera icon in the ribbon. The video frame becomes an image document and is then shown the with documents (Fig. 13.7).

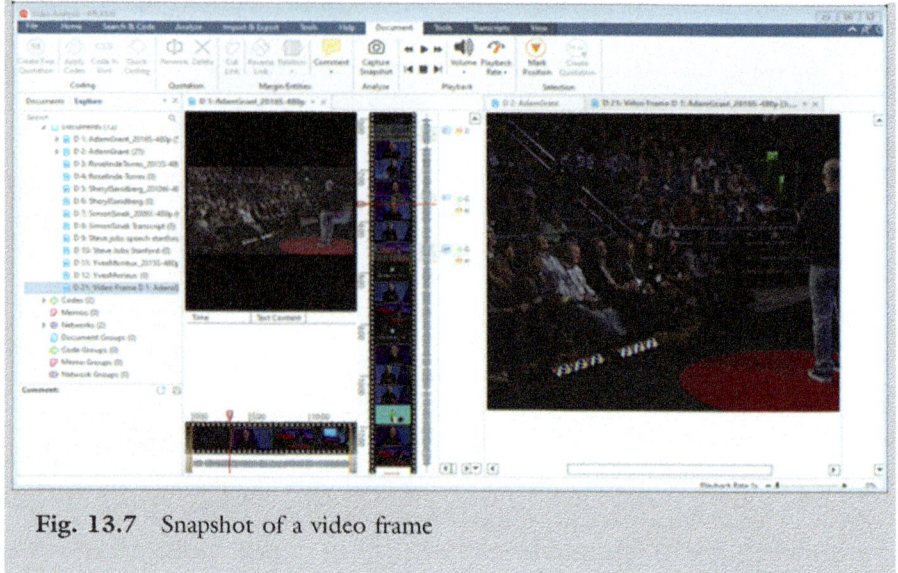

Fig. 13.7 Snapshot of a video frame

Working with Video Projects

The objectives of the study were:

1. To study the changing trends of advertisements in three brands of soap.
2. To study the influence of the endorsers' personalities in the advertisements.

The research project discussed in this section used video data. The study analysed changing trends in the advertisements of three soap brands—Vivel, Lifebuoy and Lux—and endorsers' personalities. All three videos were transcribed and imported into ATLAS.ti 22. The video was coded first. After coding, the changing trends and the endorsers' personalities were identified based on the groundedness of the quotations.

Figures 13.8, 13.9, 13.10, 13.11, 13.12 and 13.13 show the codes and themes generated. Figures 13.14, 13.15 and 13.16 are derived from the codes, code groups from video data.

Collecting Multimedia Data

In this project, the advertisements for each soap brand were downloaded from YouTube and transcribed manually. The video data was imported into ATLAS.ti 22 followed by the corresponding transcripts of each video.

13 MULTIMEDIA AND GEO DATA ANALYSIS   355

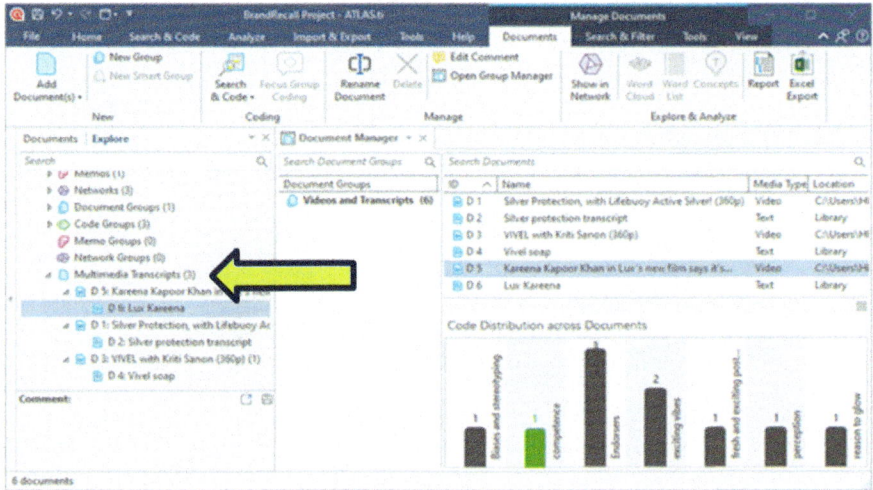

Fig. 13.8 Screenshot of the video project

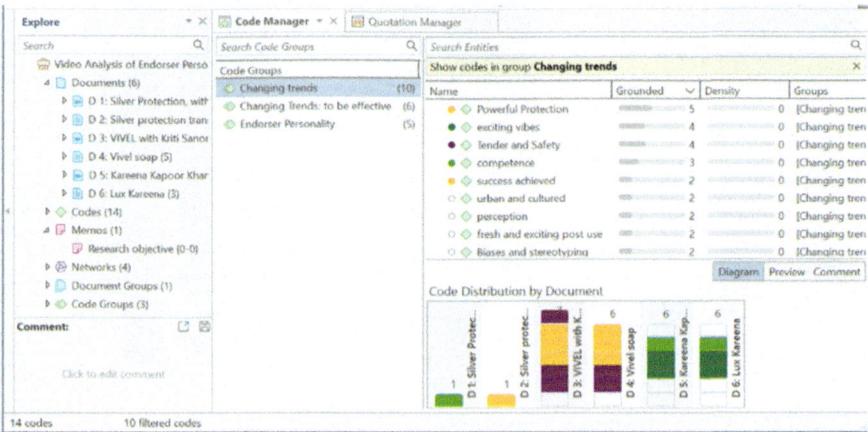

Fig. 13.9 Code groups and codes list

Figure 13.8 is the resulting display after import of the video data and the transcripts.

The multimedia transcripts are seen in the left margin, while the documents are shown on the right side in Document Manager. The association of transcripts with the videos are shown in Document Manager and Multimedia Transcript option (shown by a yellow arrow).

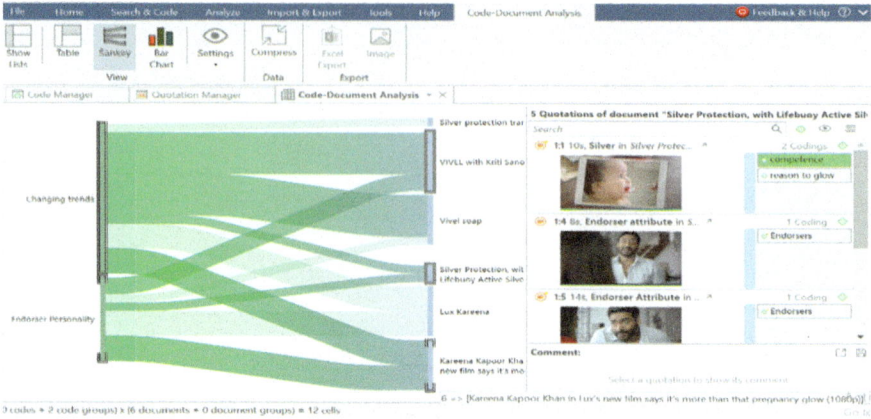

Fig. 13.10 Sankey diagram of code document table

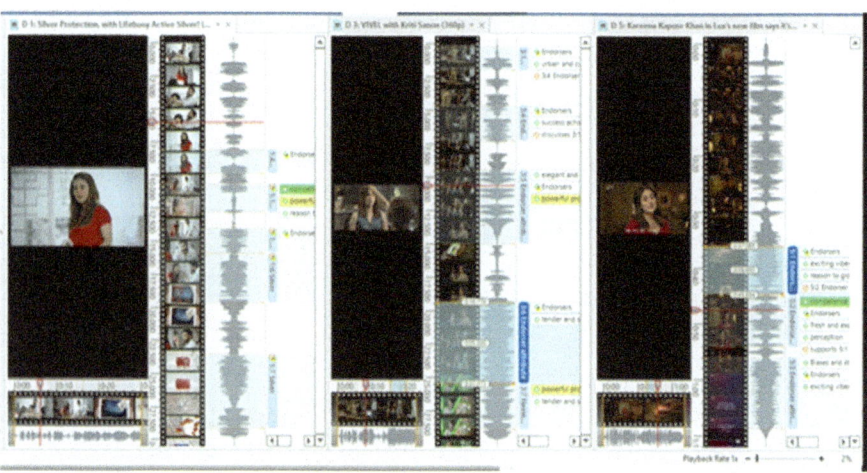

Fig. 13.11 Video coding and quotations

Working with Video Data

Data was coded by playing each video and selecting the segments of interest for the study. In the next step, the corresponding texts in the transcript were identified and the quotations of the video and text were linked. This process was repeated for all video documents. The coding sequence can also be reversed: one can also read the transcript, code the relevant segments, and then link the quotation with the corresponding segments in the video.

After coding, the code list was studied to identify patterns for creating code groups. The screenshots in Figs. 13.11 and 13.12 show, respectively, the coded video and comments.

13 MULTIMEDIA AND GEO DATA ANALYSIS 357

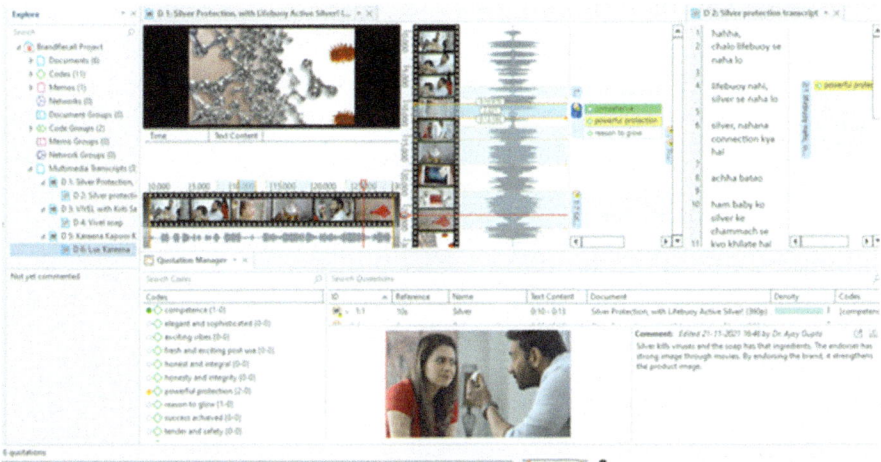

Fig. 13.12 Quotation manager and comment

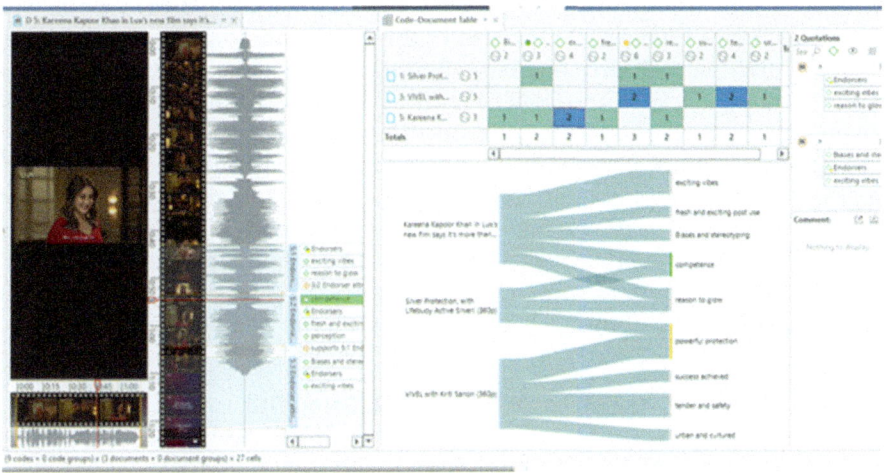

Fig. 13.13 Emerging trends, Sankey diagram and quotation

After coding, codes have been categorized and grouped. The themes—Changing trends and Endorsers' Personalities—are shown and interpreted by screenshots and charts.

A code-document table was created to study the distribution of the codes across the three brands (Fig. 13.10). The quotations are shown along with their associated codes on the right. The Sankey diagram shows the flow of concepts across documents. The figures in the table show the number of quotations associated with their respective codes. With this information, text, network, or Excel reports can be generated according to need.

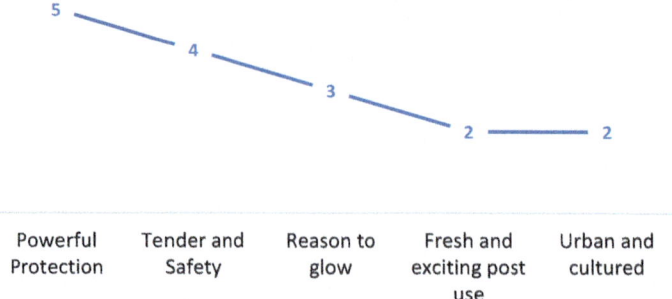

Fig. 13.14 Theme and sub-themes for changing trends

Findings and Interpretations

Two themes emerged—Changing Trends and Powerful Personality. In the first, Changing Trends, powerful protection is a significant sub-theme followed by other themes (Fig. 13.9). The size of the quotations is shown on the y-axis of the figure. In the second them, 'Endorser's Personality', 'exciting vibes' is the most significant sub-theme. Figures 13.13, 13.14 and 13.15 show the themes on a bar chart and the code-document table.

Figure 13.10 shows the code document analysis table in which the themes appear on the left, and their associated documents on the right. The flow size of the theme with documents shows the significance of the theme. Figure 13.11 shows the coding window of video data. In Fig. 13.11, quotations are video frames, and their associated codes appear on the right. Figure 13.12 shows the quotation manager and document window.

Fig. 13.15 Theme and sub-themes for endorsers' personalities

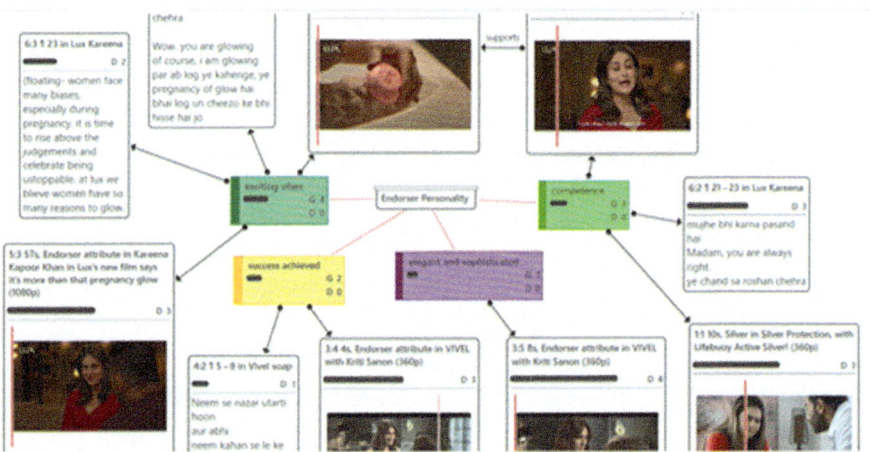

Fig. 13.16 Network of a theme

Comments about the video image are written on the right side of the video image.

Figure 13.13 shows the emerging trends in the advertisements of the three soap brands. The trends are shown on the right side, and advertisements appear left side. Figures 13.14 and 13.15 show the significance of themes based on the frequency of quotations occurrences. Figure 13.16 shows the network of a theme.

Working with Geodata

Geodata enables researchers to acquire data referenced to the earth, which is then used for analysis, modeling, simulations, and visualization. Geodata helps researchers to understand the world. It helps create a digital model or photographic interpretation of the area of study, including the spaces and their contexts.

Geodata is locative data. It includes documents, maps, images, biographical information that are spatially related to geographic locations through place-names (toponyms) and place codes (i.e., postal codes) or through geospatial referencing (Hochmair & Fu, 2006).

One need not create a separate document for importing geodata into ATLAS.ti. ATLAS.ti has three options to view geo documents in the Tools tab: Open Street map, Road map and Satellite map.

Geodata can be created in ATLAS.ti in a few simple steps:

1. Open 'Add Documents' option in Home. The option New Geo Document will appear at the bottom. Click on this option.
2. A map of the world will be displayed in the main ATLAS.ti window.

3. Use the 'Query Address' option for writing search term and enter. A new Geo Document will open on the screen and turn into a document (Fig. 13.17).

The Geo document can be renamed with a right click. Comments can also be added. Figure 13.18 shows a geo document with quotations and comments. This document has two codes and several quotations. The two codes are My Academic Journey and My Professional Journey. The process of creating codes, quotations and reporting has been explained under the section 'Creating quotations and codes in Geo Documents.

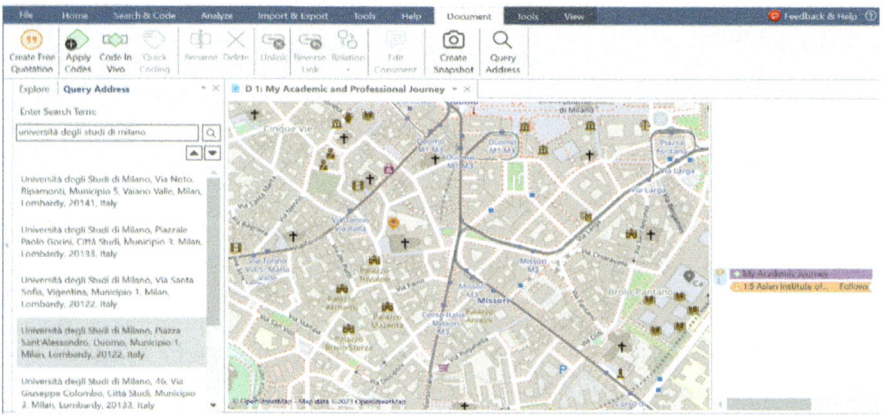

Fig. 13.17 Geo document in ATLAS.ti 23

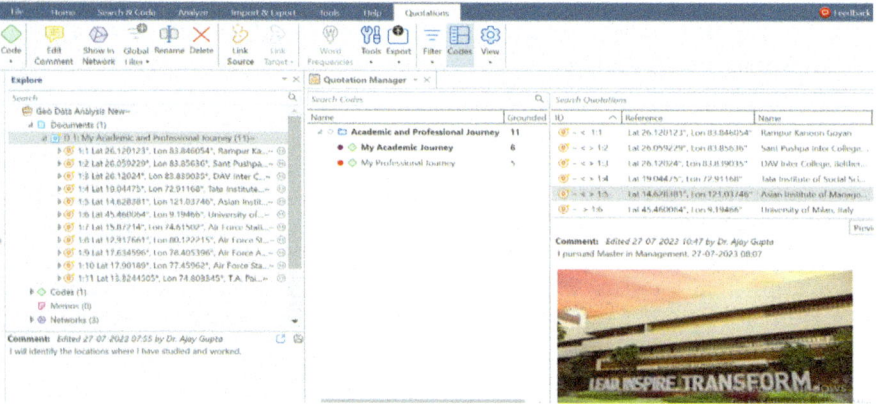

Fig. 13.18 Document and quotation manager

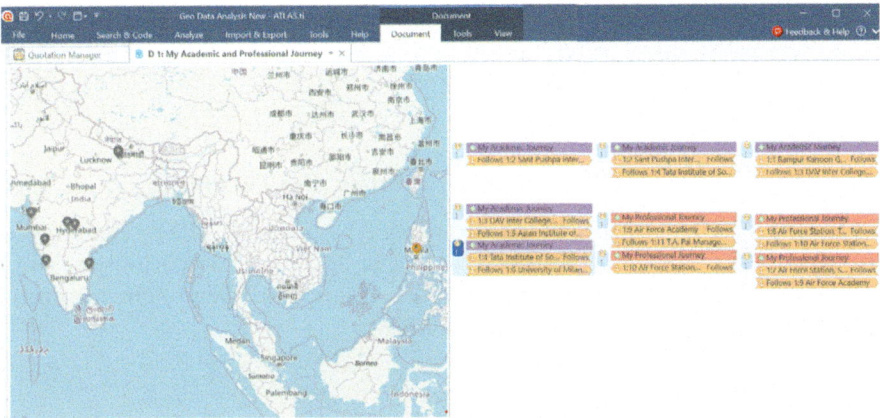

Fig. 13.19 Geo documents and snap shots

Geo Document in ATLAS.ti 23

Figure 13.17 shows the Geo Document for the queried address in the search option. The default name for any Geo document is 'New Geo Document'. With the 'Create Snapshot' option, one can take a snapshot of the location of interest for further analysis. After coding, the geo document is shown in Fig. 13.19 with codes on the right and locations in the Geo document.

The Geo document was coded by identifying the locations where the author had studied, which are coded "My Academic Journey". The locations where the author had worked are coded "My Professional Journey". Both codes are shown in the code folder named "My Academic and Professional Journey".

Creating Quotations and Codes in Geo Documents

The process of coding a Geo document is like that of textual document. One can create free quotations and code them later or create quotations and code them simultaneously (Chap. 4). After coding, one can create code groups, memos, networks, links, merge, and split codes as necessary. Reports can be generated with which the researcher can interpret the data with reference to the research's objectives.

Steps for creating quotations and assigning codes to Geodata:

1. Point the cursor on the location where the user wants to make a quotation. Right click and create a free quotation. Alternatively, the search option can be used to locate the point where a free quotation can be created. The search option can be used to locate several places in the document for creating quotations.

2. Code the Quotations by a right click on a quotation and applying the appropriate code. Alternatively, quotations can be coded in Document Manager with a right click.

Example of a Research Project in Geo Document

A Geodata-based study was conducted using ATLAS.ti 23. The research objective was to search for the Author's academic and professional journey using geodata. One Geo document was opened. Using Add Documents option, a new Geo Document was opened in ATLAS.ti.

1. Search for the addresses using the search option.
2. Right click on the location, write the name of the code.
3. Right click on quotation, open Edit Comment, write comment, insert date and time, and picture (Fig. 13.20).
4. Thus, the author identified locations where he/she studied and coded them with "My Academic Journey".
5. Next, the author identified locations where he/she was employed and coded with "My Professional Journey".
6. Then, a code folder was created consisting of the two codes. Comments were written for the quotations using the Quotation Manager option.
7. Comments were written using the Edit comment option. In the comment section, the image of a particular academic or professional institution was inserted. (This can be done by downloading the image and inserting them in the comments section for quotations.

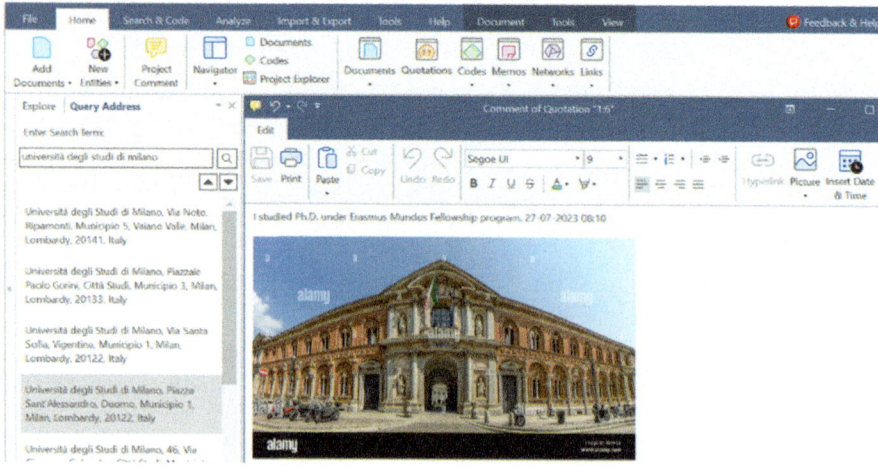

Fig. 13.20 Edit comment window

There are three options in Edit section: Hyperlink, Picture, and Insert date and time (Fig. 13.20). One must first insert date and time, write the comments, and then insert the picture and resize it. Hyperlinks should be used for mentioning the URL or website of the source of the images. Data analysis follows the same procedure as that for textual data.

Figure 13.21 shows the report option in Export menu. The report for quotations is on the right side, while the list of quotations is on the left side. A text or Excel report can be generated as required.

Figures 13.22 and 13.23 show the respective code networks for "My Professional Journey" and 'My Academic Journey" along with their associated quotations and hyperlinks. Figure 13.24 shows the network of code folder "Academic and Professional Journey", which has two codes, associated quotations with hyperlinks. A text report for each network can be created by the researcher for further analysis and interpretation.

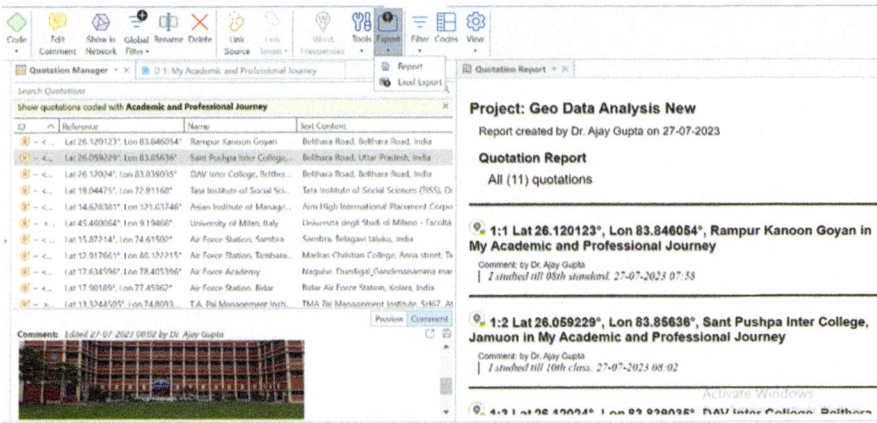

Fig. 13.21 Quotation manager and report options

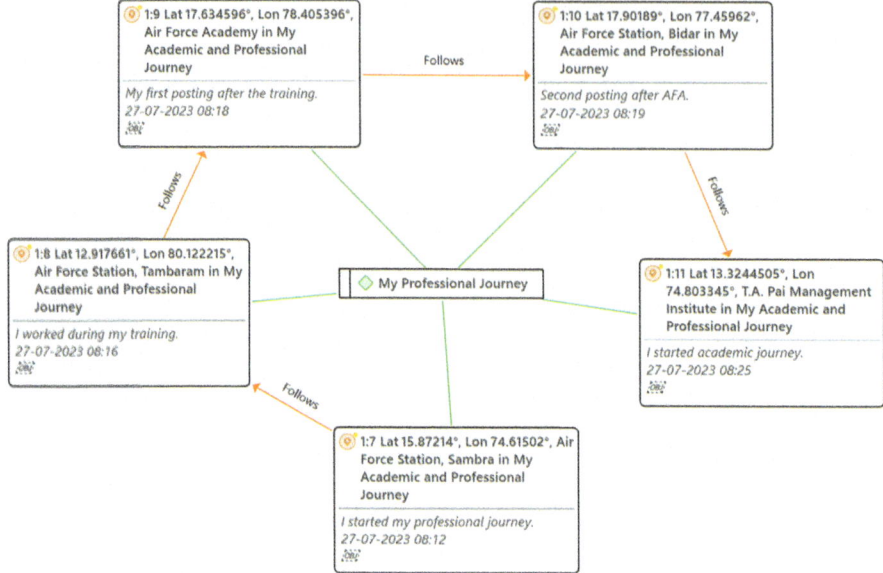

Fig. 13.22 A network of my professional journey

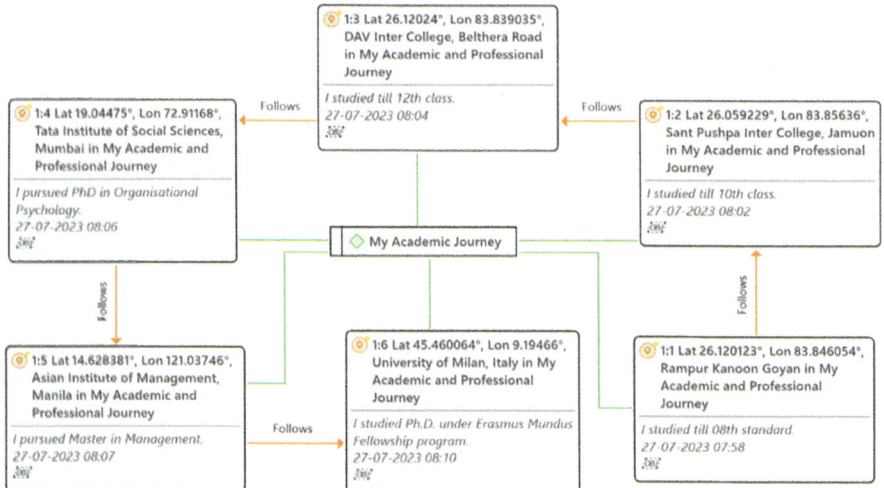

Fig. 13.23 A network of my academic journey

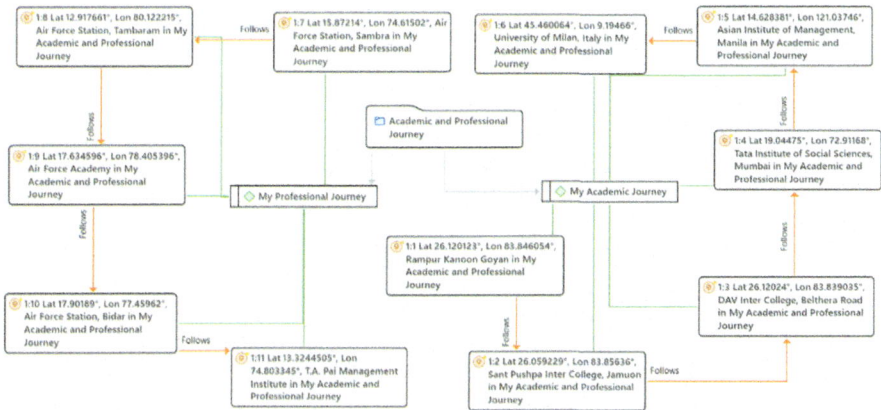

Fig. 13.24 A network of academic and professional journey

Recommended Readings

Brownlow, C., & O'Dell, L. (2002). Ethical issues for qualitative research in online communities. *Disability & Society, 17*(6), 685–694.

Bucholtz, M. (2000). The politics of transcription. *Journal of Pragmatics, 32*(10), 1439–1465.

Crichton, S., & Kinash, S. (2008). Virtual ethnography: Interactive interviewing online as method. *Canadian Journal of Learning and Technology/La revue canadienne de l'apprentissage et de la technologie, 29*(2).

Davidson, C. (2009). Transcription: Imperatives for qualitative research. *International Journal of Qualitative Methods, 8*(2), 35–52.

Easton, K. L., McComish, J. F., & Greenberg, R. (2000). Avoiding common pitfalls in qualitative data collection and transcription. *Qualitative Health Research, 10*(5), 703–707.

Eysenbach, G., & Till, J. E. (2001). Ethical issues in qualitative research on internet communities. *BMJ, 323*(7321), 1103–1105.

Heritage, J., & Atkinson, J. M. (1984). Structures of social action. *Studies in Conversation Analysis*, 346–369.

Hochmair, H. H., & Fu, Z. J. (2006). User interface design for semantic query expansion in geodata repositories.

Kvale, S. (1996). The 1000-page question. *Qualitative Inquiry, 2*(3), 275–284.

Lapadat, J. C., & Lindsay, A. C. (1999). Transcription in research and practice: From standardization of technique to interpretive positionings. *Qualitative Inquiry, 5*(1), 64–86.

McLellan, E., MacQueen, K. M., & Neidig, J. L. (2003). Beyond the qualitative interview: Data preparation and transcription. *Field Methods, 15*(1), 63–84.

Mondada, L. (2007). Multimodal resources for turn-taking: Pointing and the emergence of possible next speakers. *Discourse Studies, 9*(2), 194–225.

Ochs, E. (1979). Planned and unplanned discourse. In *Discourse and syntax* (pp. 51–80). Brill.

Poland, B. D. (1995). Transcription quality as an aspect of rigor in qualitative research. *Qualitative Inquiry, 1*(3), 290–310.

Smits, H., Wang, H., Towers, J., Crichton, S., Field, J., & Tarr, P. (2005). Deepening understanding of inquiry teaching and learning with e-portfolios in a teacher preparation program. *Canadian Journal of Learning and Technology/La revue canadienne de l'apprentissage et de la technologie, 31*(3).

Tilley, S. A. (2003). "Challenging" research practices: Turning a critical lens on the work of transcription. *Qualitative Inquiry, 9*(5), 750–773.

CHAPTER 14

Team Projects in ATLAS.ti

Abstract ATLAS.ti enables two or more researchers to use a single data set while working on different aspects of the same project. Typically, large projects require researchers to work in teams, with each member assigned a specific task or set of tasks. Usually, there is a lead researcher who has overall responsibility for the project and ensures that its objectives are fulfilled. While working with ATLAS.ti, each researcher in the project team is allowed independent access to the data set. This allows multiple approaches to data analysis simultaneously. This chapter discusses various team project scenarios and offers guidelines for teamwork. It also discusses the importance of intercoder agreement, reliability, validity, and the methods for assessing intercoder agreement (and interpretation of results). The roles of coders and lead researchers, regarding how to export, import, and merge project data have been explained. The importance and utility of the semantic domain in a collaborative project are also discussed in detail with an example.

CHAPTER OVERVIEW

This chapter is in two parts: theory and application. In the theory part, the discussions cover the concepts that the reader must become familiar with to be able to work on team projects using ATLAS.ti.

In ATLAS.ti, a large project can be broken down into several smaller projects—we can call them sub-projects—that are then assigned to different researchers. On their completion, these sub-projects can be merged into one. Each researcher is given independent access to the same data set so that different kinds of analysis can be performed simultaneously. ATLAS.ti also allows the transfer and conversion of research data while always keeping the sources of ideas identifiable.

Database management is done using the administration tool. This is an important prerequisite for collaborative work. For security and traceability, individual researchers should use the Tools and Support tab to personalize their login credentials. After the completion of the various sub-projects, the lead researcher can consolidate the outputs of each by merging them.

The usual practice in a group project is for the lead researcher to assign the coding work to a few selected team members. This is done after a discussion and agreement among the team members on the approach to coding. Such an arrangement becomes necessary where large volumes of data are involved and coding work must be distributed to save time and effort.

When coding responsibilities are distributed among team members, there must be a strong emphasis on coding discipline and the need to follow instructions. There are several quantitative and qualitative methods for testing inter-coder reliability and inter-coder agreement. Each method has its merits and limitations and therefore, it is incumbent on the researcher (or the research team) to justify the selection of a particular method, as well as explain why other methods did not fit.

Reliability and Validity in Qualitative Research

Initial conceptualizations of validity were directly applied from reliability and validity standards of quantitative or experimental research based on a positivistic philosophy (LeCompte & Goetz, 1982). Traditional definitions of reliability and validity were felt to be applicable and credible benchmarks by which the quality of all research could be judged (Popay et al., 1998). Reliability refers to the stability of findings, whereas validity represented the truthfulness of findings (Altheide & Johnson, 1994). Validity is broadly defined as "the state or quality of being sound, just, and well-founded" (Random House Webster's Unabridged Dictionary, 1999), which are certainly reasonable components of all investigations, be they qualitative, quantitative, or mixed. Validity refers to the integrity and application of the method and precision in which findings accurately reflect the data. Reliability refers to consistency within the employed analytical procedures.

The incompatibility of these terms with the underlying assumptions and tenets of qualitative research resulted in the translation of terms to be more aligned with the interpretive perspective. Many qualitative researchers argued that reliability and validity were terms pertaining to the quantitative paradigm and hence not pertinent to qualitative inquiry (Altheide & Johnson, 1998; Leininger, 1994). Kahn (1993) discussed the implications of idiosyncratic terminology associated with validity in qualitative research and emphasized that language should not obscure understanding. Because qualitative research is based on entirely different epistemological and ontological assumptions compared to quantitative research, researchers felt that validity criteria of the quantitative perspective are therefore inappropriate (Hammersley, 1992). Some suggested adopting new criteria for determining reliability and validity, and hence ensuring rigor, in qualitative inquiry (Lincoln & Guba, 1985; Leininger, 1994; Rubin & Rubin, 1995).

Guba and Lincoln (1989) substituted reliability and validity with the parallel concept of "trustworthiness," containing four aspects: credibility, transferability, dependability, and confirmability. They offered alternative criteria for demonstrating rigor within qualitative research i.e., truth value, consistency, neutrality, and applicability. Quantitative criteria: internal validity, external validity, reliability, and objectivity were substituted by qualitative counterpart: credibility (truth value), transferability (applicability), dependability (consistency) and confirmability (neutrality) respectively.

According to Lincoln and Guba (1985), credibility- refers to the truthfulness of findings. It should be viewed through the lens of respondents in their natural context. To measure the quality of credibility, the researchers should address the following questions (Miles & Huberman, 1994, pp. 278–279).

1. How does the researcher describe the context?
2. How does the researcher address negative case (deviant case)?
3. How does the researcher address triangulation?
4. How does the researcher make sense of the findings for the reader?

Transferability refers to the degree to which results can be applied to similar situations. Researchers should determine the essential context characteristics, such as detailed descriptions of surroundings, context, and individuals.

Dependability refers to the ability to produce consistent results if undertaken in similar circumstances while keeping context-specific factors in mind. To evaluate the dependability of a product or service, researchers must examine the following issues (Miles & Huberman, 1994, pp. 278–279).

1. Are research questions clearly defined and contain the features of research design?
2. Are basic paradigms and analytic constructs specified?
3. Did coding checks show adequate agreements?
4. Did researchers conduct data refraining from bias, deceit, and respondent knowledgeability? Biases inevitably exist in all social science investigations (Smith & Noble, 2014), and it is impossible to completely control or remove all social influences (Ryan, 2019).
5. How is peer or colleague review employed?
6. In teamwork data collection, do researchers follow data collection protocols?

In qualitative research, researchers must identify deviant situations that are significant for testing validity (Maxwell, 2010, p. 284). Gray (2018) asserts that researchers should explore and analyze deviant situations; otherwise, they may fail to strengthen their claim concerning findings (Anderson, 2010; Smith & Noble, 2014).

Confirmability refers to a researcher's capacity to create evidence to support their results. Respondents and circumstances should generate evidence contrary to the researcher's presumptions and prejudices. However, various scholars have different opinions about qualitative and quantitative criteria. For instance, some researchers argue for the same criteria as quantitative research (Morse et al., 2002), while others argue for different criteria (Koch & Harrington, 1998; Sandelowski, 1986), and others reject any predetermined criteria (Hope & Waterman, 2003; Johnson & Waterfield, 2004; Rolfe, 2006).

Member checking entails the researcher informally clarifying with participants, during data collection, the accuracy of their comprehension (Gray, 2018) to validate the research. They can implement member checking in interviews by repeating, paraphrasing, and clarifying responder comments. While doing so, researchers must be cognizant of respondents' tone, emphasis, and speech (Gray, 2018; Rutakumwa et al., 2020). According to Bonello (2001), verbal and non-verbal communication should be evaluated to determine if they are congruent and indicate an authentic reaction.

Respondent validation, a more rigorous member checking, allows interviewers to remark on and edit their transcribed interview records afterward (Anderson, 2010; Birt et al., 2016). However, this strategy has possible drawbacks. Respondents may reject using their comments and express the opinion that their responses have been misconstrued. They may suggest a narrow perspective on a subject (Torrance, 2012). They may recommend modifying it to represent better themselves or their organizations (Alvesson, 2012, Miltiades, 2008).

Researchers may present implicit quantitative components in qualitative study outcomes (Maxwell, 2010, p. 285). This concept can strengthen the research. To gain additional information, researchers can design qualitative questions with quantitative components. The use of simple descriptive numerical data, sometimes known as "quasi-statistics," is thus given as a beneficial supplementary type of evidence to increase validity in a primarily qualitative inquiry.

Triangulation is used to increase the rigor of a research study by suggesting the use of more diverse data collection methods and data sources (Mays & Pope, 2000; Beuving & de Vries, 2015; Fusch et al., 2018) because data from different sources may offer complementary perspectives on the same construct (Rolfe, 2006; Scott et al., 2007). Utilizing several methodologies, particularly qualitative and quantitative approaches, may provide an additional opportunity to demonstrate confirmation and completeness (McEvoy & Richards, 2006, Bekhet & Zauszniewski, 2012). Consequently, no single method will likely provide a comprehensive account of the investigated phenomenon (Torrance, 2012, p. 113).

According to Bisman (2010), replicability, which Stiles (2003) refers to as procedural trustworthiness, is one of the most important criteria for evaluating the quality of a research study. Replicability involves ensuring that the observations are repeatable and obtaining consistent results across studies that answer the same questions, each of which has obtained its data. It is the most important factor in assessing reliability in qualitative research studies.

Qualitative researchers aim to design and incorporate methodological strategies to ensure the trustworthiness of findings. They include bracketing, acknowledging biases in sampling and critical reflection, consistent and transparent data interpretation, representing different perspectives, rich and thick descriptions, reducing research biases, respondent validation, data triangulation etc. In other words, credibility can be enhanced by reflective journal, peer debriefing, data integrity, and following trustworthy research methods and reporting of findings.

Sensitivity as a validity criterion of qualitative research refers to research that is implemented in ways that are sensitive to the nature of human, cultural, and social contexts (Altheide & Johnson, 1994; Munhall, 1994). Reflexivity, open inquiry, and critical analysis of all aspects of inquiry contribute to validity in qualitative research (Marshall, 1990).

INTERCODER AGREEMENT IN TEAMWORK

Intercoder agreement is the extent to which two or more coders independently assign the same codes to the same set of data using mutually agreed-upon code definitions.

When multiple researchers code the same content, intercoder reliability ensures that they arrive at the same conclusions. The degree to which coders consistently distinguish between different responses is measured by intercoder reliability.

Intercoder agreement tests the reliability of coders to understand, interpret and code data following coding guidelines, which are based on the research objective. After ensuring an acceptable level of inter-coder agreement, the data is analyzed, and the results are interpreted. Intercoder agreement is arrived at in three steps:

1. The lead researcher and team members discuss the data, understand the research question, and approach to coding, identify relevant information segments, and finally agree on the coding decision. There must be consensus on code names, data segments, queries, code groups, semantic domains, and code categories. Once all doubts are cleared and there is an agreement, the lead researcher creates a master project on their computer.

2. Next, using the Export option under the file in the main taskbar, the project bundle is created, saved, and shared with the coders using Dropbox, OneDrive, Google Drive, or other. The lead researcher creates a snapshot of the main project before sending sub-projects to the coders. The reason is the main project remains intact, and in case of any problem in working on the project, the main project can be used to further work. On receipt of the project bundle, the coder renames and then imports the project bundle into their user account (or computer). The coder is now ready to work on the sub-project. In case of a doubt or issue, the coder writes a query in the memo and saves it in the project.
3. On completion of the sub-projects, the project bundles are returned to the lead researcher who then downloads them and imports them into the main project. After saving each sub-project, the lead researcher verifies the project's correctness and calculates the intercoder agreement. The pre-merger summary is shown in Fig. 14.8.

After pressing the 'merge option', the merge report will appear, as shown in Figs. 14.9 and 14.10. The data is ready for intercoder analysis.

ATLAS.ti measures intercoder agreement using three methods: Percentage Agreement, Holsti Index, and Krippendorff's Alpha. The author suggests the Percentage Agreement method as the results are more reliable (Fig. 14.1).

It is preferred that members should code part of the data, preferably 20 to 25% and send it to the lead researcher. The step is to ensure that coders

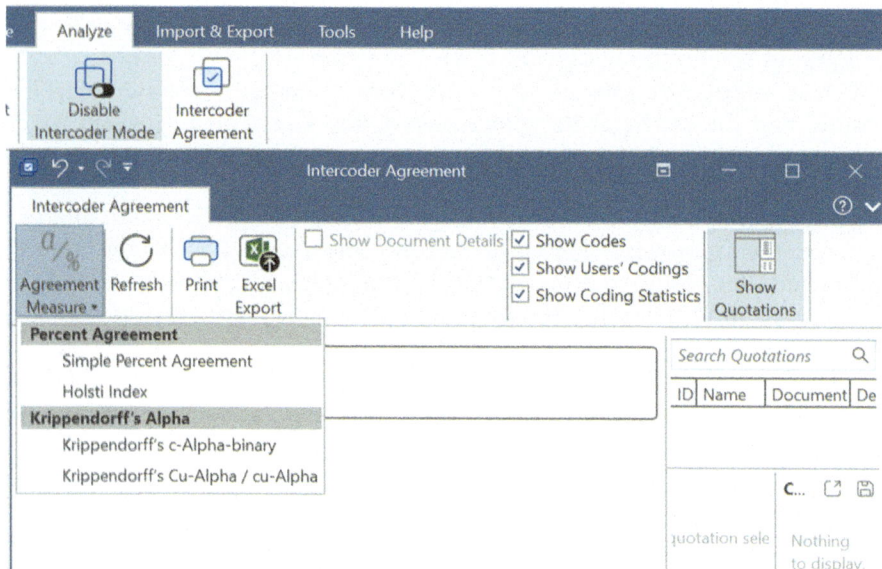

Fig. 14.1 Intercoder agreement window

understand data, coding guidelines and working knowledge of coding. The lead researcher conducts inter coder agreement and after sufficient intercoder agreement is achieved, the project bundle is sent to coders, and they are advised to code rest data for further work. The process can take two to three coding cycles. After completing the sub-projects, coders save and export the project bundle to the lead researcher to merge and analyze the data further.

Challenges in Interpretation of Intercoder Agreement

Many issues perhaps don't get much attention in the literature about the intercoder agreement. There is a need to discuss the issues that help the intercoder agreement process robust.

Silo communication. Coders should avoid discussing challenges they face during the coding, otherwise it can create chaos in the coding process. This step can also deviate from the agreed coding approach and guidelines. In addition, mutual discussions can affect the coding process and mutual coding agreement and lead to inability to reach a significant intercoder agreement.

Field notes. Every team member captures their field experience based on their understanding, experiences, and intuition. These experiences must be captured adequately, failing which can compromise the data quality, coding process, and result. A proper mechanism must be created for capturing field notes and establishing coding guidelines.

Bracketing. In qualitative research and especially in field-based data, the team members should ensure that they exercise proper care to bracket their preconceived feelings, assumptions, biases, grudges, experiences, and attitudes towards respondents, contexts, and organizations. This is important, even crucial to the research, otherwise, the quality of the data will be in question.

Sensitive research. Team members engaged in data collection should ensure that they are sensitive towards respondents, their environment, and organizations and ensure ethical protocols in documenting and reporting results. Most importantly, issues such as taboos, self-incrimination and discredit, illnesses, grief, sexual abuse, violence, drug use or homelessness should be examined from the perspective of both researchers and the participants.

Sensitizing concepts. Team members should capture jargon, local words, quotations, proverbs, and sayings to deepen the contextual understanding of the topic. Sensitization to concepts allows researchers to focus on developing empirically grounded concepts from the participants' point of view, identifying potential lines of inquiry. The process will help them to suggest future research and capture the meaning people make in their contexts. The project team members should also note that a proper coding approach should ensure capture of the meanings of words.

Deviant cases. Qualitative research often reports patterns based on significant values. However, there are deviant cases which may not fit into the pattern. The researcher should capture deviant cases and present them to show the completeness of data, analysis, and themes.

Independent coders. Independent coders can also be allowed to work in a team. While this practice has its merits, it is not free from caveats. Independent coders may not fully appreciate the context and the associated field challenges the researchers have encountered. This can influence the coding quality, understanding of the code's meaning, and effective capture of contextual components, field notes, informal communication etc.

Prerequisite Knowledge for Working in an ATLAS.ti Team Project

Team members should have a working knowledge of ATLAS.ti. The lead researcher should have experience with ATLAS.ti and be able to provide the required guidance and support to the team members.

Team members should also be familiar with, and have a good understanding of, the coding guidelines, how to identify quotations, writing up their field experiences, writing comments, creating code groups, etc. A shared understanding of requirements and approach to analysis is essential for minimizing errors and ensuring project quality.

Team Project Scenarios

There are three main scenarios in team projects:

1. A common data set that the lead researcher shares with team members at different sites.
2. A single data set that is accessed by all team members. This data is stored in the server and accessible to all team members.
3. Each team member works with only one subset of the data. The subset is usually created by the lead researcher (or team leader).

At an appropriate stage, the lead researcher (or team leader) merges the outputs of individual activities and initiates analysis.

> **Important** In most teamworking scenarios, it is essential that there is one person who has overall responsibility. This person may be designated as lead researcher or team leader and is responsible for setting up the project and assigning various tasks and activities to other team members, collecting the outputs of sub-projects, and merging the results into the main project. Project team members are provided individual user accounts in ATLAS.ti. The lead researcher (or project researcher)

exports the ATLAS.ti project, or specific components, to the team members through email or other methods. Each team member works on the component (or part) of the project that is allocated to them. On completion of each component, the output files are exported to the lead researcher for evaluation and consolidation and further processing.

Setting up a Team Project

Setting up the project is the responsibility of the lead researcher who is, usually, also the project leader (the two designations are used interchangeably in this book). The project leader is also tasked with putting together the project team. Protocols for data collection, recording field observations, information sharing, and team communications, addressing challenges and difficulties (anticipated and unforeseen), issue resolution and escalation, are discussed and finalized after meetings and discussions.

The lead researcher plays a central role which includes creating and defining the project in ATLAS.ti, data import and analysis, and the presentation of the project report. All related documents, including the codebook and coding methods, must be approved by the lead researcher before they are uploaded. Codes are decided based on mutual agreement between team members and the lead researcher.

After collection of data and its transcription, the documents are scrutinized before initiating data analysis. All team members involved in data collection must check the data for possible redundancies and ensure that the data is clean.

If the data was collected using Google forms, it is the lead researcher's responsibility to check the data and decide on further action. The documents are uploaded to ATLAS.ti only after their scrutiny by the lead researcher (or project leader).

As a good practice, by this stage, the coding scheme must be ready so that the coding activity can commence without loss of time. As with any team project, coordination and clear communications are important for preventing duplication of codes and documents. The lead researcher must establish clear guidelines for all team activities.

The lead researcher oversees the master project. The master project contains all documents, data and codes that will be shared with the team (Fig. 14.2). A master project is the complete data set that is ready for analysis. A copy of the master project, along with guidelines and work instructions, is given to the team members. It is important that every team member is accessed to complete dataset in the project. Alternatively, depending on the project scenario, only parts of its master project (the sub-projects) are shared with team members. In this case, each team member will work on a data segment independently. The lead researcher takes care that no two members will get the

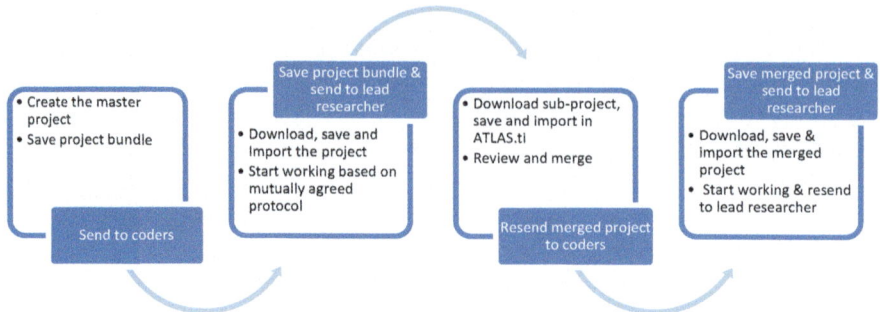

Fig. 14.2 Flowchart in team project

same document. In other words, each team member works on different set of transcripts, otherwise this may lead to duplication of codes. For example, data collected by ten respondents may be distributed among two or three team members, with each person assigned two transcripts.

After the project is saved in ATLAS.ti, the next step is to create a project bundle file which is shared with team members via emails or other workflow management tools (Fig. 14.2). Sometimes, a folder on the server is made available for access by individual members. The default name of the bundle is the project name + author and date. Depending on permission levels, individual users can rename the component of the project assigned to them. It should be remembered, however, that renaming the bundle does not change the name of the project contained within the bundle. To help understand this, the user should visualize the project bundle file as a box containing the ATLAS.ti project, including all the documents that will be analyzed. Putting a different label on the outside of the box would not change the name of the project file, which is contained *inside* the box.

On completion of all component activities, the lead researcher imports all the files and merges them into a new master file. The lead researcher should retain the original master file. The process of assigning work to team members, uploading the output files and updating the master file is an ongoing one and ends only after analysis is completed and there is agreement among the team members that their work is done.

On occasion, there may be a need for adding a new document to the project. In such a case, it is good practice to merge the project, add the new file, and then create a new master file. The process of sending subproject continues until complete data is coded by members.

Setting a up a Master Project

Speed, efficiency, and accuracy are of paramount importance in team projects and hence good practices must be adopted. To understand this need better, we will discuss with the example of a team study of "Employees' morale in public sector banks". In this project, the lead researcher has overall responsibility for the project, including monitoring its progress, and mentoring team members.

The research questions for the study are:

- What are the main components of the middle managers' morale?
- To what extent do management practices influence middle managers' morale? What can be done to improve the middle managers' morale?

The study was conducted at the country level for which data was collected from every state through interviews (telephonically or in-person) with senior managers, top management, and union directors. The subjects were selected using a purposeful sampling method.

There were ten members in the project team. Each team member conducted one interview. In all, there were twenty documents—ten transcripts and ten field notes. These documents were uploaded to ATLAS.ti by the team leader. After uploading, document groups were created according to demography, respondents' experiences, region, gender, position of the respondents (senior management, top management, union directors, gender, etc.). Formation of document groups is important for data analysis, identifying patterns and shared features, as well as differences, and while presenting results (Chap. 9).

Coding followed the formation of document groups. Prior to coding, the team leader ensured that the coding system and procedures were fully understood by all team members.

A Project with 10 Interviews

The master project consisted of 10 interviews. Members of the project team should justify the rationale of their coding decisions and show how they capture the nuances of data. Coding capabilities are about team members' ability to understand the coding process, identify appropriate segments, code, write memos, create code categories and groups, and be aware of code types and how to write codes, etc. An associated challenge is the team members' experience and intuitive understanding of the contexts in which data is collected.

Coding decisions are about working independently on the data and documenting queries, difficulties and challenges during the coding process. Members should write their understanding of encountered challenges that members can discuss and the lead researcher.

The lead researchers should make a snapshot of the project as a backup measure so that data remain intact in case anything goes wrong with the data. Next, the lead researcher sends each coder a sub-project containing five documents (25% data). Alternatively, the lead researcher can ask to code only 25% of the data.

While coding, each coder works independently of other coders. After coding activity is completed, the project bundle is returned to the lead researcher. After all sub-projects are merged, the lead researcher calculates the intercoder agreement using the appropriate method (Fig. 14.2).

> **Important** After the satisfactory intercoder agreement is met on five documents, it is understood that the team members are consistent in coding, and they can independently code the entire data. There are two ways to further complete the coding process of the remaining documents. The lead researcher can send a project containing 15 documents (75% of data) to coders. Coders complete the coding process and send it to the lead researcher after working on the sub-projects. The Lead researcher merges all sub-projects and performs the analysis.

Another way is to clear all codes from the sub-projects. It follows sending all documents, i.e., the entire project, to each coder along with coding agreement, guidelines, and protocols to work independently. The process is better because coders gain confidence in coding, and better results are likely to emerge.

After coding, the lead researcher performs the analysis based on mutual agreement, reports findings and interprets outcomes.

Creating Codes in a Team Project

Team consensus is necessary before assigning codes to data. Regardless of coding approach and the methods used, there must be a convincing rationale. After consensus is reached, the project team can use one of the two coding options in ATLAS.ti:

- The free codes option for writing definitions in Code Manager. The Code Group Manager is used for defining and describing code groups (Chap. 4).
- From the patterns identified in the word cloud, the lead researcher names the codes and code groups, and their definitions. Next, the meanings of the codes and code groups are written following which team members can write their comments. The objective of this activity is for everyone in the team to have a clear, uniform understanding of codes, code-groups, and their meanings.

This activity can also be performed offline. The names of codes and code groups are written in an Excel format. After scrutiny and agreement, the codes are imported into ATLAS.ti. It needs to be understood that such codes are in a

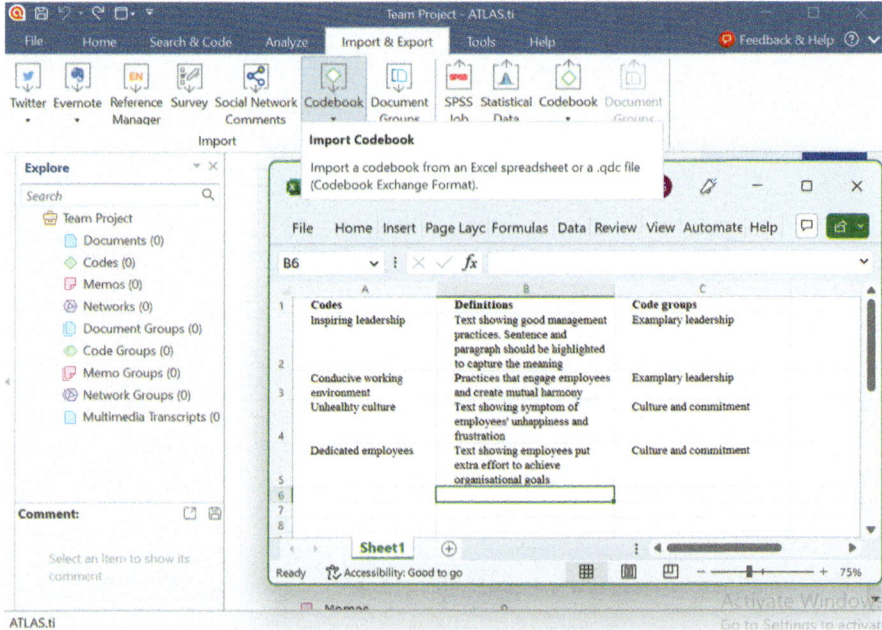

Fig. 14.3 Codes, definitions and code groups in excel format

floating form, meaning that they are not attached to any data segment. Hence, the data should be read before the codes are assigned to the segments.

Figure 14.3 shows the codes, definitions, and code groups in an Excel format which was imported into ATLAS.ti using Import Codebook in the Import & Export function. Figure 14.3 shows the codes, code groups and definitions. The code "Conductive working environment" is shown in the comment section.

After coding consensus is reached, the team leaders assign the sub-projects to individual team members. The project file is created in ATLAS.ti and then mailed to the team members who will then download it on their computers. Each team member imports the project to work on (Figs. 14.4 and 14.5).

Coding Method in a Team Project

While working on a team project, the team members are assigned different data segments for coding. Since the coding approach and methods are already decided and agreed upon, individual coders usually work independently of each other. For example, in a project with data transcripts from fifty respondents, the coding responsibilities are distributed equally. If there are five coders, each person will have the responsibility for coding ten transcripts.

On completion of the tasks, the project is returned to the lead researcher. The lead researcher then checks for redundant codes using the Tools option

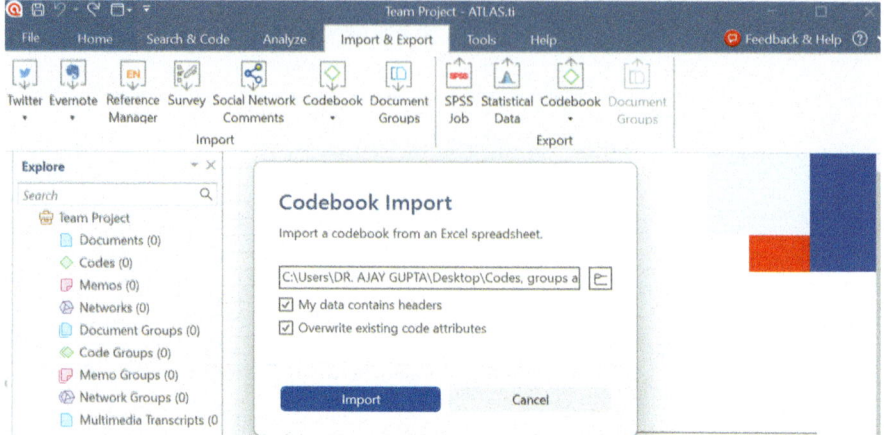

Fig. 14.4 Importing codebook into ATLAS.ti

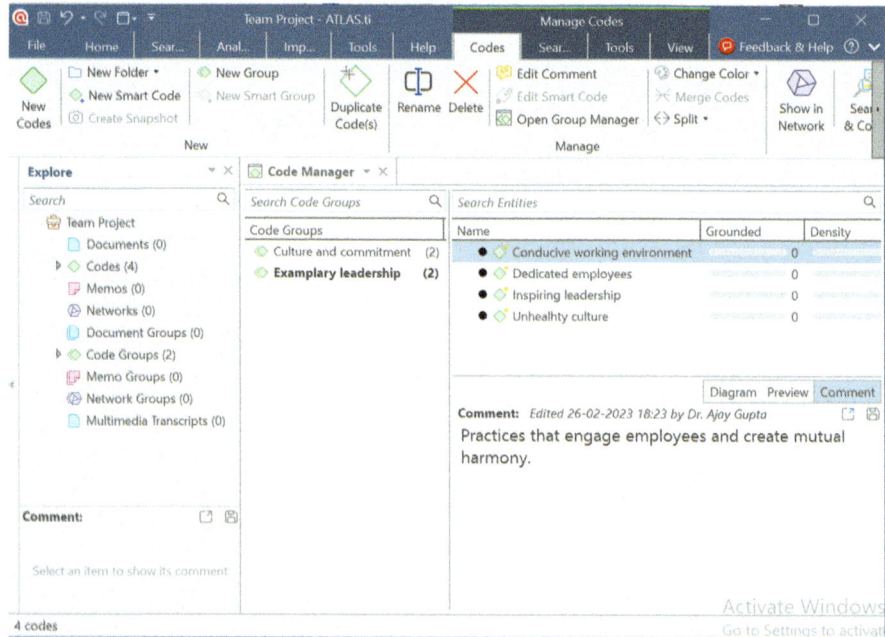

Fig. 14.5 Codes imported into ATLAS.ti

in Code Manager. In the next step, the subprojects are merged and then assigned to members for further analysis. It is good practice to check the codes for redundancy after completing the coding process and *before* exporting individual projects to the lead researcher.

Intercoder Agreement in ATLAS.ti

ATLAS.ti can test Inter-coder agreement in text, audio, and video documents. The concept of intercoder agreement is used to assess how multiple coders code a given body of data. Inter-coder agreements cannot be checked with Web versions of documents.

Considerable disagreement exists among researchers on how to ensure inter-rater consistency when teams are tasked with coding qualitative data. It is important to have a methodology for systematic coding to preserve the contextual and subjective nature of the data and ensure consistency in application of codes across varied types of data. Reaching consensus on codes and its application across the data is a challenging task in teamwork.

Visual representation of coding methods has been described below (Hemmler et al., 2022).

1. Cleaning, describing, and organizing data
2. Developing an initial codebook—inductive and deductive
3. Piloting the codebook to accommodate data
4. Developing a system of addressing uncertainty in code application
5. Calibration by co-coding one piece of each type of data after consensus
6. Completion of individual coding and writing memo and uncertainties
7. Resolving uncertainties from individual codings
8. Achieving across—and within—coder consistency

Preparing Data for Inter-coder Agreement Check

Tests for Inter-coder agreement can be performed on primary as well as secondary data (text, audio and video formats). The broad process followed is as follows (Fig. 14.6).

1. After data collection, transcription and organization, the lead researcher gives the master project a name.
2. Next, a snapshot of the master project is created and the coding system. The researcher documents the coding system, defining each code, and providing the guidelines that the coders in the project team must follow.
3. With the creation of a snapshot, the project is ready to be assigned to the coders. A snapshot is the creation of the same project so that the original document remains intact and later can be used in need arises. A snapshot is created using File option and selecting Snapshot. Figure 14.6 shows the snapshot of a project.
4. Snapshots can be given a separate name for easy identification, such as, in this example, 'Sub-project for ICA'. ATLAS.ti creates two projects: a snapshot of the master project, and a sub-project for ICA.

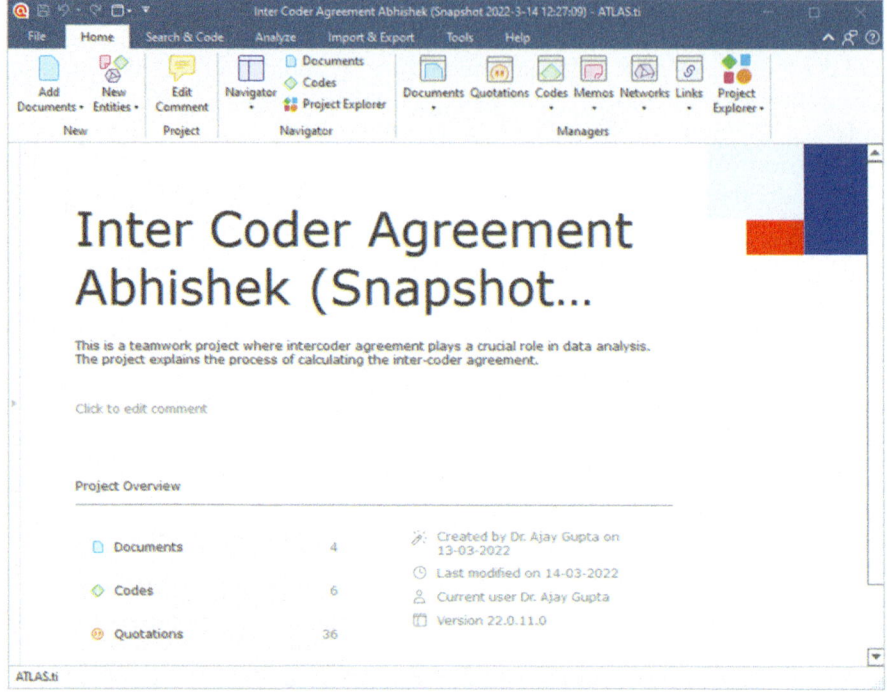

Fig. 14.6 Snapshot of a project

The lead researcher may send only 10–20% of the project materials that require coding are shared with the coders. The purpose of is to ensure that the coders have understood the data, follow appropriate coding approach based on consensus. Once the required level of intercoder agreement is achieved, the remaining data is shared with the coders. Coding can be done in two ways:

1. Using the pre-defined codes and instructing the coders to apply these codes to the data segments, or
2. Instructing the coders to code the selected quotations.

The Lead Researcher's Role

After the coders have completed their work, the files are imported by lead researcher who then:

1. Reviews each coder's work and comments. Coding errors, if any, are corrected and discrepancies, and issues resolved.

2. Merges the projects using Merge Project option. The project bundles of coders can also be merged directly. However, the codes must be checked for errors and discrepancies.

Figures 14.7, 14.8, 14.9, 14.10, 14.11, 14.12 and 14.13 show the steps in the analysis for inter-coder agreement. The flow chart is shown in Fig. 14.2.

After opening the merged file of the project, a snapshot of the project is created. The snapshot file is opened. Before merging the projects, the inter-coder mode is enabled in the Analyse option. Open the project of coder A and select merge under file option. Select the project to be merged and proceed. After the merging of projects, the following screen appears, and the data is ready for inter-coder analysis.

Figure 14.9 shows the resulting display after selecting the Merge option.

Then, the inter-coder analysis function is applied to the merged project in the following steps:

1. Open Inter-coder Agreement in the Analysis section
2. Click on the Add Coder button and select Coders (Fig. 14.11).
3. Click on the Add Documents button and select the documents for inter-coder analysis.
4. Click on the Add Semantic Domain button and add codes to the domain (Fig. 14.11).

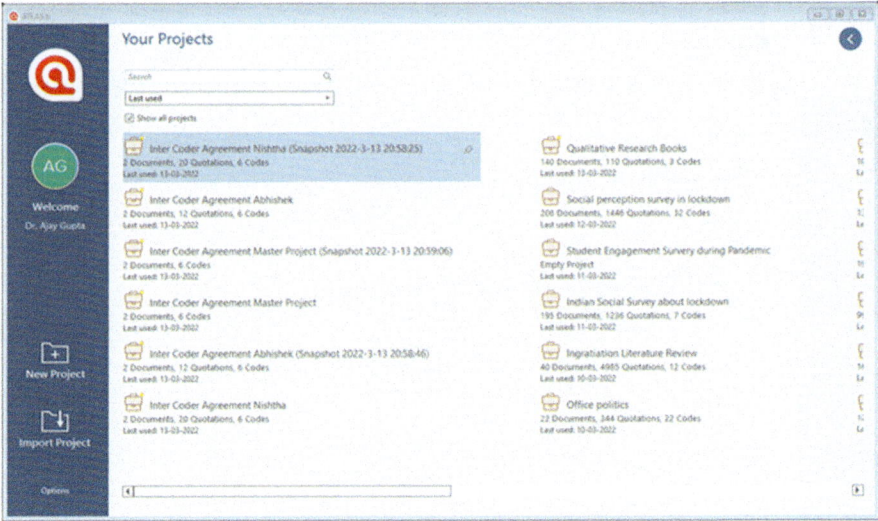

Fig. 14.7 Snapshot of projects in ATLAS.ti 22

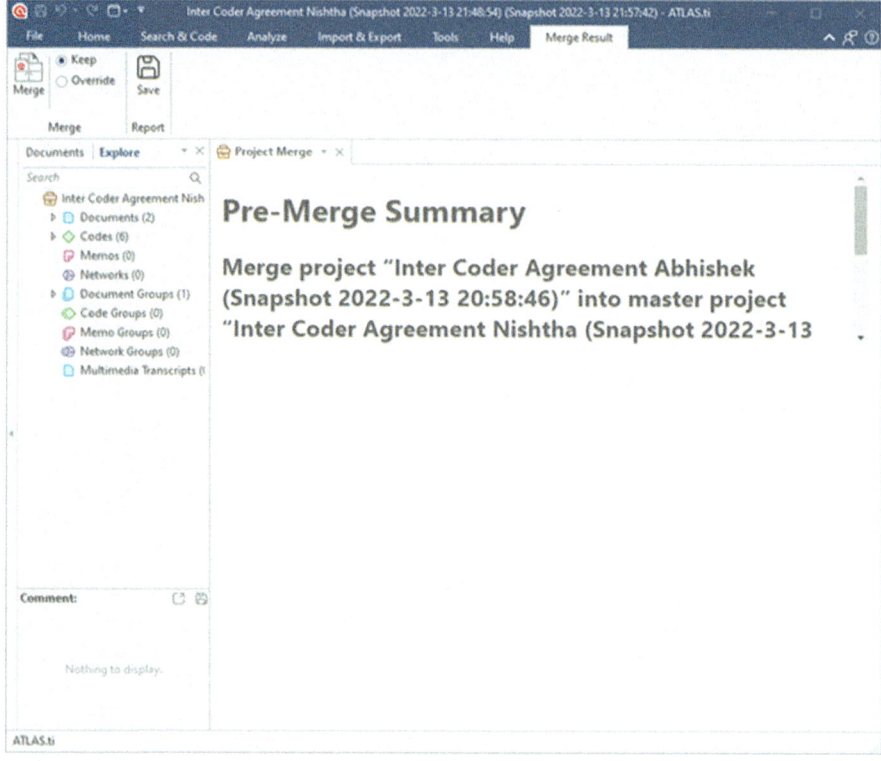

Fig. 14.8 Pre-merge summary for inter-coder agreement

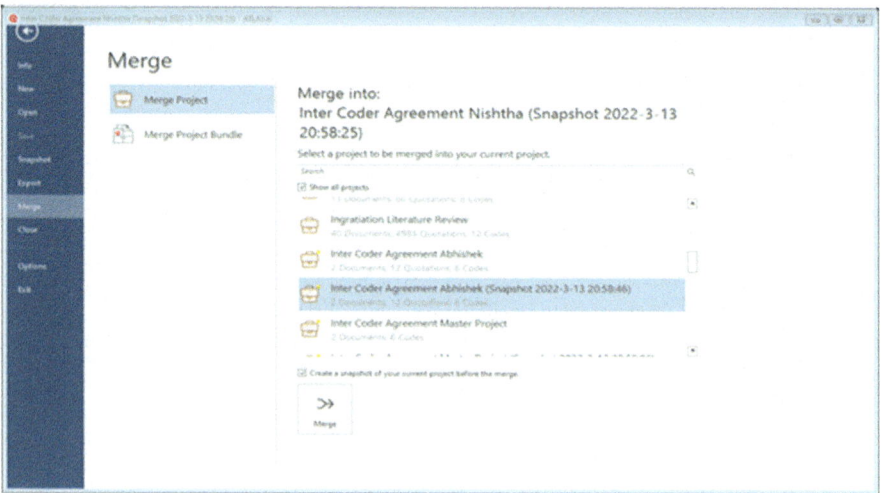

Fig. 14.9 Project merge option

Fig. 14.10 Merge report for two coders

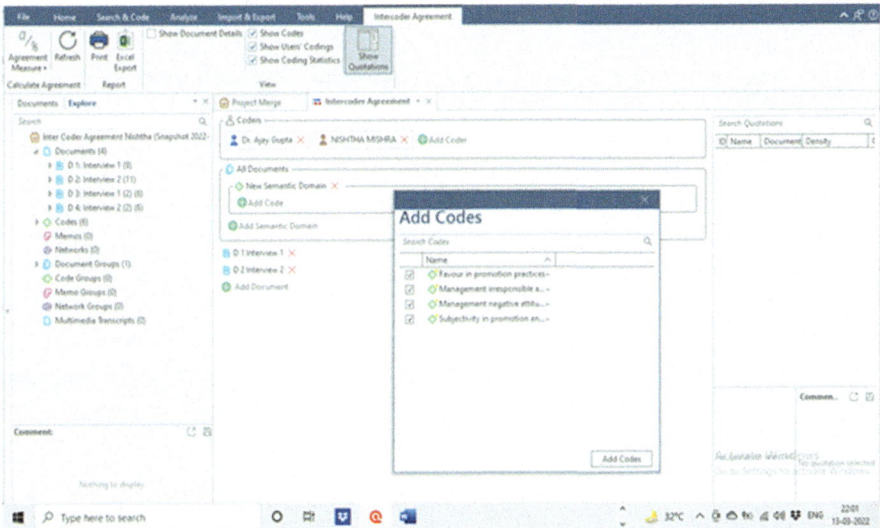

Fig. 14.11 Adding coders, documents and codes for Ica

On selecting Show Quotations, the list of quotations will be displayed on the right. The report on the results of Inter-coder analysis in Excel format is shown in Fig. 14.13.

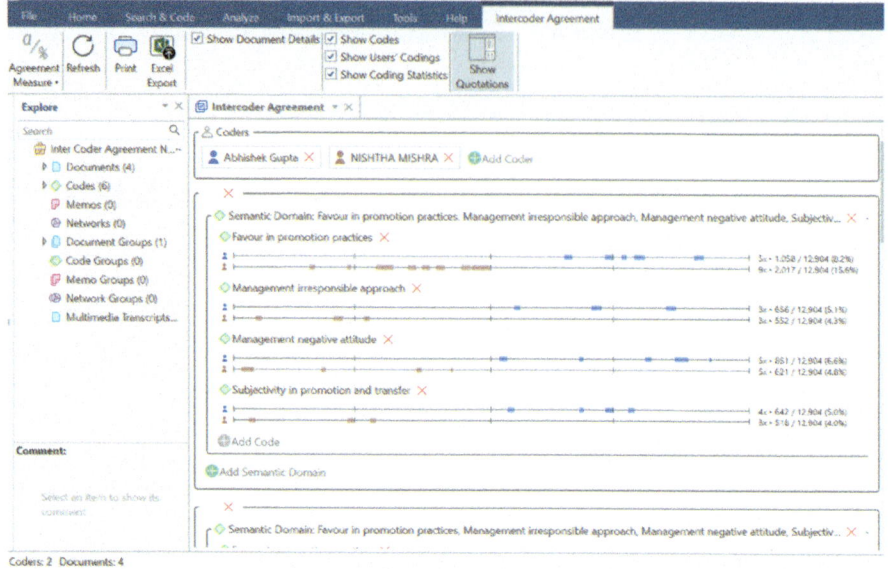

Fig. 14.12 Inter-coder analysis results

Intercoder Agreement for Project Inter Coder Agreement Abhishek (Snapshot 2022-3-14 12:27:09)					
Agreement Coefficient:	Krippendorff's Cu-α/cu-α				
Legend					
Applied*	Number of times the code has been applied				
Units*	Number of units* the code has been applied				
Total Units*	Total number of units* across all selected documents				
Total Coverage*	% Coverage within the selected documents				
Coders					
	Abhishek Gupta				
	NISHTHA MISHRA				
All Documents					
Semantic Domain:	Favour in promotion practices Management irresponsible approach Subjectivity in				
Code	Coder	Applied*	Units*	Total Units*	Total Coverage*
Favour in promotion practices					
	Abhishek Gupta	5	1058	12904	8.20%
	NISHTHA MISHRA	9	2017	12904	15.63%
Management irresponsible					
	Abhishek Gupta	3	656	12904	5.08%
	NISHTHA MISHRA	3	552	12904	4.28%
Subjectivity in promotion and					
	Abhishek Gupta	4	642	12904	4.96%
	NISHTHA MISHRA	3	518	12904	4.01%
Management negative attitude					
	Abhishek Gupta	5	851	12904	6.59%
	NISHTHA MISHRA	5	621	12904	4.81%

Fig. 14.13 Excel report for inter-coder analysis

Percentage Agreement

ICR is more commonly calculated as a percentage of agreement (Cohen, 1960; Hallgren, 2012; Lombard et al., 2002) as compared to formal statistical tests like Krippendorff's alpha (Feng, 2014). Many qualitative analyses draw on quantitative information, such as frequency counts of the number of interviews that contain a given code (Maxwell, 2010). While ICR can increase the rigor and transparency of the coding frame and its application to the data (Hruschka et al., 2004; Joffe & Yardley, 2003; MacPhail et al., 2016; Mays & Pope, 1995), it challenges the interpretative agenda of qualitative research (Braun & Clarke, 2013; Hollway & Jefferson, 2013; Vidich & Lyman, 1994; Yardley, 2000). In other words, percentage method may not provide space for interpretation.

The most common method is percentage agreement of coded data (Feng, 2014; Kolbe & Burnett, 1991). Miles and Huberman (1994) suggest reliability can be calculated by dividing the number of agreements by the total number of agreements plus disagreements. However, percentage-based approaches are almost universally rejected as inappropriate by methodologists because percentage figures are inflated by some agreements occurring by chance (Cohen, 1960; Hallgren, 2012; Lombard et al., 2002).

For percentage agreement approaches, there is no universally accepted threshold for what indicates acceptable reliability, but Miles and Huberman (1994) suggest a standard of 80% agreement on 95% of codes. Most of the commonly used statistical tests of ICR present results on a scale between -1 to $+1$, with figures closer to 1 indicating greater correspondence. Neuendorf (2002) reviews "rules of thumb" that exist for interpreting ICR values, observing ICR figures over 0.9 are acceptable by all, and over 0.8 acceptable by many, but considerable disagreement below that. Researchers often cite Landis and Koch's (1977) recommendation of interpreting values less than 0 as indicating no, between 0 and 0.20 as slight, 0.21 and 0.40 as fair, 0.41 and 0.60 as moderate, 0.61 and 0.80 as substantial, and 0.81 and 1 as nearly perfect agreement.

There is no universal agreement regarding how to manage low-performing codes. Some researchers may opt to discard codes below a certain ICR threshold. Others may modify poorly performing codes and with the revised coding frame, repeating this process until an acceptable ICR is attained. Researchers must also decide how to treat instances of inter coder disagreement when finalizing the definitive coded data set. Some research teams may introduce a third coder and adopt a "majority rules" decision. Others may decide that the judgments of one coder (usually the PI or more experienced researcher) outweigh those of the others. Finally, some research teams may adopt a consensus approach where disagreements are discussed, and joint decisions reached. Campbell et al. (2013) describe such a strategy of "negotiated agreement" (Campbell et al., 2013, p. 305), which ultimately increased reliability from 54 to 96%. Once the coding frame is finalized, it should be

Table 14.1 Percentage method of inter-coder agreement

Segments	1	2	3	4	5	6	7	8	9
Coder A	Coded	Not coded	Coded	Not coded	Coded	Coded	Not coded	Coded	Coded
Coder B	Not coded	Not coded	Coded	Coded	Coded	Coded	Not coded	Coded	Coded

systematically applied to the entire data set. This typically involves the recoding of data originally coded during the ICR process (MacQueen et al., 1998).

The percentage represents the proportion of the total number of codes that the coders have agreed on. Therefore, the total agreement is the number of coded and uncoded segments that match between coders, that is, inter-coder agreement is calculated by adding the matched coded and uncoded segments between coders (Table 14.1). The ratio is multiplied by 100 to get the percentage agreement.

Table 14.1 shows coder agreement for nine segments in a document. The document was coded by Coder A and Coder B. Both coded some segments and left a few segments uncoded. There is agreement between the two coders over the segments that were coded and not coded. The number of 'agreements' for coded segment is 5, that for uncoded segments is 2. Thus, the number of agreements, coded and uncoded, is 7. The percentage agreement, PA, is calculated from the expression.

PA = number of agreements/total number of segments = 7/9 = 0.77 or 77%.

Establishing Intercoder Percentage Agreement

Intercoder agreement is the iterative process of unitizing, coding, discussing coding discrepancies, and refining codes and code definitions. To ensure high inter-coder agreement,

1. The coders in the team should use some sample of transcripts to develop a coding scheme having as high a level of intercoder reliability as possible.
2. Disagreements over coding should be resolved through discussion among coders.
3. Deploy coding scheme on the full set of transcripts after acceptable levels of intercoder agreement.

According to Miles and Huberman (1984: 63), intercoder reliability is calculated as the "number of times that all coders used it in the same text unit/the number of times that any coder used it in the transcript. For example, with two coders, if 40 text units were coded with "motivation" by at least one of them and in 25 of those cases both had invoked the code on the same text unit, then the level of intercoder reliability would be 25/40 = 62.5% for the "motivation" code. Using the same method, intercoder reliability is calculated as follows.

Intercoder reliability = total number of agreements for all codes/the total number of agreements and disagreements for all codes combined.

There is no universal agreement on an acceptable level of intercoder reliability. Literature shows considerable variation according to the standards of different researchers, as well as the method of calculation (Dunn 1989: 37). Fahy (2001) held that an intercoder reliability range of 70–94% was "acceptable" to "exceptional" for his analysis of transcripts from conference discussions. Kurasaki (2000) reported intercoder agreement scores of 70% during coder training and 94% during actual coding—both of which were deemed acceptable.

Calculation of Intercoder Agreement

According to O'Connor and Joffe (2020), statistical software can be used to calculate ICR. Further, ICR is calculated on just a subset of the data which is a about 10–25% of data units would be typical. It is good practice to double code a small amount of data i.e., one interview. It can lead to coding framework refinement before commencing the formal independent double-coding with the larger subset of data. Some researchers recommend that coders should be external to the research team who had no role in designing the coding frame (Kolbe & Burnett, 1991). However, ethical and data protection considerations should be taken care of.

A meeting should be organized after the first round of coding to address a difference among coders before second round of coding begins (Campbell et al., 2013; Hruschka et al., 2004). Training for coders depends upon the nature of data and degree of interpretations. Coded segments differ across studies. In other words, there is no protocol that decides what segment of data is coded. However, Kurasaki (2000) proposed coding strategy for the segment whereby allowing coders to select their own segments, randomly picking certain lines in the document, and comparing codes recorded withing a radius of five lines.

Regarding the number codes, MacQueen et al. (1998) suggest researchers concerned with achieving satisfactory reliability should work with an upper limit of 30–40 codes. Hruschka et al. (2004) recommend a limit of approximately 20 and further suggest that for semi structured interview data, codes should be specific to the interview questions.

The suggested procedure for ICR assessment (O'Connor & Joffe, 2020) is given below.

1. Researchers must make a priori decisions regarding the number of coders, amount of data to be coded in duplicate, the unit of coding, the

conceptual depth, the reliability measures, the threshold that will indicate acceptable reliability.
2. The research team develops the first draft of a coding frame after an intensive reading of the data. The first coder applies the codes to the data, saves the file in duplicate. From the duplicate file, he removes the code, saves the file, and sends it to the second coder to code and resends to the first coder.
3. Both coded files should be sent to SPSS, merged and reliability should be measured as per consensus reached, as mentioned in the first point. Further codes that fall short of the threshold can be evaluated to identify potential reasons for inconsistency of interpretation and removed or revised in accordance with the team's best judgement.
4. Further, the same process can be repeated using different subsets of data. Once the research team is satisfied with the overall reliability of the coding frame, the entire data set can be coded by a single coder or team of coders.

Interpretation of Results of Intercoder Analysis

ATLAS.ti treats documents as a textual continuum. Each character of the text is a unit of analysis in the test for inter-coder agreement. For audio and video documents, the unit of analysis is a second. Therefore, inter-coder analysis measures characters or seconds that have been coded and not the quotations themselves. The parts of the quotations that overlap are treated in analysis as agreement and the those that do not overlap as disagreements.

In Figs. 14.11 and 14.12, two coders, Abhishek Gupta and Nishtha Mishra independently coded the project data. Their outputs were merged for intercoder analysis. Figure 14.12 shows a line for each coder and code. A statistical calculation is shown to the right of every code. The intercoder agreement in this example is explained as follows:

The codes assigned by two users are shown in different colors. On each coder line, there are marked areas which show the number of times the coder has applied a code and the number of characters coded. The Total shows the total number of characters in the selected documents. Thus, the code 'Favour in promotion practices' was coded five times in 2 documents by Abhishek Gupta. The total number of characters coded is 1058. The total number of characters in both documents is 12,904. Using descriptive statistics, we get 8.2%. The percentage was obtained by multiplying 1058 by 5 and dividing the product by 12,904 and expressing the result as a percentage.

Recommended Readings

Altheide, D. L., & Johnson, J. M. (1994). Criteria for assessing interpretive validity in qualitative research.

Altheide, D. L., & Johnson, J. M. (1998). Criteria for assessing interpretive validity in qualitative research. In N. K. Denzin & Y. S. Lincoln (Eds.), *Collecting and interpreting qualitative materials* (pp. 283–312). Sage.
Alvesson, M. (2012). Understanding organizational culture. *Understanding Organizational Culture*, 1–248.
Anderson, C. (2010). Presenting and evaluating qualitative research. *American Journal of Pharmaceutical Education, 74*(8) 1–7, Article 141.
Bekhet, A. K., & Zauszniewski, J. A. (2012). Methodological triangulation: An approach to understanding data. *Nurse researcher, 20*(2).
Beuving, J., & De Vries, G. (2015). *Doing qualitative research: The craft of naturalistic inquiry*. Amsterdam University Press.
Birt, L., Scott, S., Cavers, D., Campbell, C., & Walter, F. (2016). Member checking: A tool to enhance trustworthiness or Merely a Nod to validation? *Qualitative Health Research, 26*(13), 1802–1811.
Bisman, J. (2010). Postpositivism and accounting research: A (personal) primer on critical realism. *Australasian Accounting, Business and Finance Journal, 4*(4), 3–25.
Bonello, M. (2001). Perceptions of fieldwork education in Malta: Challenges and opportunities. *Occupational Therapy International, 8*(1), 17–33.
Braun, V., & Clarke, V. (2013). *Successful qualitative research*. Sage.
Campbell, J. L., Quincy, C., Osserman, J., & Pedersen, O. K. (2013). Coding in-depth semistructured interviews: Problems of unitization and intercoder reliability and agreement. *Sociological Methods & Research, 42*(3), 294–320.
Cohen, J. (1960). A coefficient of agreement for nominal scales. *Educational and Psychological Measurement, 20*(1), 37–46.
Dunn, G. (1989). *Design and analysis of reliability studies: The statistical evaluation of measurement errors*. New York: Oxford University Press.
Fahy, P. J. (2001). Addressing some common problems in transcript analysis. *The International Review of Research in Open and Distributed Learning, 1*(2).
Feng, G. C. (2014). Intercoder reliability indices: Disuse, misuse, and abuse. *Quality & Quantity, 48*, 1803–1815. https://doi.org/10.1007/s11135-013-9956-8
Fusch, P., Fusch, G. E., & Ness, L. R. (2018). Denzin's paradigm shift: Revisiting triangulation in qualitative research. *Journal of Sustainable Social Change, 10*(1), 2.
Gray, D. (2018). *Doing research in the real world* (4th ed.). Sage.
Guba, E. G., & Lincoln, Y. S. (1989). *Fourth generation evaluation*. Sage.
Hallgren, K. A. (2012). Computing inter-rater reliability for observational data: An overview and tutorial. *Tutorials in Quantitative Methods for Psychology, 8*(1), 23.
Hammersley, M. (1992). Some reflections on ethnography and validity. *Qualitative Studies in Education, 5*(3), 195–203.
Hemmler, V. L., Kenney, A. W., Langley, S. D., Callahan, C. M., Gubbins, E. J., & Holder, S. (2022). Beyond a coefficient: An interactive process for achieving inter-rater consistency in qualitative coding. *Qualitative Research, 22*(2), 194–219.
Hollway, W., & Jefferson, T. (2013). *Doing qualitative research differently: Free association, narrative and the interview method* (2nd ed.). London: Sage.
Hope, K. W., & Waterman, H. A. (2003). Praiseworthy pragmatism? Validity and action research. *Journal of Advanced Nursing, 44*(2), 120–127.
Hruschka, D. J., Schwartz, D., St. John, D. C., Picone-Decaro, E., Jenkins, R. A., & Carey, J. W. (2004). Reliability in coding open-ended data: Lessons learned from HIV behavioral research. *Field Methods, 16*(3), 307–331

Joffe, H., & Yardley, L. (2003). Chapter four: Content and thematic analysis. In Marks, D., & Yardley, L. (Ed.), *Research methods for clinical and health psychology* (pp. 56–68). Sage Publications.

Johnson, R., & Waterfield, J. (2004). Making words count: The value of qualitative research. *Physiotherapy Research International, 9*(3), 121–131.

Kahn, D. L. (1993). Ways of discussing validity in qualitative nursing research. *Western Journal of Nursing Research, 15*(1), 122–126.

Koch, T., & Harrington, A. (1998). Reconceptualizing rigour: The case for reflexivity. *Journal of Advanced Nursing, 28*(4), 882–890.

Kolbe, R. H., & Burnett, M. S. (1991). Content-analysis research: An examination of applications with directives for improving research reliability and objectivity. *Journal of Consumer Research, 18*(2), 243–250.

Kurasaki, K. S. (2000). Intercoder reliability for validating conclusions drawn from open-ended interview data. *Field Methods, 12*(3), 179–194.

Landis, J. R., & Koch, G. G. (1977). The measurement of observer agreement for categorical data. *Biometrics, 33*, 159–174.

LeCompte, M. D., & Goetz, J. P. (1982). Problems of reliability and validity in ethnographic research. *Review of Educational Research, 52*(1), 31–60.

Leininger, M. (1994). Evaluation criteria and critique of qualitative research studies. In J. Morse (Ed.), *Critical issues in qualitative research methods* (pp. 95–115). Sage Ltd.

Lincoln, Y. S., & Guba, E. G. (1985). *Naturalistic inquiry*. Newbury Park, CA: Sage.

Lombard, M., Snyder-Duch, J., & Bracken, C. C. (2002). Content analysis in mass communication: Assessment and reporting of intercoder reliability. *Human Communication Research, 28*(4), 587–604.

MacPhail, C., Khoza, N., Abler, L., & Ranganathan, M. (2016). Process guidelines for establishing intercoder reliability in qualitative studies. *Qualitative Research, 16*(2), 198–212.

MacQueen, K. M., McLellan, E., Kay, K., & Milstein, B. (1998). Codebook development for team-based qualitative analysis. *Cam Journal, 10*(2), 31–36.

Maxwell, J. (2010) Chapter 17. Validity. How might you be wrong? In Luttrell, W. (Ed.), *Qualitative educational research*. Routledge.

Mays, N., & Pope, C. (1995). Qualitative research: rigour and qualitative research. *Bmj, 311*(6997), 109–112.

Mays, N., & Pope, C. (2000). Assessing quality in qualitative research. *Bmj, 320*(7226), 50–52.

Marshall, J. E. (1990). *An investigation of the construct validity of the test of basic process skills in science: a multitrait-multimethod analysis*. University of South Florida.

McEvoy, P., & Richards, D. (2006). A critical realist rationale for using a combination of quantitative and qualitative methods. *Journal of Research in Nursing, 11*(1), 66–78.

Miles, M. B., & Huberman, A. M. (1984). *Qualitative data analysis: A sourcebook of new methods*. Beverly Hills, CA: Sage.

Miles, M. B., & Huberman, A. M. (1994). *Qualitative data analysis: An expanded sourcebook*. Sage.

Miltiades, H. B. (2008). Interview as a social event: Cultural influences experienced while interviewing older adults in India. *International Journal of Social Research Methodology, 11*(4), 277–291.

Morse, J. M., Barrett, M., Mayan M., Olson, K., & Spiers J. (2002). Verification strategies for establishing reliability and validity in qualitative research. *International Journal of Qualitative Methods, 1*(2), Article 2.

Munhall, P. L. (1994). *Revisioning phenomenology: Nursing and health science research* (No. 41). Jones & Bartlett Learning.

Neuendorf, K. A. (2002). *The content analysis guidebook.* Thousand Oaks: Sage.

O'Connor, C., & Joffe, H. (2020). Intercoder reliability in qualitative research: Debates and practical guidelines. *International Journal of Qualitative Methods, 19*, 1609406919899220.

Popay, J., Rogers, A., & Williams, G. (1998). Rationale and standards for the systematic review of qualitative literature in health services research. *Qualitative Health Research, 8*(3), 341–351.

Random House Webster's Unabridged Dictionary. (1999). Validity [Def. 1]. New York, NY: Random House.

Rolfe, G. (2006). Validity, trustworthiness and rigour: Quality and the idea of qualitative research. *Journal of Advanced Nursing, 53*(3), 304–310.

Rubin, H. J. & Rubin, I. S. (1995). *Qualitative interviewing: The art of hearing data.* Thousand Oaks, CA: Sage.

Rutakumwa, R., Mugisha, J. O., Bernays, S., Kabunga, E., Tumwekwase, G., Mbonye, M., & Seeley, J. (2020). Conducting in-depth interviews with and without voice recorders: A comparative analysis. *Qualitative Research, 20*(5), 565–581.

Ryan, G. (2019). Postpositivist critical realism: Philosophy, methodology and method for nursing research. *Nurse Researcher, 27*(3), 20–26.

Sandelowski, M. (1986). The problem of rigor in qualitative research. *Advances in Nursing Science, 8*(3), 27–37.

Scott, M. L., Spear, M. F., Dalessandro, L., & Marathe, V. J. (2007, September). Delaunay triangulation with transactions and barriers. In *2007 IEEE 10th International Symposium on Workload Characterization* (pp. 107–113). IEEE.

Smith, J., & Noble, H. (2014). Bias in research. *Evidence-Based Nursing, 17*(4), 100–101.

Stiles, W. B. (2003). Qualitative research: Evaluating the process and the product. *Handbook of Clinical Health Psychology*, 477–499.

Torrance, H. (2012). Triangulation, respondent validation and democratic participation in mixed methods research. *Journal of Mixed Methods Research, 6*(2), 111–123.

Vidich, A. J., & Lyman, S. M. (1994). Qualitative methods: Their history in sociology and anthropology. In N. K. Denzin & Y. S. Lincoln (Eds.), *Handbook of qualitative research* (pp. 23–59). London: Sage.

Yardley, L. (2000). Dilemmas in qualitative health research. *Psychology and Health, 15*(2), 215–228.a2.

CHAPTER 15

Literature Review and Referencing

Abstract ATLAS.ti supports literature review and referencing. It helps researchers coding articles, organize references, extract in-text citations, and identify patterns in published studies on the topic of interest from which the researcher can draw inferences and present their insights. One can perform a literature search and review of multiple documents in chronological or any logical order. Literature review in ATLAS.ti involves three stages: data preparation, import, and conversion. Reference Manager tools are used to generate the files of articles, which are then imported to ATLAS.ti for literature review and referencing. Using the export and import functions, ATLAS.ti imports all the information associated with the data. It then automatically generates document groups and references in the comment sections, which aids data analysis and concept discovery, coding, followed by review, analysis, and reporting. The discussions in this chapter use an example of qualitative content analysis to explain the literature review process in ATLAS.ti, and the steps for reporting references in text, network, and Excel formats.

AN OVERVIEW OF THE LITERATURE REVIEW

Literature review is crucial to the research process. Literature review helps researchers to ground their studies, explain the rationale of their approach, and establish the scope and objectives. A literature review requires the researcher to cite related studies and provide references. ATLAS.ti supports the import of articles from reference manager tools and online resources. The studies referred to by the researcher can be organized and presented in textual, network, and Excel formats.

Figure 15.1 shows how references are managed in ATLAS.ti. The researcher can use either the manual or auto reference options for importing references. In the manual method, Google Scholar (or any other digital resource) is used to copy the references and paste them into the comment section of the document in ATLAS.ti. For auto-referencing, the Reference Manager tools in ATLAS.ti are used. ATLAS.ti supports Endnote, Mendeley, Zotero, and Reference Manager (Figs. 15.2, 15.3 and 15.4).

To import, one must first create an export file in Endnote XML format. It is good practice to prepare the data in the Reference Manager tool so that the export only contains a pre-selected set of documents. Figure 15.4 shows data preparation for ATLAS.ti using Zotero. The documents can be added manually or with the help of the Reference Manager tools to import research papers or articles. One can also use the Document Manager to add articles and documents. Depending on user preferences, either option—Add File or Add Folder Contents—can be used. Alternatively, the documents and articles, along with comments, can also be imported using the Reference Manager tool.

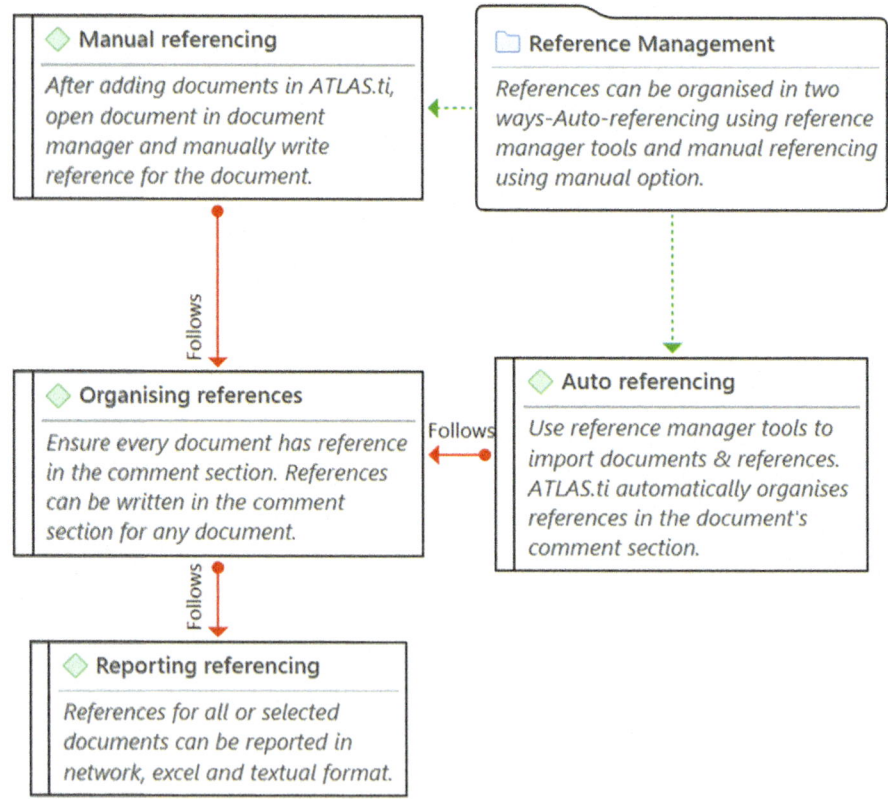

Fig. 15.1 Flow diagram for referencing

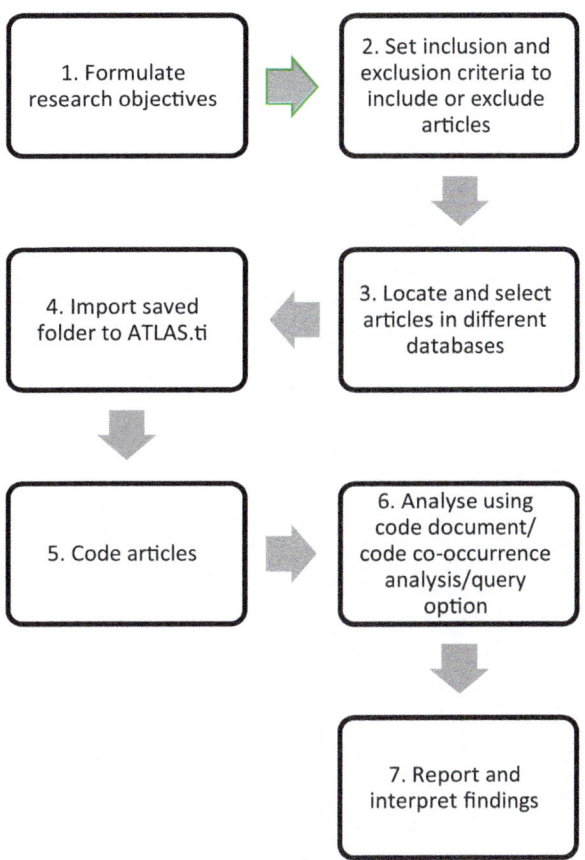

Fig. 15.2 Literature review in ATLAS.ti

The Literature Review Process

Literature review in ATLAS.ti starts with a clearly defined research objective. The researcher selects and reviews research articles and other published studies to ground their study. These articles should be included in the reference manager tool. While identifying and collecting articles, it is important to ensure that the articles are referenced in the reference manager tool. If this has not been done, references should be inserted manually. Researchers using different reference manager tools should prepare articles in the proper format for import into ATLAS.ti for literature review and referencing.

Approach to coding should be aligned with the objectives of the study. If the researcher is interested in identifying concepts or words from articles year- or author-wise, they can search for concepts or words from the articles. By using code-document and co-occurrent analysis options, concepts can be identified in the articles spanning authors and years of publication. Researchers

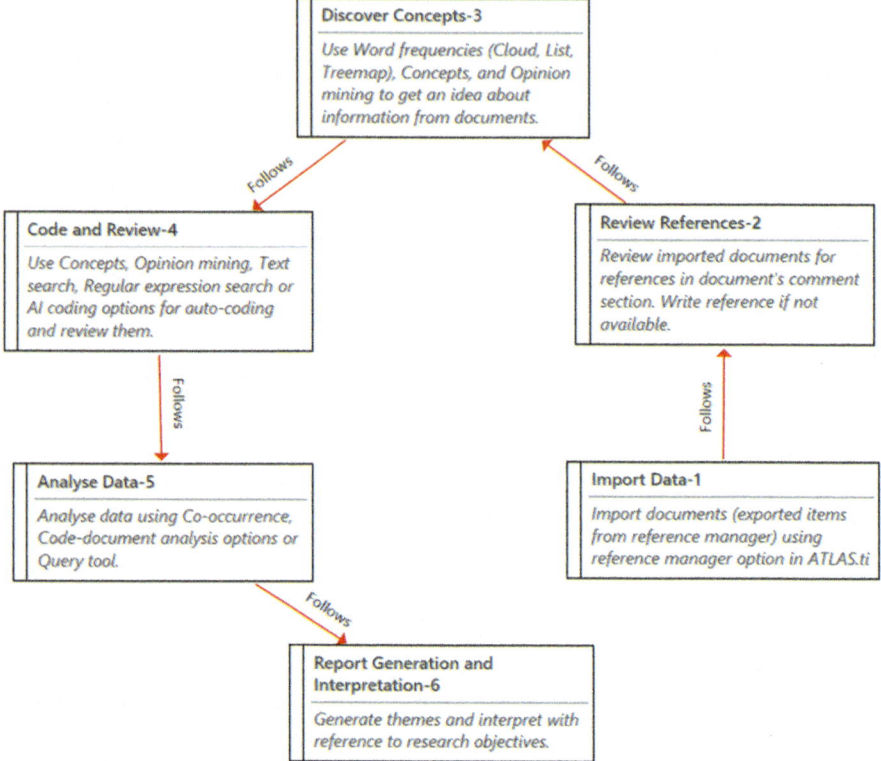

Fig. 15.3 Steps for conducting a literature review in ATLAS.ti

can code, group, and organize them for further reading and interpreting the literature. Figure 15.2 shows the literature review process in ATLAS.ti.

Based on research objectives, articles should be selected, downloaded and imported to reference manager tools. Articles should be exported in Endnote XML format (Fig. 15.3) and saved on computer. Using reference manager under 'Import & Export' option in ATLAS.ti, the saved folder should be imported to ATLAS.ti. The next step starts with coding articles and analysing using code co-occurrence, code document and query tool appropriately. The last step is generating and interpreting findings.

Export Literature from Reference Manager Tool to ATLAS.ti

1. Open the Reference Manager tool which contains the articles for the literature review. These articles are taken from several research databases.
2. Export the folder containing these articles and save. The default name of the folder will be "Exported Items" (Fig. 15.4).

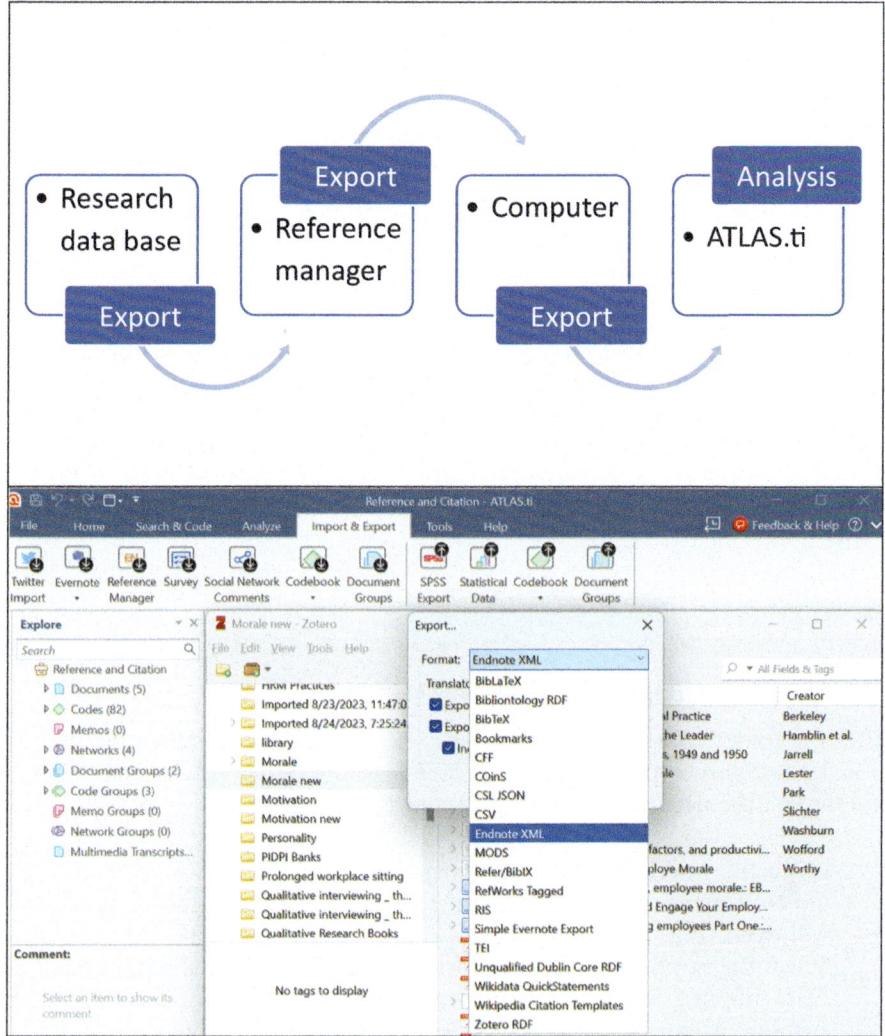

Fig. 15.4 Reference manager option

3. Open ATLAS.ti and import the "Exported Items" folder using export and import function of ATLAS.ti.

Figure 15.3 shows the steps followed for literature review in ATLAS.ti. Figure 15.4 shows data preparation using Zotero.

In this example, the user used Zotero and selected the 'Morale New' folder in 'My Library'. After selecting all the articles, with a right click, the 'Export Items' was displayed with a list (Fig. 15.4). Then, the researcher selected "Endnote XML" format, which is compatible with ATLAS.ti, for exporting data. Articles should be downloaded from a research database and exported reference manager tool. Selected articles from the reference manager should be exported to the computer and finally exported to ATLAS.ti for analysis to begin.

The user can select the articles that one wants to export to ATLAS.ti. After selection of the articles, with a right-click, select Export (display in Fig. 15.4). The user then selects the Endnote XML format for saving the file.

> *Important* Zotero offers several file export formats of which the user should select the Endnote XML option. It is also possible, whenever required, to export notes and the full contents of the article. While exporting from Zotero, one must open the file to locate the export library function. The user should select both options 'Export Notes and Export Files. The comments and notes will be imported as an article comment in ATLAS.ti. The file is then saved in a location decided by the user. Once saved, the files can be imported to ATLAS.ti.

Importing a Saved Folder into ATLAS.ti

After the folder containing the articles is exported from reference manager, it must be saved in computer. The next step is to open the ATLAS.ti application and import the data using the Import and Export function:

1. Open the Import and Export function in ATLAS.ti.
2. Go to Reference Manager, open, and locate the saved folder. A window will open (Fig. 15.5).
3. Select the saved folder (the exported items) and click to import the files. After import, the data can be sorted in the alphabetical order of the authors' names, and the years of publication of the articles.
4. If the full article is not available, select import Abstract. ATLAS.ti will treat the abstract as a document. The researcher's comments on the original article will be imported into the comment section.
5. On opening Document Manager, details of the article can be seen on the left, while comments are shown at the right in the comments section (Fig. 15.6).
6. Document groups are created based on the selections made in Reference Manager import (Fig. 15.5). Users need to check the box shown under Create Document Groups under Reference Manager Import.
7. If there are no references for the article, a URL for the article can be added in the comments field.

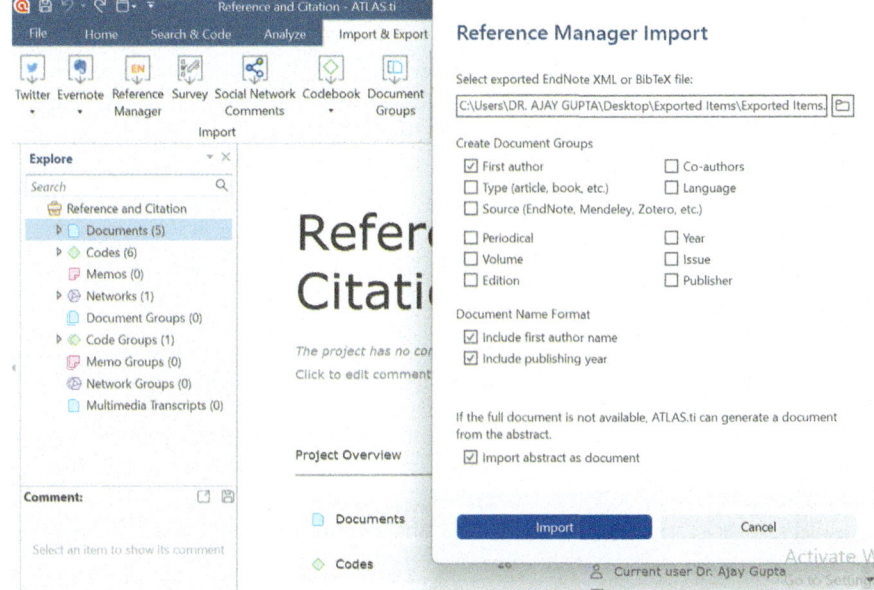

Fig. 15.5 Reference manager import selection list

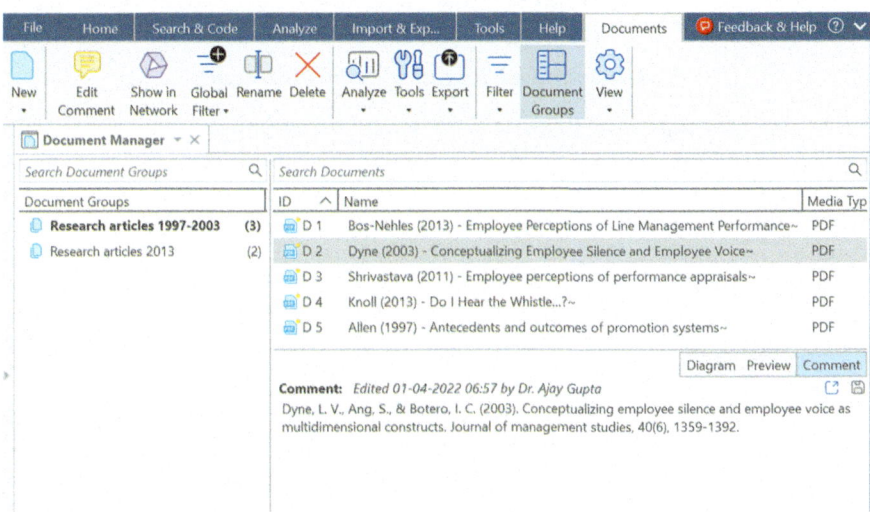

Fig. 15.6 Imported research project from Zotero

> *Important* ATLAS.ti does not carry out literature reviews. ATLAS.ti helps researchers in identifying, organizing, and retrieving information to facilitate review of the selected information, concepts, authors, year, and even region-wise to gain insights from the literature review. Based on the objectives of their study, researchers can make the appropriate interpretations and present their insights, as well as organize reference based on researcher's preferences.

Auto-Coding Options for Literature Review

Manual reading of sources is the best approach for a literature review because the researcher can directly engage with the concepts, contexts, scope, and the insights presented by various authors. The value of using ATLAS.ti lies in its ability to sort the information and sharpen researcher focus to help the researcher to gain deep insights into the topic.

Using the Text Search or the Regular Expression Search options, in-text citations can be generated for the year-range selected by the users. Table 15.1 compares the auto-coding options that are available to the user.

Text Search and Regular Expression Search

Either the Text Search or Regular Expression Search options (Figs. 15.7, 15.8 and 15.9) can be used to code the selected quotations with the key words in the word list. In the Text Search Option, the user writes the word and checks 'Include Inflected Forms'. Thus, the search for 'motivate' will include 'motivation', motivated', etc. When the Regular Expression Search is used, the word should be written exactly in the way it is mentioned in the quotation. Thus, if both 'silence' and 'Silence' are mentioned, the user should write silence|Silence in the search option. Both options will provide identical search results.

Whether using Text Search or Regular Expression Search, the documents and document groups can be selected for which the text search is needed. ATLAS.ti will search Sentences/Paragraphs/Words/Exact Matches only in the selected documents. After coding, the search option can be used to search the text within quotations. For this, the codes should be selected from the list and the box checked (Fig. 15.7). On clicking 'Continue', a search window will open. The search text should be written in the search option. On pressing the 'Show Results' button, all quotations containing the searched text will be shown. The process is used to refine the search within quotations.

Table 15.1 Auto-coding options

Options	Descriptions
Text search	Searches word/s and synonyms in "Paragraphs, Sentences, Words, Exact matches" option for documents. Using "And" and "OR" several combinations of search can be performed. Search can be performed from documents and codes. From documents text segment can be produced and quotation can be coded while using codes option, search for a word can be performed from quotations. Users can check the "Include Inflected Forms" option to search all inflected forms of a word (i.e., a search for motivation in a sentence will include motivated, motivating as well
Regular expression search option	Searches word/s in "Paragraphs, Sentences, Words, Exact matches" option from documents. It uses (GREP) globally search for a regular expression and print matching lines. Further using codes option, search for the word/s can be performed from quotations
Named entity recognition	Finds sentences or paragraphs in four entity categories—person, location, organisation, and miscellaneous from documents. By checking any of four entities, the information in a sentence or paragraph is shown. Using Code Name and Create Code Groups functions, a default code name and category can be generated. Using the Codes option, the search for the word/s can be performed from quotations
Sentiment analysis	Finds and classifies text segments by sentiment and suggests appropriate codes from documents. Positive, negative, and neutral sentiments can be shown in paragraphs and sentences. Sentiments are coded using Apply Proposed Code function
Concepts	Extracts concepts from the selected documents and codes then Using the Apply Proposed Codes function. The concepts can be saved as memos using the Save as Memo option. The memo can be treated as a document for analysis. In addition, the concepts can be visualized in cloud, list and treemap form
Opinion mining	Extracts positive and negative opinion terms from the selected documents and makes bar charts of words with their associated components. Coding decisions can be made to code opinions
Word frequencies	Finds occurrences of words in Cloud, List and Treemap forms. By checking 'Show Details' and 'Show Percent', more information about the words can be shown
AI coding	Extracts relevant information and codes contained in one or more selected documents
Search project	Finds occurrence of words in documents, codes, memos, networks, and groups based on the search term. This works like a regular expression search option

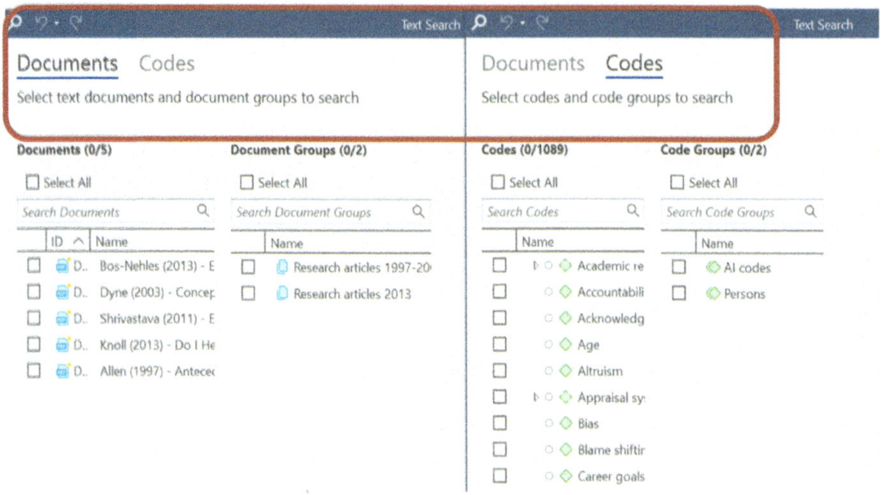

Fig. 15.7 Text search options

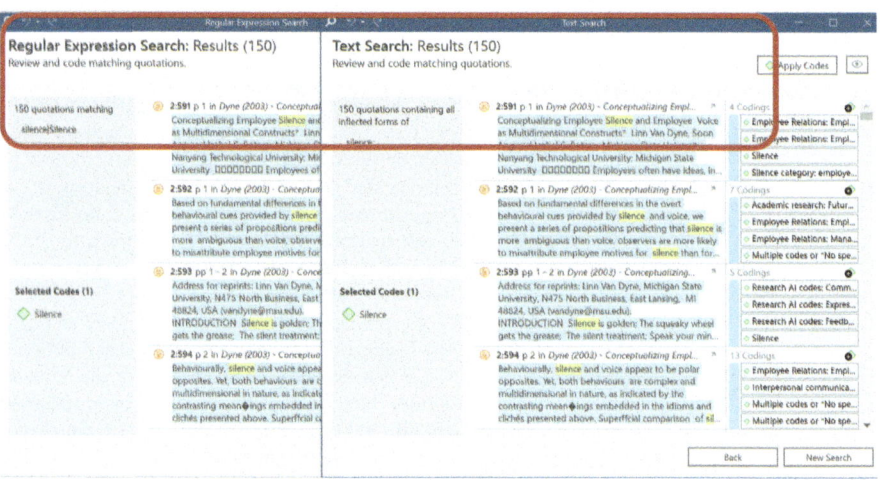

Fig. 15.8 Regular expression and text search result

Named Entity Recognition

Like Search Text and Regular Expression Search Option, the "Named Entity Recognition" function is used to identify the key information from articles on four parameters: Person, Location, Organization, and Miscellaneous (Fig. 15.10). There are two search options that researchers can apply: Paragraph and Sentence. Sentence selection is usually the preferred option as it helps the researcher to read and capture information about the key concept. Searching for Person will show result of all authors present in the article.

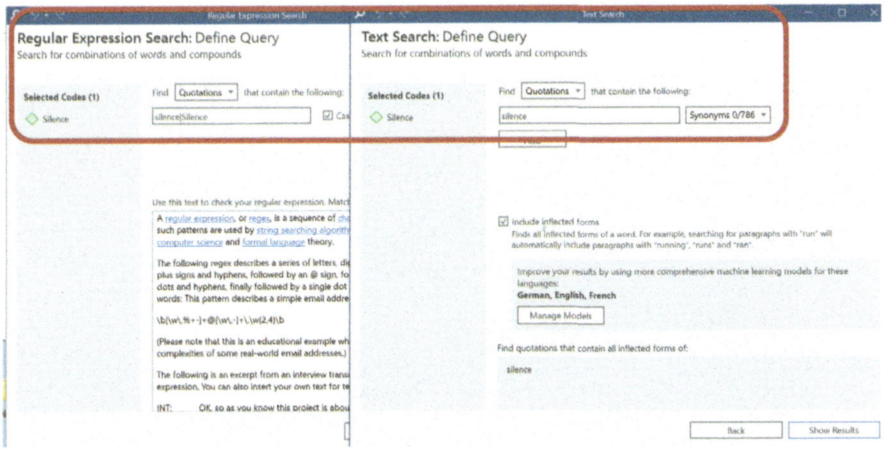

Fig. 15.9 Regular expression and text search window

Researchers can code paragraph/sentences using the "Apply Proposed Code" option which will automatically code all selected quotations. The users can select "Apply All Codes" to create the codes automatically. If the user selects 'Person', 'Person' becomes the category and the name of the person becomes the code.

Selection of "Category Only" will result in a single code for all quotations. If user selects 'Person', the person becomes the code. Similarly, depending on the selection, location, organisation or miscellaneous become codes.

Checking "Creating Code Groups from Categories" will create a code group in which each member of the group is a code.

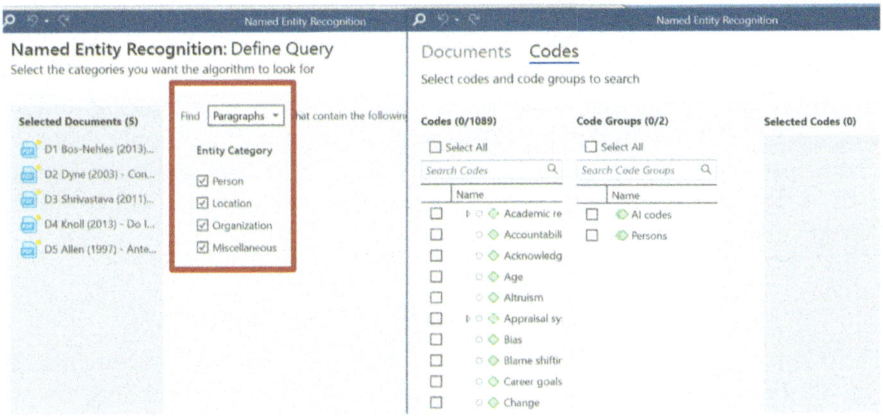

Fig. 15.10 Named entity recognition option

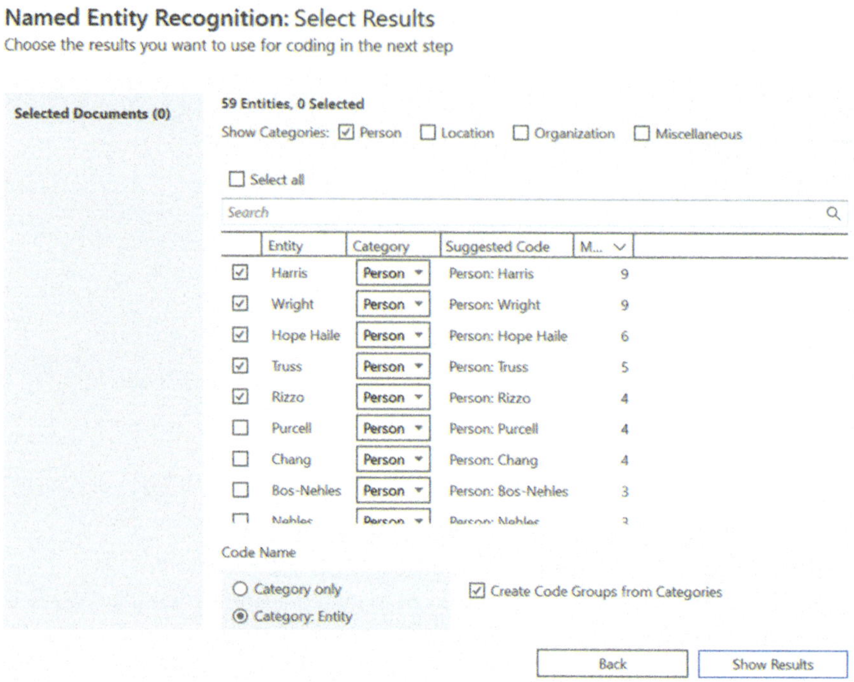

Fig. 15.11 Named entity recognition window

Selection of "Category: Entity" from code name and "Create Code Groups from Categories" will create a code category and a code group (Fig. 15.11). In Fig. 15.12, the number 109 under 'Person' shows the occurrence of the entity in the selected document/s.

Five articles were selected to identify authors appearing in articles. 'Person' was selected under "Named Entity Recognition" (Fig. 15.12). Ten persons were identified as having high grounded. In other words, the number of quotations that appeared for the person is shown under grounded (Fig. 15.13). All the quotations (109) were coded using the 'Apply Proposed Codes' option. All persons were coded with their names and were categorized as 'Person'. A code group was also automatically created, which was the Person's name (Fig. 15.13).

Concepts

The Concepts option extracts concepts from the selected documents. In Fig. 15.14, the word 'silence' is mentioned 500 times. Depending on the context, the silence was of various types: employee silence, acquiescent silence, quiescent silence, prosocial silence, opportunistic silence, organisational silence, and so on. This can be determined by clicking once on any

15 LITERATURE REVIEW AND REFERENCING 407

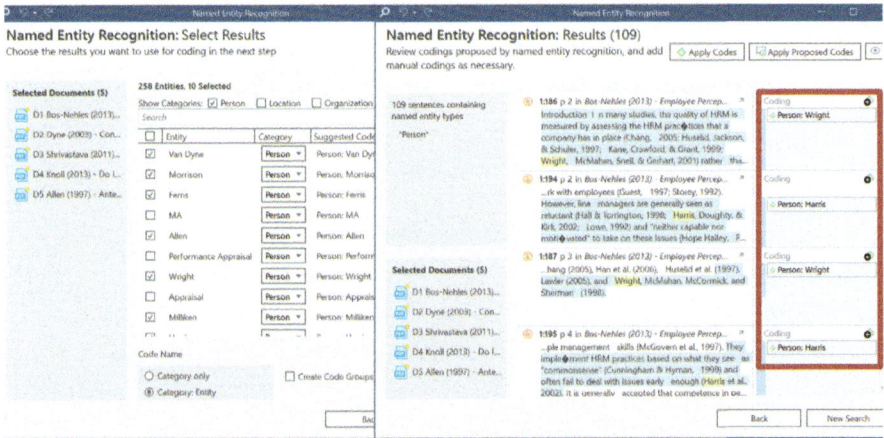

Fig. 15.12 Named entity recognition result window

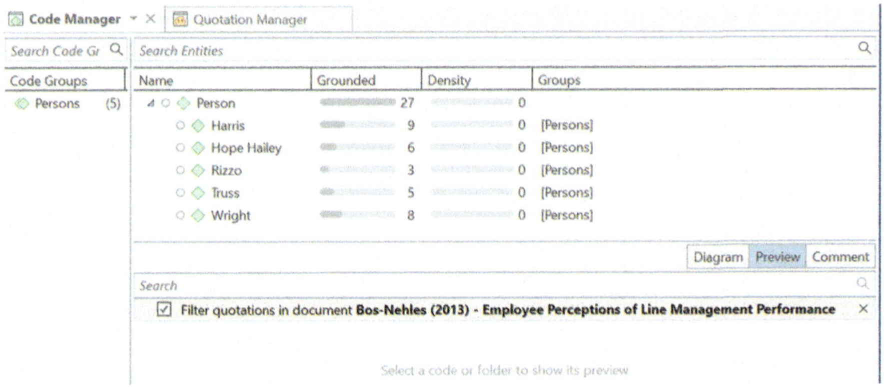

Fig. 15.13 Named entity recognition coding window

expression containing the word silence after which the associated quotations will be shown on the right side. The concept option is used to create codes, code categories, smart codes and code groups using the auto-coding facility. This can save time and help in faster data analysis.

Figures 15.14 and 15.15 show the use of words and concepts in documents. The Word List shows the percentage weight of words in individual documents. Here, the weighted percentage shows the usage of the word across documents. The Word Cloud shows word frequencies in spiral and typewritten forms. The Concepts window shows the concepts and sub-concepts used in documents (ATLAS.ti 22). By hovering the cursor over a concept, a flag will show the sub-concepts and associated quotations (Fig. 15.14).

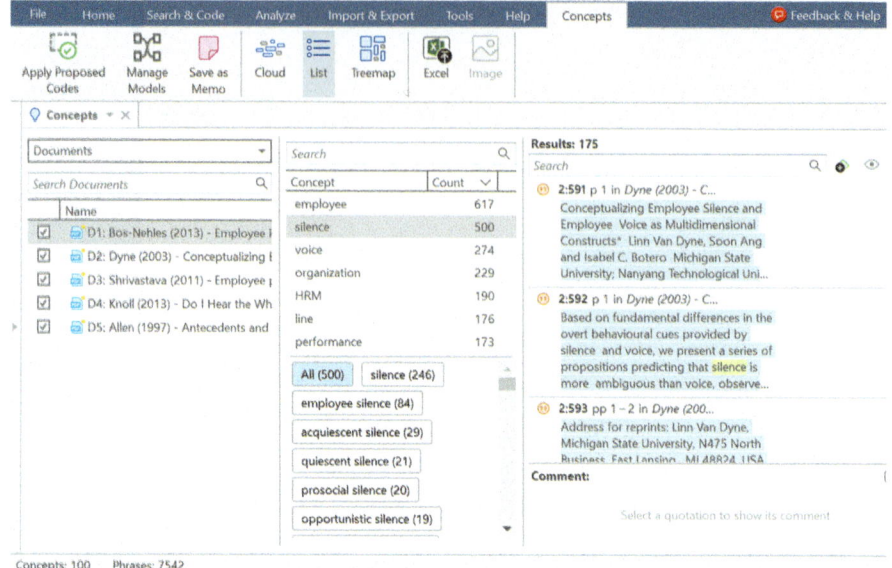

Fig. 15.14 Concepts window

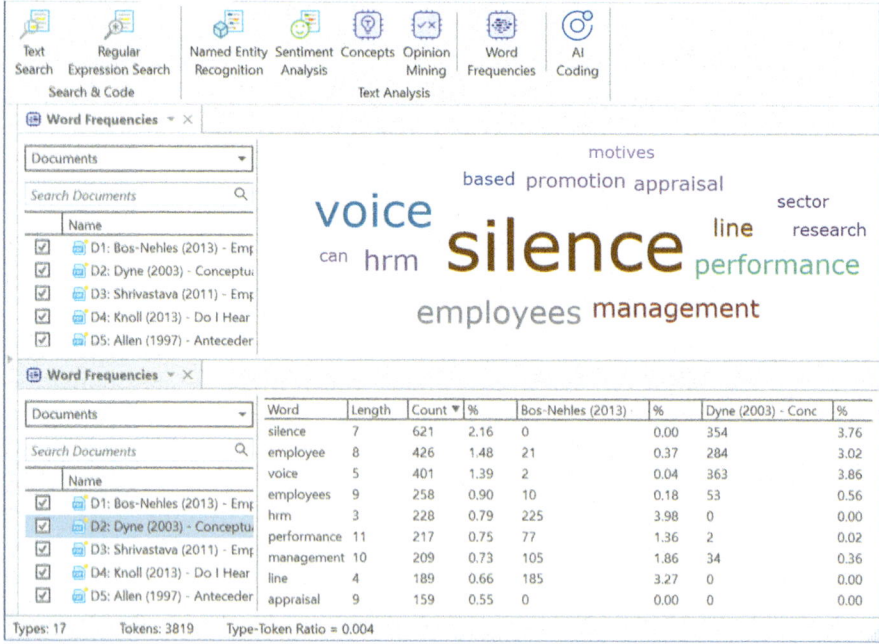

Fig. 15.15 Word cloud and word list window

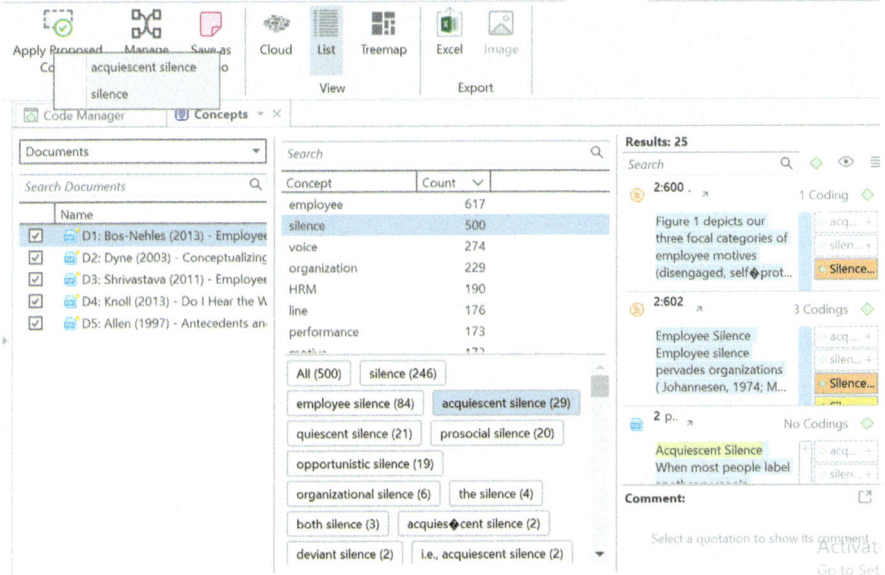

Fig. 15.16 Concept and apply proposed code window

Auto-coding of concepts is done using 'Apply Proposed Codes' (Fig. 15.14). For example, by clicking on 'Employee Silence' all the associated quotations will be shown (on the right in Fig. 15.14) after which all or the selected quotations for 'Employee Silence' are coded using the 'Apply Proposed Codes' option (Fig. 15.16). The code name can be retained or modified as desired by the researcher.

Coding Preference

The Named Entity Recognition and Concepts approaches are preferred for identifying key information in a document and then extracting it. With Named Entity Recognition, the user can create categories of entities (persons, locations, organisations and miscellaneous).

The concepts option is useful for identifying concepts, counts and number of quotations to create the core concepts and their associated sub-concepts. Using the "Apply Proposed Code" function, the quotations appearing on right side (Fig. 15.16) can be coded. The default code will be the words appearing in the concept list:

1. Click once on a concept in the concept list. The use and context will be shown below the list.
2. In this example, 'Acquiescent silence (29)' was selected. The associated quotations are shown on the right side (Fig. 15.16).

After coding, one can form code categories for each concept. Code document analysis can be carried out to identify person/ organization/ location/ miscellaneous in all the selected articles. For example,

- Location helps to understand the context of the study. Users can select 'Context' to understand the contextual pattern for the concepts.
- 'Organisation' helps the researcher to understand settings of the context.
- Qualitative and quantitative content, thematic, sentiment, opinion mining can be conducted using Concepts, Sentiment analysis, opinion mining, text search, regular expression search, and named entity recognition options. Researchers can gain insight about the literature using these methods.
- Co-occurrence analysis can be performed to identify the significant concepts based on c value.

Literature Review in ATLAS.ti

The objective of the literature review project is to extract literature about ethical issues for the period 1990 to 2009. This includes extraction of in-text citations and organizing references on the topic of internet ethics. Four articles were identified for the literature review. These were imported into ATLAS.ti from reference manager tool.

With the help of word cloud/treemap, and after removing redundant words, four words were identified based on the frequency of their occurrences. They are online, internet, privacy and public. These were coded using the 'Text Search' option and selecting Sentences (Fig. 15.17).

The year of publication was coded as shown in the Figure. In the figure, the search item 199[0–9] will select sentences with years 1990 to 1999 mentioned in them. By using the Apply Codes option, the sentences were coded '1990–1999'. Similarly, sentences containing the year 2000 to 2009 were coded '2000–2009' (Fig. 15.18).

The code lists for the documents appear in the Fig. 15.19. The groundedness value shows the number of quotations associated with the code in the sentences. The next step is to use co-occurrence analysis to find the code-to-code relations using the C value.

Figure 15.20 shows the c values in brackets and the code-to-code strength. The c value for the codes 'Online New' and '2000–2009' is 0.08, which is significant with 82 co-occurrences. In other words, eighty-two sentences were coded by both codes. Finally, a report for the quotations can be generated and interpreted (Fig. 15.21).

Code document analysis shows the year-wise usage pattern of the words in the documents from 1999 to 2009. For comparison, the data was normalized. The percentage value shows the weightage of the codes in the documents. From these values, the relative importance of the concepts (codes) in the

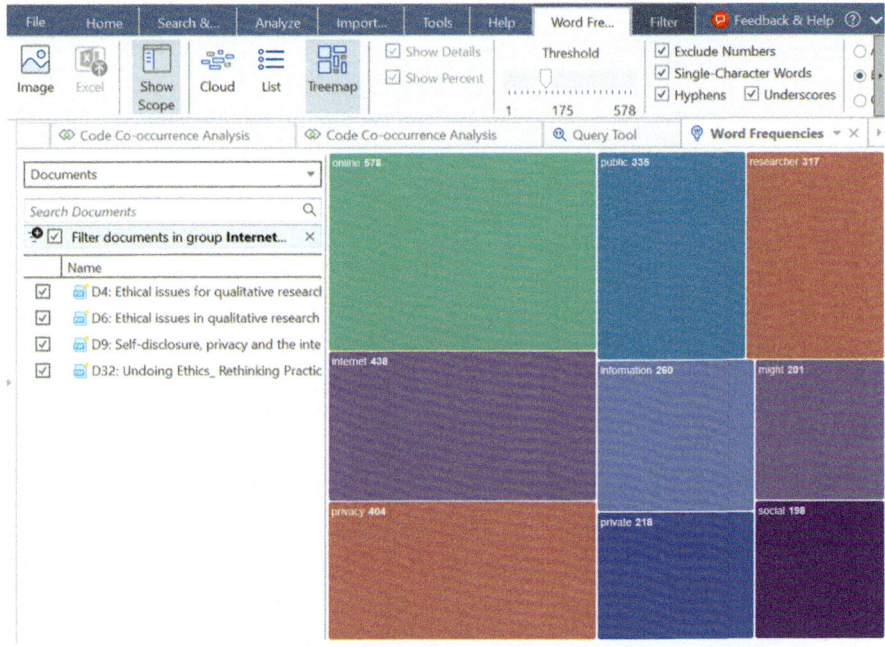

Fig. 15.17 Treemap for selected documents

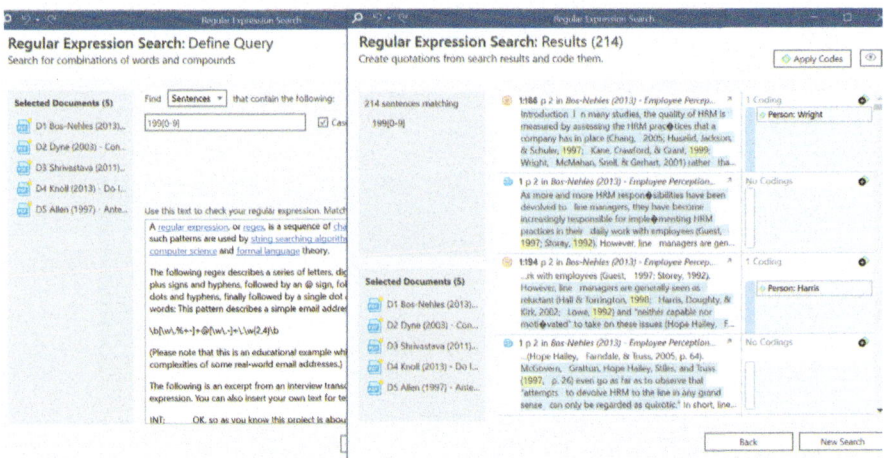

Fig. 15.18 Regular expression search

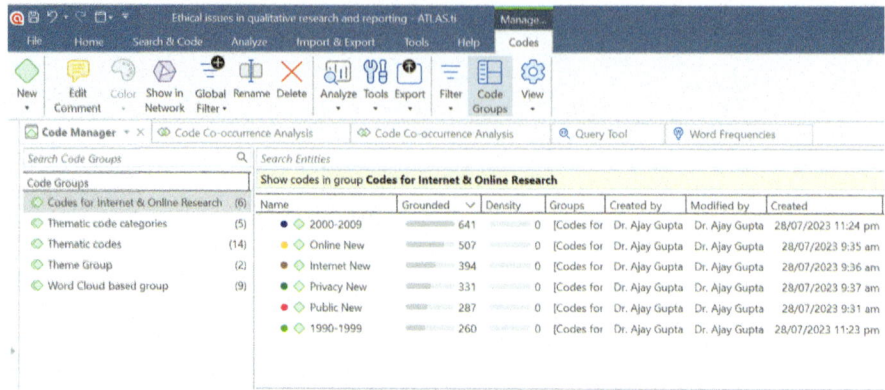

Fig. 15.19 Code lists for documents

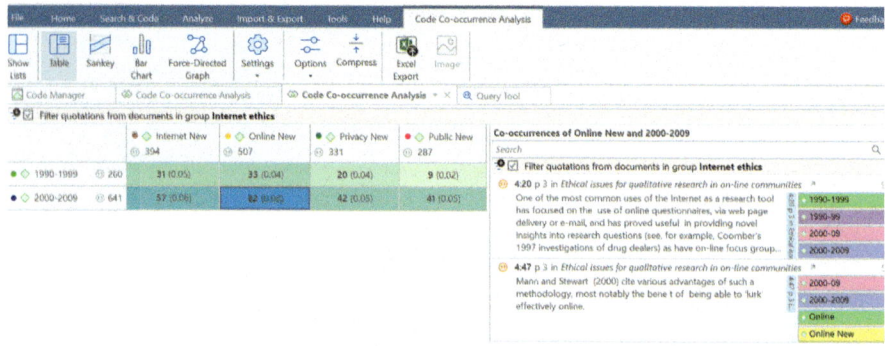

Fig. 15.20 Code co-occurrence analysis

Fig. 15.21 Sankey diagram for code co-occurrence analysis

documents can be ascertained, which is then interpreted by the researcher. The Sankey diagram visualizes the flow of concepts and their relationships with other concepts (Figs. 15.22 and 15.23).

Using Query Tool, reports for concepts (codes) can be generated for the years 1990–1999 and 2000–2009. Figure 15.24 shows the results for 1990–1999. A smart code can be created for the selection. The text report is shown in Fig. 15.25.

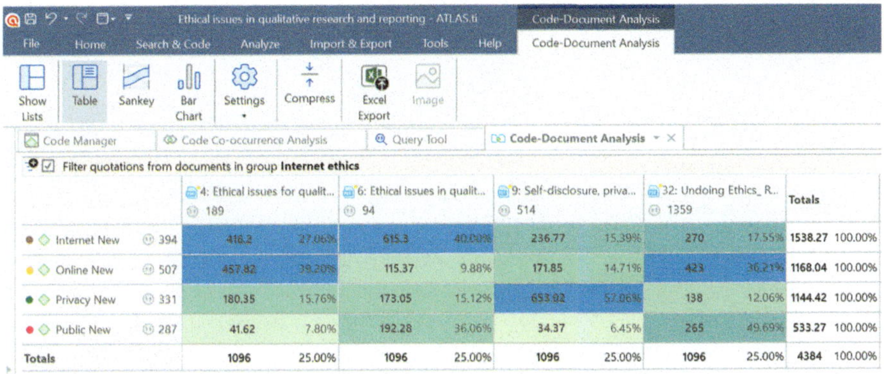

Fig. 15.22 Code document analysis

Fig. 15.23 Sankey diagram for code document analysis

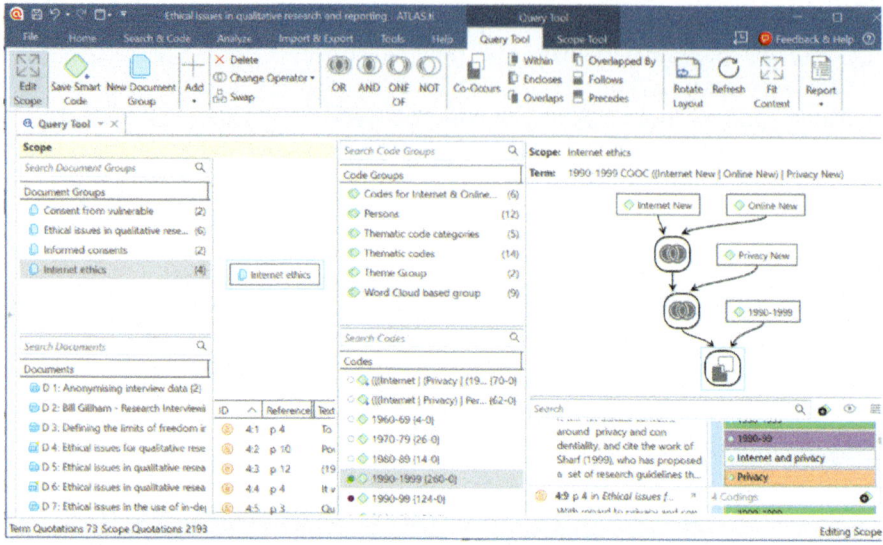

Fig. 15.24 Query tool

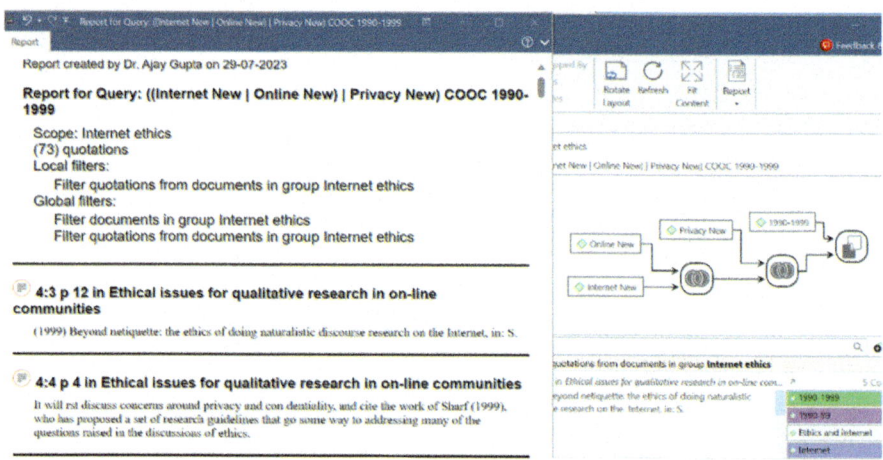

Fig. 15.25 Query tool and report for query

A report for the year 2000–2009 has been created as created for the year 1990–1999 (Fig. 15.26). References appear under the comment section. The number of quotations appears under the Quotation Count (Table 15.2).

Using 'Named Entity Recognition,' researchers can identify authors for the concepts in different articles. The code co-occurrence analysis table (Fig. 15.26) shows the quotations co-occurring with authors and concepts. The number outside the bracket shows the number of quotations coded by

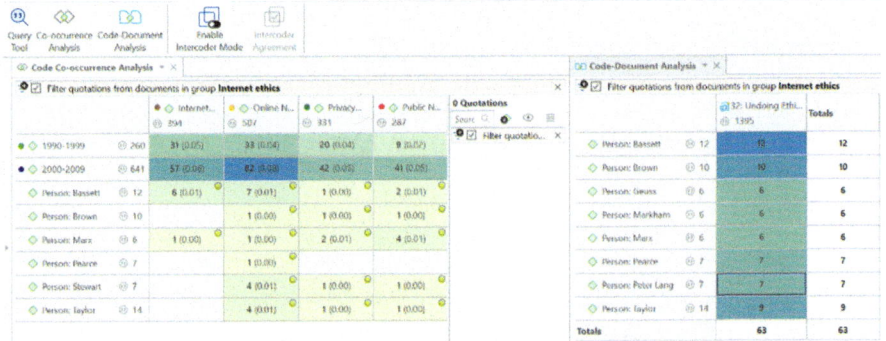

Fig. 15.26 Code co-occurrence and code-document analysis

Table 15.2 References for documents

Document	Comment	Document groups	Quotation count
Ethical issues in qualitative research on on-line communities	Brownlow, C., & O'Dell, L. (2002). Ethical issues for qualitative research in on-line communities. Disability & Society, 17(6), 685–694	Internet ethics Ethical issues in qualitative research	189
Ethical issues in qualitative research on internet communities	Eysenbach, G., & Till, J. E. (2001). Ethical issues in qualitative research on internet communities. Bmj, 323(7321), 1103–1105	Internet ethics Ethical issues in qualitative research	94
Self-disclosure, privacy and the internet	Joinson, A. N., & Paine, C. B. (2007). Self-disclosure, privacy and the Internet. The Oxford handbook of Internet psychology, 2,374,252, 237–252	Internet ethics	514
Undoing Ethics_ Rethinking Practice in Online Research-Springer US (2012)	Whiteman, N., & Whiteman, N. (2012). Undoing ethics (pp. 135–149). Springer US	Internet ethics	1359

both codes appearing in columns and rows. Code-document analysis table shows the frequency of occurrence of the authors' names in the document. Both options help researchers to identify and refine information based on need.

Important Using in-text citations, researchers can select references from the reference list of the document, and code them to create a reference list. They can identify documents appearing in in-text citations and select appropriate references from the document.

Reporting In-Text Citations and References

After selecting the documents of interest in Document Manager, the researcher can generate reports in text and Excel formats, or even in the form of a network for the document group. All references in the comment section of the documents will be reproduced in the report. In the network, by selecting comments in the View option, the references for the documents will be shown.

The researcher can generate in-text citations for the code/concept for the particular year or range of the year. For example, all in-text citations, articles with the code 'Acquiescent Silence', 'Employee Motive', 'Employee Silence', and 'Employee Voice', which are mentioned in articles between years 1980 to 2019, can be retrieved by using co-occurrence option in the query tool (Fig. 15.27).

In Fig. 15.27, the codes 'Employee Silence' and '2000–09' have 37 co-occurrences, meaning that 37 quotations were coded with both codes. The c value is 0.11. The next step is to read all the quotations and categorize the information associated with 'Employee Silence':

1. Click on the number 37. All quotations will be displayed on the right.
2. Selecting the code symbol appearing above quotations and write the code name. All quotations will be added to the code.

'Employee Silence' and other codes can be interpreted with reference to the comments and their use in the contexts. The comments can include definitions, antecedents, consequences, triggers, synonyms, usage, and the possible effects of silence.

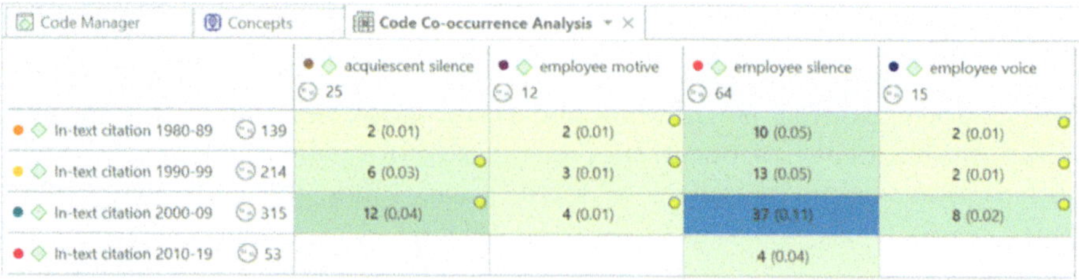

Code Manager | Concepts | Code Co-occurrence Analysis

	acquiescent silence 25	employee motive 12	employee silence 64	employee voice 15
In-text citation 1980-89 139	2 (0.01)	2 (0.01)	10 (0.05)	2 (0.01)
In-text citation 1990-99 214	6 (0.03)	3 (0.01)	13 (0.05)	2 (0.01)
In-text citation 2000-09 315	12 (0.04)	4 (0.01)	37 (0.11)	8 (0.02)
In-text citation 2010-19 53			4 (0.04)	

Fig. 15.27 Code co-occurrence table

15 LITERATURE REVIEW AND REFERENCING

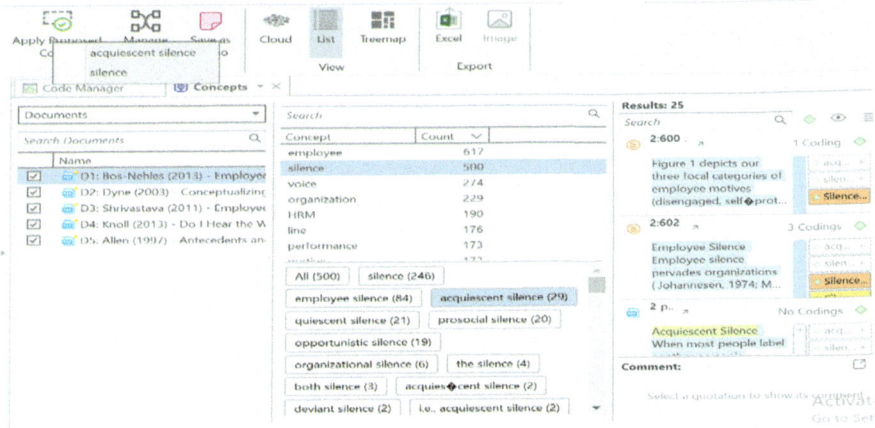

Fig. 15.28 Concepts and apply proposed code window

After identifying the contexts, the next step is to analyse the usage and relationships of the key words/concepts with reference to their contexts. Researchers can also use the concepts window to analyse the usage of words in their contexts. For example, Fig. 15.28 shows the use of 'Silence' in various contexts, which reflects different attitudes, such as acquiescence, a prosocial silence, opportunistic silence, and others. Next, after reading the quotations for each context, the researcher can give more focus to the meanings of the words.

LITERATURE REVIEWS IN ATLAS.TI—GOOD PRACTICE

The literature review is a dialogue between ideas and evidence (Ragin, 1994). According to Phelps et al. (2007), and Silverman (2010), researchers should follow good practices while reading articles for their literature reviews.

> *Important* Never leave an article without noting your thoughts about it. Make a habit of reading an article with a keen eye, and for a specific purpose.
>
> 1. Highlight the segment that draws your attention; write about it, describing your thoughts about the segment. Summarize your thoughts about the article.
> 2. Your notes and summaries must stand alone so that you do not need to go back to the original articles.

Source: adapted from Phelps et al. (2007: 175–6).

Strauss and Corbin (1990) suggest that researchers can use existing literature for five purposes.

1. To stimulate theoretical sensitivity, 'providing concepts and relationships that [can be] checked out against (your) actual data'.
2. To provide secondary sources of data to be used for initial trial runs of your own concepts and topics.
3. To stimulate questions during data gathering and data analysis.
4. To direct theoretical sampling to 'give you ideas about where you might go to uncover phenomena important to the development of your theory'.
5. To be used as supplementary validation to explain why your findings support or differ from the existing literature.

The literature review must be connected to the topic of the research. The researcher must remember to segregate information on situations, settings, methodological issues, ideas, etc. Each segregated part of the information can be identified with a suitable description, such as important, quotable, remarkable, etc. The objective is to capture the relevant segment that hooks the researcher.

Multiple papers can be grouped, with each document group reflecting one perspective about the topic. This helps the researcher to retrieve information systematically.

While importing or adding documents (research articles) to ATLAS.ti, for easy identification, the researcher can rename the documents based on the focus and purpose of the documents. A document naming scheme should be developed for easy classification and retrieval. The scheme should also demonstrate that the researcher has followed convention consistently.

Consistency and coherence are of vital importance while naming codes. Codes should be created and named such that their nature and meanings are clear, and they can be easily distinguished from one another. Then, based on user need and preferences, the codes can be grouped, merged, or split. If needed, smart codes or smart groups can also be created.

An Innovative Method for Retrieving Literature

The literature review for research must be comprehensive; otherwise, the study will lack substance and credibility. A comprehensive literature review, which evaluates and critiques existing knowledge, reflects the researcher's understanding of the topic. The literature review is also essential for justifying the need for the present study. It must demonstrate to the audience that the researcher has reviewed all scholarly works relevant to the topic of study. Thus, it is essential for the researcher to access a substantial body of reports and peer-reviewed studies for the literature review. Hence, the researcher must spend as much time and effort as is necessary to gather and organize the information required for the literature review.

Every article, paper, report, and book accessed for the literature review must be read and reread as often as is required for the researcher to gain diverse perspectives and insights into the topic of the study. Literature search and review is tedious and time-consuming work. There are no short-cuts.

ATLAS.ti helps researchers save time and effort by searching the material collected relevant to the topic of study, and sort them according to user preferences. To give an indication of its time-saving potential, the software can search and identify, using various search options, the relevant information in a complete set of documents within seconds. The researcher then only needs to focus on critiquing and review.

For example, in this study, the researcher had collected articles published from the years 1980 to 2019 for the literature review. Using the GREP tool option in ATLAS.ti, in-text citations were extracted year-wise, decade-wise, as well as for the entire period 1980–2019. To understand how this was done, one must first understand the utility of the GREP tool.

The GREP tool uses symbols to extract information based on user preferences. It is of crucial importance to understand the correct use of symbols to retrieve information of interest. In this example, the researcher searched for in-text citations from the years 1980 to 2019. The search was done in four segments—1980 to 1989, 1990 to 1999, 2000 to 2009, and 2010 to 2019. Figure 15.29 shows how the search option is written. For citations from the years 1980 to 1989, the search requirement was written as 198[0–9]. ATLAS.ti extracted the citations for the periods specified. A vertical line separates each year range (the vertical line will include all the years only when the user checks the GREP option).

IMPORTANT: There must be no space between the years and vertical line |.

The search counts each year as one hit. Therefore, in a sentence that mentions three different years, the search will show three hits. The complete sentence is counted as one hit. Thus, even if a sentence mentions four different years, auto-coding will show only one hit. The number of results from a search will depend on whether the search is for words or sentences.

This approach is especially useful when it is necessary to understand the meaning of a word. Then, the user can select a paragraph. The auto-coding process will highlight the paragraph for review by the researcher who, after reading, can decide whether to select the sentence or the complete paragraph. But the reader must also be aware that when a paragraph is selected, the number of quotations will be less than when the sentence option is selected. It also goes without saying that words will get more hits than sentences. This

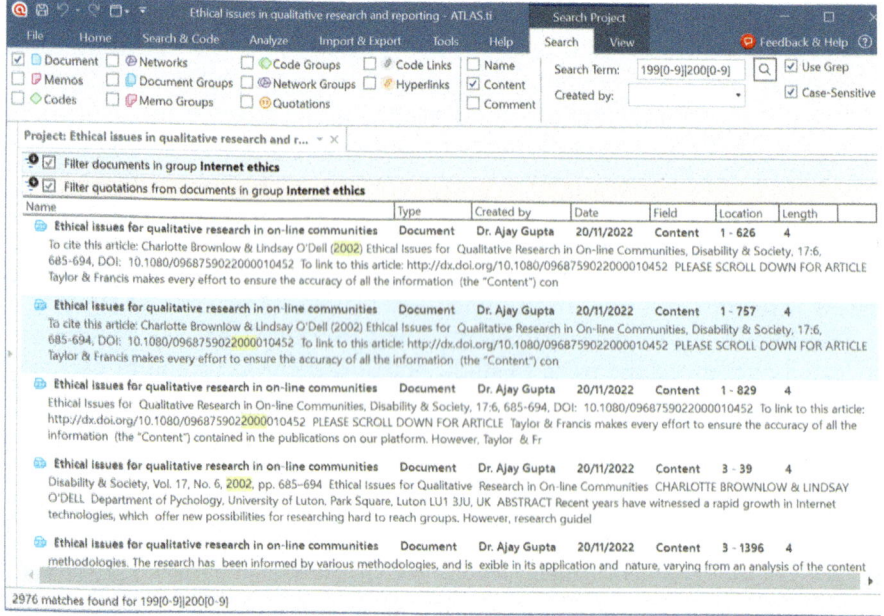

Fig. 15.29 Project search and regular expression search view

feature in ATLAS.ti is of great value when the researcher wants to perform quantitative content analysis, where interest is in the frequency of use of certain words.

Using Search Project

1. Open the project.
2. Decide whether the search should cover the complete document group or a specific document.
3. For working with the document or document groups, highlight the documents, and open the Search Project option.
4. Write the search preference, and then check the box 'Use GREP Cell'.
5. Select Search and obtain the result.
6. A double-click on the result will open the document for that result.
7. Write search option, e.g., 199[0–9]|200[0–9]
8. Select Search. The result will be displayed (see Fig. 15.29).

Reporting and Interpretation

Coding is followed by code review, code grouping, creation of code categories and smart codes. The user can analyse the findings based on research objective/research questions. Data analysis can be conducted on a concept-to-concept basis, where the linkages between concepts are determined and then interpreted in the context of the study. With code co-occurrence tables and code coefficients, significant concepts can be discussed. The same table can also be used to study the relationships between concepts and author/organisation/location and interpret them. Data analysis can also be conducted using single on many documents.

In the example discussed in this chapter, qualitative content analysis was done to show concept-author and concepts-document relations for interpretation of significant concepts across the selected articles. If required sentiment analysis of the contents can be done using the Sentiment Analysis option.

In this example, one significant concept which emerged from analysis is "Voice Category & Silence Category". These are discussed in the two articles shown in Fig. 15.30.

In Fig. 15.31, the Query Tool was used for generating a report for the concept (Voice Category & Silence Category) for all documents. There were 82 quotations in the report. The researcher selected two articles from documents using OR operators (D4 and D2). The meaning of the query is to find out information under the code 'Voice category & Silence category' within two articles. By using the report option, 82 quotations appeared that is shown by an arrow.

The quotations are then read to understand how 'silence' and 'voice' are discussed by various authors in different contexts. The user should identify and categorize the-sub concepts in 'Voice' and 'Silence', and then make interpretations that are consistent with theory.

Code-Document Analysis		Voice category & Silence category 82	Totals
1: Bos-Nehles (2013) - Employee Perceptions of Line Management Performance	33		0 / 0.00%
2: Dyne (2003) - Conceptualizing Employee Silence and Employee Voice	157	67 / 81.71%	67 / 81.71%
3: Shrivastava (2011) - Employee perceptions of performance appraisals	47		0 / 0.00%
4: Knoll (2013) - Do I Hear the Whistle...?	122	15 / 18.29%	15 / 18.29%
5: Allen (1997) - Antecedents and outcomes of promotion systems	19		0 / 0.00%
Totals		82 / 100.00%	82 / 100.00%

Fig. 15.30 Concept and articles association window

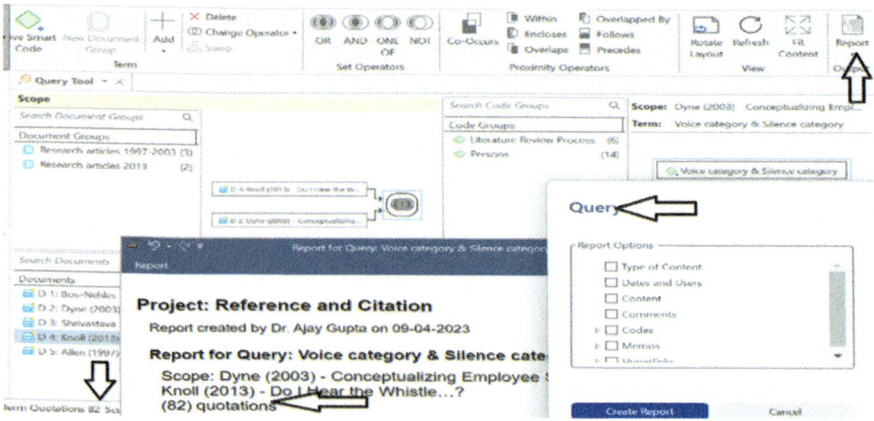

Fig. 15.31 Concepts, query tool, scope and text report

Recommended Readings

Phelps, R., Fisher, K., & Ellis, A. (2007). *Organizing and managing your research: A practical guide for postgraduates*. Sage.

Ragin, C. C. (1994). Qualitative comparative analysis. The comparative political economy of the welfare state, 320.

Silverman, D. (2010). Doing qualitative research. *Doing qualitative research*, 1–100.

Strauss, A., & Corbin, J. (1990). *Basics of qualitative research: grounded theory procedures and techniques*. Newbury Park: Sage Publications.

Index

A
Abductive coding, 100, 102
Advocacy, 4, 5, 19
AI Coding, 274, 287, 313
Anonymity, 36, 65–69, 76
Appreciation, 80, 92
Artifacts, 46, 47
Artificial intelligence, 118
Attribute coding, 115
Audio, 343–347, 349, 350
Authenticity, 49, 90, 92
Auto-coding, 125, 130, 132, 135, 141–144, 152, 275, 276, 400, 405, 417
Autonomy, 68
Avoidance, 79, 84, 90
Axial codes, 104–106, 109, 111, 117, 134, 135, 173
Axial coding, 42–45, 104, 105, 107, 108

B
Beneficence, 65, 68
Biases, 11, 20–22, 26, 34, 47, 120, 159, 320, 321, 367, 369, 371
Binarized, 201
Bracketing, 20, 21, 371

C
CAQDAS, 97, 118–120
Case Study, 13
Categories, 41–45
Categorization, 108, 111, 112, 114, 117
Causation coding, 116
C-coefficient, 205, 206, 211
Challenges, 371–373, 375, 385
Classification, 244, 266
Clipping, 345, 346
Closed Coding, 107
Clues, 32, 52
Code categories, 97, 103–105, 107–109, 111, 112, 116, 204, 220, 250, 274, 282, 288, 301, 314, 345, 405, 407, 418
Code coefficient, 304
Code co-occurrence, 204, 210, 301, 303, 304, 306, 335
Code co-occurrence analysis, 278, 412
Code document, 201, 218, 269, 276, 279, 283, 285, 288
Code document analysis, 328, 333, 407, 409
Code groups, 223–228, 234, 235, 237, 238, 240, 241, 243, 245, 249, 251, 253–255, 264, 266, 284, 285, 297, 299, 301, 308, 345, 352, 354, 359
Codes, 43–45, 53, 56, 97–99, 127, 130, 132, 134–140, 143, 153, 165, 167,

172, 173, 175, 201–207, 209–212, 214, 215, 217, 218, 220
Code-theme relationships, 173
Code to code relations, 254, 255
Coding, 98, 343, 345, 348, 351, 352, 354–356, 359
Coding agreement, 371, 376
Coding methods, 373, 379
Coding option, 274, 333, 337
Comments, 201, 220, 394, 398, 399, 413, 414
Concepts, 301, 302, 310, 311, 395, 399–401, 403, 405–407, 409, 411, 412, 414, 415, 418, 419
Conceptual coding, 105, 108, 117, 118
Confidentiality, 33, 50, 63, 66–69, 71, 76, 82
Conscious, 84, 88, 91
Consent, 32, 33, 35, 61, 63, 66–69, 72, 74, 78, 79, 92
Consistency, 366, 367, 379
Constructivism, 4, 7, 9
Constructivist, 5, 42, 44, 45, 70, 71
Convenient, 61, 78, 82
Conversations, 35, 36, 321, 325
Co-occurring Codes, 184, 194
Core category, 105, 107, 109, 118
Credibility, 24, 26, 47, 367, 369
Cues, 82, 87, 90, 91
Culture, 64, 81, 85, 92
C-value, 286

D
Data integrity, 369
Deductive coding, 101, 107
Deductive disclosure, 67
Density, 225
Descriptive responses, 279
Deviant cases, 371
Disagreements, 320
Disclosure, 33, 67, 73, 84
Discourse, 322, 331, 335
Document groups, 243–247, 266
Main and Taxonomic coding, 116

Emotions, 9, 11, 32, 53, 62, 68, 79, 84, 87, 114, 158, 162, 331
Epiphany, 37
Ethical, 50, 61–66, 68, 71, 75, 76, 87, 92, 321, 371, 387, 408, 413
Ethical concerns, 31, 32
Ethnographic, 45–47, 49
Ethnographic research, 45
Ethnography, 1, 11, 15, 31, 37, 82
Ethnomethodology, 7–10
Exploration, 1, 16, 17
Export, 393, 394, 397, 398
External validity, 367

F
Field diary, 50
Field notes, 31, 33, 35, 43–47, 49, 50, 52, 53, 56, 88, 125, 127, 129, 156–160, 170
Field observations, 160
Fieldwork, 46, 48
First-class link, 181
Flowchart, 327
Focused coding, 105, 108, 117
Focus group, 32, 34, 36, 41, 319, 320, 322, 325–327, 331
Focus group discussion, 319–322, 325, 328, 331, 332, 335
Free codes, 131, 132

G
Generalizability, 34
Geo documents, 358, 359
Gestures, 40, 52
Global filters, 195, 196, 198
Good code, 99
Good practice, 276, 293, 344, 345, 347, 349
Google forms, 70, 279
Groundedness, 225, 232
Grounded theory, 1, 15, 16, 31, 37, 41–45, 51, 57, 103, 105, 108
Group dynamics, 320–323, 330

H
Hermeneutic, 31, 39

Hesitant, 34, 51, 82
Heuristic, 98
Hybrid coding, 102
Hyperlinks, 179–181, 185, 187–191

I
Images, 343
Importing data, 298, 301
Independent coders, 372
Inductive, 66, 71, 81
Inductive coding, 100, 102, 103
Informal conversations, 61, 63, 71, 76, 87, 91, 162
Information retrieval, 278
Informative network, 254
Informed consent, 61, 63, 65, 68, 69, 71, 78, 84, 92
Initial codes, 99, 117
Insights, 63, 65, 75, 76, 78, 80, 82, 87, 89, 91, 92, 274, 290
Interactions, 32, 35, 41, 52, 61, 62, 65, 68, 71, 73, 76, 78, 79, 88, 91, 92
Intercoder agreement, 365, 370, 371, 376, 379, 380, 386–388
Intermediate coding, 105
Internal validity, 367
Internet communities, 68
Interpretations, 31, 40, 41, 195, 218, 238, 261, 274, 293, 303, 330, 343, 357, 361, 387
Interpreting, 396
Interpretivist, 108
Interviews, 319, 320, 326, 328
In-text citations, 393, 400, 408, 413, 416, 417
In vivo codes, 100

J
Justice, 68

L
Lead researcher, 365, 366, 369, 370, 372–377, 379, 380
Literature review, 393, 395
Lived experience, 40, 174
Local filters, 195

Lumping, 115

M
Medicine, 14, 126
Member checking, 368
Memo groups, 243, 245, 257, 259, 260, 264, 266
Memos, 40, 44, 49, 52–54, 56, 57, 126–129, 140, 155–163, 167–170, 173, 175, 201, 202, 220
Mergers, 237
Merging, 125, 139–141
Methodological, 71, 83
Methodology, 42, 55, 57
Moderators, 320–322, 325, 327, 328
Multimedia data, 343–345, 347, 350, 351

N
Narrative inquiry, 37
Narrative research, 12
Natural coding, 107
Naturalistic inquiry, 78
Neighbors, 184, 194
Network groups, 262, 263, 266
Networks, 126, 134, 158, 160, 161, 163, 168, 169, 175, 179, 182–184, 194, 223, 226, 238, 240, 241, 243, 245, 246, 262–266, 274
Nodding, 84, 85
Nodes, 179–181, 184, 186, 194
Nonmaleficence, 68
Nonverbal, 32, 51, 52, 322, 323, 337
Normal codes, 223, 224
Normalized, 201, 217, 218

O
Observations, 31, 32, 35–37, 41–43, 45–47, 49–53, 56, 57, 69, 74, 78–81, 87–93, 129, 155–160, 162, 168–170, 320, 322
Observer, 34, 35
Open codes, 134
Open coding, 107
Operators, 201–203, 209, 211, 212, 220, 276, 289
Opinion mining, 407

P

Paradigm, 1, 7
Pattern, 3, 15, 99, 114, 169, 171, 276
Pattern coding, 117
Peer debriefing, 369
Percentage Agreement, 370
Perceptions, 64, 65, 72, 78, 92
Phenomenological, 1
Phenomenological research, 37, 39
Phenomenology, 6–8, 10, 13–15
Piggyback focus groups, 325
Pragmatist, 5
Privacy, 69
Probing techniques, 84
Proximity operators, 223–226, 231–233, 235
Psychological, 39, 40

Q

Query tools, 202, 223, 224, 226–228, 233, 278, 301, 303
Quotations, 128, 134, 136, 138–141, 146, 148, 151, 154, 155, 157, 161, 163, 168, 169, 202–208, 210–212, 214, 215, 217, 218, 220, 248–250, 254, 255, 259, 262, 264, 327, 328, 334, 337, 339, 343–345, 348, 349, 351, 352, 354–361

R

Recording, 49
Redundant coding, 139
Reference manager tools, 393, 395, 396
Reflections, 36, 41, 48, 49, 62, 88, 102, 114, 156, 168, 369
Reflexivity, 4, 20, 21
Regular Expression Search, 274, 275
Regular Expression Search Option, 400
Relative significance, 278, 285, 288, 289, 294
Reliability, 34, 57, 365–367, 369, 385–388
Reporting, 393
Reports, 223, 224, 226, 234–238, 241, 243, 245, 247, 248, 259, 265, 266, 297, 301, 303, 306
Research objectives, 301, 306

Research paradigm, 1, 7
Research questions, 278, 279
Rigor, 20, 50, 127, 129, 170, 367, 368, 385

S

Sankey diagram, 207, 218, 288
Saturation, 42, 57
Schema, 41
Scope, 201, 203, 220, 276, 290, 293
Search Project, 418
Search Text, 400
Second-class links, 182
Selective coding, 105, 107, 108, 118
Sensitive, 32
Sensitive issues, 62, 82, 84
Sensitive research, 371
Sensitizing concepts, 20, 21, 371
Sentiment analysis, 131, 132, 145, 274, 297, 310, 314, 401, 407, 418
Set operators, 229
Silo communication, 371
Smart codes, 171, 202, 223–226, 228, 229, 232, 234–238, 241, 276, 301, 304, 328, 345, 405, 416, 418
Smart groups, 127, 132, 157, 160, 171, 223, 226–229, 234–236, 238, 240, 241, 276, 416
Social context, 3, 19, 21
Social interactions, 4, 10
Social media, 297, 306, 307
Social networks, 306
Social phenomenon, 1
Social reality, 2, 3, 7, 8, 10
Social relationships, 3, 10, 11
Social setting, 1
Source, 188, 190
Split coding, 136
Splits, 237
Splitting, 98, 115
Stop and Go List, 277
Sub-codes, 105, 250
Subjectivity, 40, 51
Survey data, 269, 270
Survey questions, 270, 274, 277, 278, 280, 282, 287, 288
Symbolic, 1
Symbolic interactionism, 8

T
Tab groups, 191, 193
Tabs, 191
Target, 181, 186, 188–191
Team project, 372
Text data, 343, 345, 348
Text reports, 293
Thematic, 25, 37, 56, 100, 103, 106, 109, 114, 115, 127, 128, 148, 182, 201, 223, 225, 238, 240, 256, 270, 279, 289, 303, 304, 306, 307, 311, 319, 322, 332, 335, 407
Themes, 3, 4, 10–12, 14, 15, 17, 19, 25–27, 37, 39–41, 45, 46, 53, 54, 56, 57, 74, 97, 98, 101, 107, 109, 111, 114, 115, 117, 119, 125–130, 132–134, 148, 160, 162, 169–175, 201, 206, 207, 214, 220, 221, 223, 224, 238, 240, 241, 269, 270, 274, 278, 283, 285, 287–290, 294, 343, 352, 355–357
Theoretical coding, 105, 106, 118
Thick description, 1, 11
Thin description, 1, 11
Timestamp, 351
Transcendental, 40
Transcription, 31, 49–54, 56, 322, 327
Transcripts, 344–346, 348–351
Transferability, 367

Treemap, 275–277, 282
Triangulation, 56, 57
Trustworthiness, 367, 369
Truth value, 367
Twitter data, 297, 298, 301, 303, 304, 307, 310
Two-level Analysis, 330

U
Unit of analysis, 323, 331

V
Validation, 41
Validity, 34, 57, 365–369
Verbatim, 51, 56, 69, 81
Video clips, 54
Video data, 343–346, 348, 352, 354, 356
Vulnerable populations, 62, 63, 67

W
Word Cloud, 275, 278, 281

Y
YouTube comments, 297, 313

The manufacturer's authorised representative in the EU is Springer Nature Customer Service Centre GmbH, Europaplatz 3, 69115 Heidelberg, Germany. If you have any concerns regarding our products, please contact ProductSafety@springernature.com

Printed and bound by CPI Group (UK) Ltd, Croydon, CR0 4YY
27/03/2026
02079485-0002